Transactions on Computer Systems and Networks

Transactions on Computer Systems and Networks is a unique series that aims to capture advances in evolution of computer hardware and software systems and progress in computer networks. Computing Systems in present world span from miniature IoT nodes and embedded computing systems to large-scale cloud infrastructures, which necessitates developing systems architecture, storage infrastructure and process management to work at various scales. Present day networking technologies provide pervasive global coverage on a scale and enable multitude of transformative technologies. The new landscape of computing comprises of self-aware autonomous systems, which are built upon a software-hardware collaborative framework. These systems are designed to execute critical and non-critical tasks involving a variety of processing resources like multi-core CPUs, reconfigurable hardware, GPUs and TPUs which are managed through virtualisation, real-time process management and fault-tolerance. While AI, Machine Learning and Deep Learning tasks are predominantly increasing in the application space the computing system research aim towards efficient means of data processing, memory management, real-time task scheduling, scalable, secured and energy aware computing. The paradigm of computer networks also extends it support to this evolving application scenario through various advanced protocols, architectures and services. This series aims to present leading works on advances in theory, design, behaviour and applications in computing systems and networks. The Series accepts research monographs, introductory and advanced textbooks, professional books, reference works, and select conference proceedings.

Nonita Sharma · Monika Mangla ·
Subhash K. Shinde
Editors

Big Data Analytics in Intelligent IoT and Cyber-Physical Systems

Springer

Editors
Nonita Sharma
Department of Information Technology
Indira Gandhi Delhi Technical University for Women
New Delhi, Delhi, India

Monika Mangla
Department of Information Technology
Dwarkadas J. Sanghvi College of Engineering
Mumbai, Maharashtra, India

Subhash K. Shinde
Lokmanya Tilak College of Engineering
Navi Mumbai, India

ISSN 2730-7484 ISSN 2730-7492 (electronic)
Transactions on Computer Systems and Networks
ISBN 978-981-99-4517-7 ISBN 978-981-99-4518-4 (eBook)
https://doi.org/10.1007/978-981-99-4518-4

This Springer imprint is published by the registered company Springer Nature Singapore Pte Ltd.
The registered company address is: 152 Beach Road, #21-01/04 Gateway East, Singapore 189721, Singapore

Paper in this product is recyclable.

Acknowledgements

First and foremost, editors would like to thank to the almighty who has inspired us and been our strength during the journey of this book. This book is the result of inspiration, guidance, encouragement, and support several close knits. The editors would like to acknowledge the contribution to all the authors from various countries for submission of their research work in this book. The editors would also like to give a special thanks to reviewers across the globe whose efforts towards providing constructive feedbacks have helped to accomplish this book as a quality product.

Dr. Nonita Sharma, Dr. Monika Mangla, and Dr. Subhash K. Shinde would like to express their earnest thanks to their workplace namely Indira Gandhi Delhi Technical University for Women, New Delhi, Dwarkadas J. Sanghvi College of Engineering, Mumbai and Lokmanya Tilak College of Engineering, Navi Mumbai, respectively, for providing a conducive research environment to complete this project. Also, editors would like to thank all the people around us including college administration and authorities, colleagues, and students who have been a constant source of motivation and guidance throughout this journey. They would like to thank everyone enough for their involvement to take on this project beyond their comfort zones.

Editors wish to thank their parents for their love and encouragement during the journey of this project. Not to mention, editors would like to thank their family who has always been the pillar of strength and instrumental in successful completion of this book.

Last but not at all the least, we acknowledge the publishing team at Springer Pearson for supporting our work and continuously guiding us through the process. We are very grateful to the entire team members of Springer involved in the publishing of this book. Without their continuous and extensive support, this work would never have gone this far.

Nonita Sharma
Monika Mangla
Subhash K. Shinde

Contents

Editors and Contributors

About the Editors

Dr. Nonita Sharma is working as Associate Professor at Indira Gandhi Delhi Technical University for Women, New Delhi. She has more than 15 years of teaching experience. Her major area of interest includes data mining, bioinformatics, time series forecasting, and wireless sensor networks. She has published several papers in the International/National Journals/Conferences and book chapters. She received several best paper awards for her research work at renowned international conferences. She has been awarded the Best Teacher Award in view of recognition of her contributions, achievements, and excellence in Computer Science & Engineering at NIT Jalandhar. She has been awarded Best Content Guru Award by Infosys twice. She is a member of IEEE and has been shortlisted in the Top 5 for IEEE Women Achiever Award. She is the reviewer of many peer-reviewed journals and contributed to academic research in terms of projects, papers, and patents.

Dr. Monika Mangla received her Ph.D. from the Thapar Institute of Engineering and Technology, India, in 2019. She is working as an Associate Professor in the Department of Information Technology at Dwarkadas J. Sanghvi College of Engineering, Mumbai. Her interest areas include IoT, cloud computing, algorithms and optimization, location modelling, and machine learning. She has published several research papers and book chapters (SCI and Scopus Indexed) with reputed publishers. She has also been associated with several journals of national and international repute as a reviewer. She has two patents applied to her credit. She is a life member of CSI and IETE.

Dr. Subhash K. Shinde has earned his Ph.D. in Computer Science and Engineering from Shri Guru Gobind Singhji Institute of Engineering and Technology, India. He has over 24 years of academic and research experience. His research areas include machine learning, computer networks, network security, data warehousing, and mining. He has also been guiding research scholars at the University of Mumbai, India. Currently, he is working as a Professor and vice-principal at Lokmanya Tilak College of Engineering (LTCoE), India. He has published numerous research papers at various reputed national/international conferences and journals (SCI and Scopus Indexed).

Contributors

Alankrita Aggarwal Panipat Institute of Engineering and Technology, Panipat, India;
Computer Science & Engineering—Apex Institute of Technology, Chandigarh University, Mohali, India

Rakhi Akhare CSED, Lokmanya Tilak College of Engineering, Navi Mumbai, India

Shefali Bajaj B. R. Ambedkar National Institute of Technology, Jalandhar, India

Sujay Bansal B. R. Ambedkar National Institute of Technology, Jalandhar, India

Noel Job Benzer School of Computer Science and Engineering, Digital University Kerela, Thiruvananthapuram, India

Saswati Chatterjee Center for AI & ML, SOA University, Bhubaneswar, Odisha, India

Shivani Gaba Panipat Institute of Engineering and Technology, Panipat, India

Anju Gandhi Panipat Institute of Engineering and Technology, Panipat, India

Uttam Ghosh Department of Computer Science & Data Science, Meharry Medical College Nashville, Nashville, TN, USA

Nitin Goyal Department of Computer Science and Engineering, Central University of Haryana, Mahendragarh, Haryana, India

I. Kala Department of Computer Science and Engineering, PSG Institute of Technology and Applied Research, Coimbatore, India

Shahnawaz Khan Design and Information and Communications Technology, Bahrain Polytechnic, Isa Town, Bahrain

Chandana Kuntala Department of Information Technology, Indira Gandhi Delhi Technical University for Women, New Delhi, India

Monika Mangla Dwarkadas J Sanghvi College of Engineering, Mumbai, India

Stuti Mehla Panipat Institute of Engineering and Technology, Panipat, India

Mukesh Mishra Massey University, Palmerston North, New Zealand

Ritunsa Mishra FET, Sri Sri University, Cuttack, Odisha, India

Shally Napgal Panipat Institute of Engineering and Technology, Panipat, India

Sangeeta Oswal Research Scholar, Computer Engineering LTCE, Navi Mumbai, India

Wasique Haleem Pandit Dr. B. R. Ambedkar National Institute of Technology, Jalandhar, India

Bibudhendu Pati Department of CS, Rama Devi Women's University, Bhubaneswar, Odisha, India

Monika Poriye Department of Computer Science and Applications, Kurukshetra University, Kurukshetra, Haryana, India

Manik Rakhra Department of Computer Science and Engineering, Lovely Professional University, Phagwara, Punjab, India

Reena Computer Science, Edge Hill University, Ormskirk, Lancashire, England

Ruchi School of Computer Science and Engineering, Lovely Professional University, Phagwara, Punjab, India

Pritika Sabharwal Department of Electronics and Communication Engineering, Netaji Subhas University of Technology, (Formerly known as Netaji Subhas Institute of Technology), New Delhi, India

Rahul Sachdeva Indira Gandhi Delhi Technical University for Women, Delhi, India

Harish Kumar Saini Department of Computer Science and Applications, Kurukshetra University, Kurukshetra, Haryana, India

Soumya Samarpita FOS, Sri Sri University, Cuttack, Odisha, India

Rabinarayan Satpathy FET, FOS, Sri Sri Univeristy, Cuttack, Odisha, India

Suneeta Satpathy Center for AI & ML, SOA University, Bhubaneswar, Odisha, India

Aditi Sharma Department of Computer Science, Nazarbayev University, Astana, Kazakhstan

Deepak Kumar Sharma Department of Information Technology, Indira Gandhi Delhi Technical University for Women, New Delhi, India

Divyansh Sharma Bow Valley College, Calgary, AB, Canada

Krishna Pal Sharma Computer Science and Engineering, Dr. B R Ambedkar National Institute of Technology, Jalandhar, India

Nonita Sharma Department of Information Technology, Indira Gandhi Delhi Technical University for Women, Delhi, India

Ravi Sharma Computer Science and Engineering, Dr. B R Ambedkar National Institute of Technology, Jalandhar, India;
Department of Computer Science and Engineering, NIT Jalandhar, Jalandhar, India

S. Sherine Computer Science and Engineering, Karunya Institute of Technology and Sciences, Coimbatore, India

Subhash K. Shinde Professor and Vice Principal, Lokmanya Tilak College of Engineering, Navi Mumbai, India

Shivam Computer Science and Engineering, Dr. B R Ambedkar National Institute of Technology, Jalandhar, India

Arun Singh School of Computer Science and Engineering, Lovely Professional University, Phagwara, Punjab, India

Dalwinder Singh School of Computer Science and Engineering, Lovely Professional University, Phagwara, Punjab, India

Janpreet Singh School of Computer Science and Engineering, Lovely Professional University, Phagwara, Punjab, India

Kartik Singhal Department of Manufacturing Process and Automation Engineering, Netaji Subhas University of Technology, (Formerly known as Netaji Subhas Institute of Technology), New Delhi, India

Sristi Department of Computer Science and Engineering, Indira Gandhi Delhi Technical University for Women, New Delhi, India

Puneet Thapar Department of Computer Science and Engineering, Lovely Professional University, Phagwara, Punjab, India

Krishna Pal Singh Tiwari Cognizant Technology Solutions Pvt. Ltd, Howrah, India

Priyanka Tomar Indira Gandhi Delhi Technical University for Women, Delhi, India

Preeti Vats Department of Information Technology, Indira Gandhi Delhi Technical University for Women, Delhi, India

M. Vijayalakshmi AI and Data Science VESIT, Mumbai, India

Aditya Vishnu School of Computer Science and Engineering, Digital University Kerela, Thiruvananthapuram, India

Javaid Ahmad Wani Qualcomm India Private Limited, New Delhi, India

Sourabh Yadav Mobile Computing Lab at UNT, University of North Texas, Denton, TX, United States

Jothimanivannan Yuvanesh Department of Computer Science and Engineering, PSG Institute of Technology and Applied Research, Coimbatore, India

Part I
Cyber Physical System

Chapter 1
An Outline of Cyber-Physical System: Issues and Security Risks

Noel Job Benzer, Aditya Vishnu, Saswati Chatterjee, and Suneeta Satpathy

1.1 Introduction

1.1.1 Overview

A Computer security Framework is a platform that successfully combines cyber and physical components using modern sensor, computation, and communication networks (Zeadally and Jabeur 2016; Ghazani et al. 2012). A new cloud computing known as cyber-physical-social computing or physical-cyber-social computing has emerged as a result of CPS and the cyber-social system (Sheth et al. 2013). A development of CPSs that incorporates social space and telltale signs of human interaction is known as cyber-physical-social systems (CPSSs) (Zeng et al. 2016). The widespread adoption of CPS is related to the concept of "Industry 4.0" (Zeng et al. 2016), which depicts the method of combining knowledge and technology to give independence, dependability, coherency, and management requiring user intercession. The primary technology breakthroughs that support CPS include IoT technologies (IoT), intelligent systems, cloud-based services, etc. The development of smart manufacturing, smart healthcare, smart infrastructure, smart cities, smart cars, wearable technology,

N. J. Benzer · A. Vishnu
School of Computer Science and Engineering, Digital University Kerela, Thiruvananthapuram, India

S. Chatterjee (✉) · S. Satpathy
Center for AI & ML, SOA University, Bhubaneswar, Odisha, India
e-mail: cshiva68@gmail.com

S. Satpathy
e-mail: suneeta1912@gmail.com

N. Sharma et al. (eds.), *Big Data Analytics in Intelligent IoT and Cyber-Physical Systems*, Transactions on Computer Systems and Networks,
https://doi.org/10.1007/978-981-99-4518-4_1

mobile systems, defense technology, meteorology, etc., is based on CPSs. Applications for CPS are expanding quickly, which causes a lot of security and confidentiality issues information must be kept private and must not be shared with unauthorized parties, organizations, or procedures. Integrity is defined as accuracy and thoroughness. Being reachable and usable at the request of an authorized entity is the quality of availability. Additional qualities that could be important are integrity, transparency, non-repudiation, and dependability. Due to the extensive usage of communication devices for data collection, transmission, and control devices, when a detection system is employed, information security systems are becoming more and more important in business. The independence and remote location of CPS devices raise the danger of trespassing and attacks. When using several devices at once, some of them may get compromised. CPS security creates a lot of many additional challenges. (Alguliyev et al. 2018) These systems are more vulnerable to cyberattacks (like DDoS) due to the proliferation of the Internet of Things, security threat modeling, the development of a structured approach for assessing CPS security problems, and the layout of reliable and fault-tolerant frameworks for the management of swiftly evolving cyber and physical threats. New methodologies and technologies must be developed in order to meet CPS standards for the security, dependability, and confidentiality of personal data (Fig. 1.1).

Multiple networks of fixed or moving actuators and sensors make up a CPS, which is controlled by an intelligent decision-making system. These systems are renowned for their multi-information flow, cross-domain sensor cooperation, and smart choice abilities. A CPS's specific features and attributes depend on the application for which it is intended. To track and control the material reality, CPSs mix computing, networking, and physical processes. Sensing Components (SC) and Controlling Components are two divisions of the CPS components (CC). While CCs

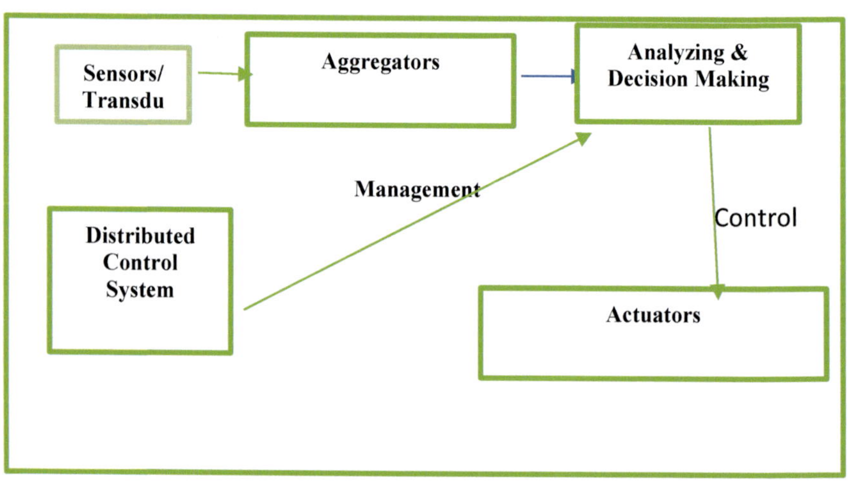

Fig. 1.1 Infrastructure of CPS

watch and manage signals, SCs collect and perceive data. Sensors, aggregators, and actuators are the major sensing components used in a CPS. A CPS may employ a variety of control methods, including Programmable Logic Controllers, Distributed Control Systems, and Remote Terminal Units (Yaacoub et al. 2020).

1.2 Challenges in Cyber-Physical Systems

Cyber-physical system (CPS) implementation and management can be difficult. These challenges can be complex due to the dynamic and interconnected nature of CPS, as well as the need to integrate and work effectively with a variety of different technologies and systems.

1.2.1 Security

A secure system must be capable to guard against unlawful resource restriction, unauthorized modifications to information, and improper sharing of private information. Because of their complexity, scalability, and dynamic properties, cyber-physical systems (CPS) are particularly susceptible to errors and attacks. Cyber-physical systems (CPS) are susceptible to malicious attacks such as eavesdropping, man-in-the-middle attacks, denial-of-service attacks, compromised key assaults (Wang et al. 2010), and the injection of phony sensor measurements or activation requests. These assaults may target the physical components of the system or the digital infrastructure (such as the information management layer, telecommunications network, or decision-making mechanisms) with the intention of interfering with system performance or obtaining confidential data (Gunes et al. 2014).

Cyber threats can affect various aspects of cyber-physical systems (CPS), including (Alguliyev et al. 2018; Wang et al. 2010):

1. Confidentiality: By "packet sniffing" on forms of communication, cyberattacks can attempt to influence the status of the real object and compromise the integrity of private user data stored in CPS.
2. Integrity: Cyber threats can allow unauthorized changes to data or resources.
3. Availability: Breakdowns in digital technology, organization, connectivity, and infrastructure can be brought on by cyberattacks.
4. Reliability: Cyber threats can undermine the ability to confirm the identities of parties involved in CPS.
5. Authenticity: Threats from the internet might make it difficult to locate a subject or source.
6. Non-repudiation: Cyber threats might make it more difficult to show that certain things happened.

7. Accountability: Cyber threats can make it difficult to trace the actions of an entity to that entity.

Due to their reliance on expansive networks usage of protocols for unprotected transmission, incorporation of existing systems, and quick embrace of widely viable technologies, CPSs are also in danger. It is important to address these security threats in order to ensure the dependability and trustworthiness of CPS (Anwar and Ali 2012; O'Reilly 2013).

1.2.2 Latency

Practical applications must have high throughput since data transmission latencies might have negative effects. For instance, in the healthcare industry, telemonitoring cycles of malware systems can be disrupted by delays in the transmission of patient data, which has an adverse effect on the prompt administration of medications and primary care. Similarly, to this, it is crucial for application scenarios that hazard delay, or the interval between a defect occurring and being recognized, be kept to a minimum in order to raise system stability. In general, low latency and low fault latency are critical requirements for real-time applications in order to ensure that systems operate effectively and efficiently (Verma 2022; Dowdeswell et al. 2020).

1.2.3 Sustainability

A sustainable system is one that is able to continue operating effectively over a prolonged period of time without experiencing significant degradation in performance or outcomes. To achieve this, the system should have self-healing and dynamic tuning capabilities that allow it to adapt and recover from changes in its environment. This includes the ability to maintain its functionality and efficiency under evolving circumstances. A system that is extremely sustainable can therefore survive a long time and be flexible enough to function continually. Policies for energy provision and management must include sustainability from an energy standpoint (Gunes et al. 2014). For instance, by adding renewable sources of energy generated from the natural environment, the Smart Grid facilitates power distribution, management, and customization from the perspective of customers or service providers. However, maintaining the long-term functionality of the Smart Grid is hampered by intermittent energy supply and unclear or poorly defined load characterization. The Smart Grid needs dynamic energy optimization approaches, real-time performance assessments, environment-aware duty cycling of computer units, planning and operation under uncertain conditions, the identity of energy distribution networks.

1.2.4 Heterogeneity

Heterogeneity is the characteristic of a system that involves the combination of various types of interconnected components that work together to form a complex entity. These components may be diverse in terms of their functions, capabilities, or technologies, and they may interact with each other in a variety of ways. The resulting system may be highly complex, with many different parts working together to achieve a common goal. Based on their physical characteristics, cognitive components, instruction set, and variety of communication methods, cyber-physical systems (CPS) are intrinsically complex. As a result, most parts of the system must be varied for CPS (Gunes et al. 2014). For instance, the combination of cyber-physical systems (CPS) is necessary for the communication and data storage services offered by outside service providers in a smart city healthcare ecosystem. For effective and productive operation, these CPS must integrate and encapsulate a multitude of devices with various data types. The requirement to communicate with equipment that has varied technology and software resources presents a significant difficulty in this heterogeneous environment. Data exchange and collaborative functionality are required for the integration and higher levels of healthcare cyber-physical systems (HCPS), which necessitate an approach that has the following, architecture, testing, and evaluation. To ensure the secure storage of data that can be accessed by medical professionals, insurance providers, clients, and academics, it is necessary to collaborate with the following providers of cloud services (Verma 2022; Mosterman and Zander 2016).

Heterogeneous control networks and closed-loop control of connected devices are essential because future medical equipment with diverse processing and communication capabilities are likely to be coupled in increasingly sophisticated open systems using a connector method. Based on the unique medical requirements of each patient, the setup of these devices may be highly dynamic. Future medical systems are anticipated to be far more complicated and competent than existing systems, enabling scenario element autonomy, cooperation management, actual assurances, and varied tailored configurations (High Confidence Software and Systems Coordinating Group 2009).

1.2.5 Scalability

A scalable system should be able to handle changes in size or workload without experiencing significant degradation in performance. In order to do this, the system must be equipped with tools for effectively allocating and collecting responsibilities in order to balance the load, as well as adequate communication channels. These strategies help the system maintain good performance and take full advantage of changes in size or workload, ensuring that it can continue to function effectively even as demand for its services or products increases. The hundreds of advanced electronics, sensors, and actuators that make up cyber-physical systems (CPS) may need to cooperate well. To satisfy rising computational demands, CPS may leverage numerous topologies with network cable networks to provide scalability (Gunes and Givargis 2014). Cyber-physical systems (CPS) demand a high, flexible system that allows CPS entities to access and depart the system constantly if they are to be able to adapt to changing conditions and fully utilize the resources that are available. When data is often shared among CPS entities, dynamic software updates, which entail upgrading the software program in actuality, can assist in improving the efficiency of CPS resources. This can help to ensure that CPS applications are able to adapt to changing circumstances and make the most effective use of available resources (Park et al. 2010).

1.2.6 Maintainability

For a system to be maintainable, it should be easy to repair, not cause additional problems during maintenance, and be able to be fixed at a low cost. While sustaining software systems entails finding and fixing errors, maintaining traditional engineering systems entails fixing problems that develop over time. For malware systems, both sorts of operations are necessary. Furthermore, these systems' interconnectedness provides greater options for updating and upgrading them (Broy and Schmidt 2014). The architecture of cyber-physical systems (CPS) can be continually tested and monitored using autonomous predictive and corrective diagnostic procedures, allowing for the detection of broken components. Due to the tight contact between system components, such as detectors, operators, cyber elements, and tangible things, CPS can make good use of these mechanisms. In order to increase the general maintainability of the CPS, it is possible to identify components that are prone to recurring failures by regularly testing and monitoring the infrastructure (Ruiz-Arenas et al. 2014; Germany Trade and Invest 2013). These components can then be redesigned or replaced with higher-quality components.

1.3 Security Issues in Cyber-Physical Systems

The rapid expansion of CPS's application base exposes it to a variety of security and confidentiality issues. Since virtual and physical technologies are used in cyber-physical systems, security should be taken extremely carefully. This is a result of the CPS's integration of physical and computational processes (Alguliyev et al. 2018). Their physical environment is impacted by changes in the calculation process and vice versa. Additionally, due to their distant place, CPS devices are subject to hacker and invasion risks. Experts find it considerably more challenging to build an attack model and identify the attack paths as there are more linked devices. A breach in CPS' security casts doubt on the system's confidentiality, integrity, and availability, just like it does with any cyber system. Information is kept private and not disclosed to unapproved parties. Integrity refers to the quality and comprehensiveness of the data. Availability is the quality of being reachable at any time by a legitimate entity (Alguliyev et al. 2018). Other qualities like accountability, integrity, non-repudiation, and durability are also important in addition to these three.

1.3.1 Types of Attacks in CPSs

(1) The following is a list of some of the main security risks that CPSs and other cyber systems must deal with:

(2) **Denial of service attack**
In denial-of-service attacks, requests are flooded into the system from non-legitimate traffic sources and thereby taking up all the network and computational resources for use by legitimate users. These types of attacks transmit a huge amount of data into the network making it difficult for the server to handle any more requests. The network won't be able to operate normally as a result of this. As soon as the attacker has access to the CPS's network, they can:

- Overburden the system with load until it collapses from overload.
- Block bandwidth to only allow authorized parties access.
- Send invalid data to the controller and system network which can cause abnormal behavior in the system.

(3) **Eavesdropping**
Eavesdropping is a type of attack in which the adversary can listen to any information that is communicated through the system. In this assault, the attacker merely watches the system in a passive manner without interfering with its operation. Consequently, it is a form of a passive attack, and we might never be aware that one has occurred. Eavesdropping can affect data privacy rules and expose critical data such as patient health data in the case of CSP systems in healthcare.

(4) **Communication attacks**
Communication attacks can affect system traffic by hindering the routing of system packets. This can disrupt the resource allocation mechanism attacks on information can support attackers. collect sensor information and maybe sometimes make the loss of data packets which can affect the quality of the dataset.

(5) **Man-in-the-middle attack**
In an operation of this nature, the adversary stands in the middle of the communication link relaying the information being sent. The attacker can modify the commands or data being sent and make the system act abnormally. In a CPS, such an attack can cause undesirable events to happen since the system interacts directly with its physical environment.

The fact that cyber-physical dangers come from cyberspaces and have the potential to affect the workspaces of the system is one of their most crucial aspects. The number of digital threats is growing daily. They are easily replicated and distributed freely through many sources. Attackers are exploring newer techniques to exploit vulnerabilities in cyberspace. Thus, a robust security framework is very essential in safeguarding the devices which are connected to the network. Especially due to the heterogeneous nature of the CPS which includes different architectures and communication protocols they are more vulnerable to such attacks. Different CPS vulnerabilities are as follows:

1.3.2 CPS Vulnerabilities

Susceptibility is an unidentified major flaw in the network system that can be used by enemies to conduct either passive or active espionage, including reconnaissance. As a result, a security study entails analyzing the CPS systems to discover their flaws and then implementing the necessary corrective, preventative, and mitigating actions to safeguard them against potential attacks (Yaacoub et al. 2020). The three main categories into which vulnerabilities are divided are:

(1) **Network vulnerability**
Vulnerabilities in wired or wireless communication channels that make the system susceptible to many sorts of attacks, including eavesdropping, man-in-the-middle attacks, replay, sniffing, spoofing, DoS/DDoS attacks, etc.
(2) **Platform vulnerability**
Technology, software, system control, and database risks fall under this category.
(3) **Management vulnerability**
Security flaws are brought on by an absence of suitable frameworks, rules, and regulations (Zeng et al. 2016).

1.4 Conclusion

The actual world could be dramatically affected by cyber-physical systems (CPS), which have the ability to change both present and future engineering systems. By linking them with the digital world and facilitating greater contact with the physical environment, CPS systems are designed to enhance the quality, accessibility, and performance of products and systems. They are an important part of Industry 4.0 and have already begun transforming how we interact with the world. They are increasingly being used in various industries, offering numerous benefits such as improved efficiency, safety, and reliability. However, CPS also poses challenges and security issues that can affect their safety, efficiency, and reliability. These issues need to be addressed in order to ensure the successful deployment of CPS systems. In this work, some of these difficulties and security concerns are covered. Understanding the issues and security concerns that CPS is facing, as well as putting plans into place to solve them, is crucial for addressing these challenges and enhancing CPS security. By doing so, industries can better plan for and address these issues, leading to more successful and secure implementation of CPS, maximizing their benefits and minimizing their risks.

References

Alguliyev R, Imamverdiyev Y, Sukhostat L (2018) Cyber-physical systems and their security issues. Comput Ind 100:212–223

Anwar RW, Ali S (2012) Trust based secure cyber physical systems. In Proceedings of workshop proceedings: trustworthy cyber-physical systems, tech report series, computing science, Newcastle University

Broy M, Schmidt A (2014) Challenges in engineering cyber-physical systems. Computer 47(2):70–72

Dowdeswell B, Sinha R, MacDonell SG (2020) Finding faults: a scoping study of fault diagnostics for Industrial cyber-physical systems. J Syst Softw 168:1–16

Germany Trade & Invest (2013) Industrie 4.0–smart manufacturing for the future

Ghazani SHHN, Lotf JJ, Alguliev RM (2012) A study on QoS models for mobile ad-hoc networks. Int J Model Optim 2(5):634–636

Gunes V, Peter S, Givargis T, Vahid F (2014) A survey on concepts, applications, and challenges in cyber-physical systems. KSII Trans Internet Inf Syst 8(12):4242–4268

Gunes V, Givargis T (2014) XGRID: a scalable many-core embedded processor. In: Proceedings of the 11th IEEE International conference on embedded software and systems (ICESS), Paris

High Confidence Software and Systems Coordinating Group (2009) High-confidence medical devices: cyber-physical systems for 21st century health care. In: A Research and Development Needs Report, NCO/NITRD

Mosterman PJ, Zander J (2016) Industry 4.0 as a cyber-physical system study. Softw Syst Model 15(1):17–29

O'Reilly P (2013) Designed-in cyber security for cyber-physical systems. In: Workshop report by the cyber security research alliance (CSRA) and co-sponsored with NIST

Park MJ, Kim DK, Kim W-T, Park S-M (2010) Dynamic software updates in cyber-physical systems. In: Proceedings of the IEEE International conference on information and communication technology convergence (ICTC), pp 425–426

Ruiz-Arenas S, Horvath I, Mejia-Gutierrez R, Opiyo EZ (2014) What is with the maintenance principles of cyber-physical systems? J Mech Eng

Sheth A, Anantharam P, Henson C (2013) Physical-cyber-social computing: an early 21st-century approach. IEEE Intell Syst 28(1):78–82

Verma R (2022) Smart city healthcare cyber physical system: characteristics technologies and challenges. Wireless Pers Commun 122:1413–1433

Wang EK, Ye Y, Xu X, Yiu SM, Hui LCK, Chow KP (2010) Security issues and challenges for cyber physical system. In: IEEE/ACM international conference on cyber, physical and social computing, Hangzhou, China

Yaacoub J-PA, Salman O, Noura HN, Kaaniche N, Chehab A, Malli M (2020) Cyber-physical systems security: limitations, issues and future trends. Microprocess Microsyst 77:103201

Zeadally S, Jabeur N (2016) Cyber-physical system design with sensor networking technologies. The Institution of Engineering and Technology, London UK

Zeng J, Yang LT, Lin M, Ning H, Ma J (2016) A survey: cyber-physical-social systems and their system-level design methodology. Future Gener Comput Syst. https://doi.org/10.1016/j.future.2016.06.034

Chapter 2
Classification of Vulnerabilities in Cyber Physical Systems: Approach, Security and Challenges

Anju Gandhi, Stuti Mehla, Shivani Gaba, Alankrita Aggarwal, and Shally Napgal

2.1 Introduction

Cyber Physical System integrates cyberspace and real space in a dynamic environment. A CPS is a feedback loop mechanism involving a set of physical devices (sensors and actuators) controlled by computer-based algorithms. With the help of CPS, service providers demonstrate their products to their customers and gain a better understanding. It is used to model many real-time applications such as automotive, factory, healthcare, agriculture, and my monitoring. The main goal of CPS is to maximise the implementation of large systems by improving their adaptability, flexibility, performance, functionality, reliability, protection, and accessibility. CPSs have the following two main elements.

- Actual Time data collecting from the internet intelligence feedback and real world are made possible by advanced technologies.
- Cyberspace relies on intelligent data processing, analysis, and computing power.

Cyber physical systems use IoT as its foundational or enabling technology. Cyber physical systems are the IoT's advancement in terms of full conception and perception, and they have a significant capacity for physical world control. Traditional embedded and control techniques are also a part of cyber physical systems,

A. Gandhi (✉) · S. Mehla · S. Gaba · A. Aggarwal · S. Napgal
Panipat Institute of Engineering and Technology, Panipat, India
e-mail: anjugandhi.cse@piet.co.in

S. Mehla
e-mail: stutimehla.cse@piet.co.in

A. Aggarwal
Computer Science & Engineering—Apex Institute of Technology, Chandigarh University, Mohali 140413, India

N. Sharma et al. (eds.), *Big Data Analytics in Intelligent IoT and Cyber-Physical Systems*, Transactions on Computer Systems and Networks,
https://doi.org/10.1007/978-981-99-4518-4_2

which have evolved them into cutting-edge techniques. For dependable transmission and information processing, IoT links information acquiring devices including sensors, Cloud Computing and RFID (Radio Frequency Identification) wireless sensor networks technology. In contrast, CPS is a control technique that combines computation, communication, and IoT control. It is scalable and reliable. IoT, on the other hand, focuses on information processing and transmission, whereas CPS not only has the capacity to perceive but also has a potent ability to control. Cyber and physical aspects are related to one another in CPS on both a geographic and temporal scale, revealing a variety of distinct behavioural processes and cooperating with one another in a variety of ways that change the context.

Next generation engineered systems are referred to as CPS. In 2006, Helen Gill at the NSF (National Science Foundation) introduced the term "Cyber-Physical System". The terms "cyber-physical system" and "cyber-security," which have no connection to physical processes, are frequently used interchangeably. The close integration of computations, algorithms, and physical devices is known as CPS. The technologies are seen as connecting the information world with the real world. CPS communicates via well-known technologies including the Industrial Internet of Things (IoT), Industry 4.0, Intelligent Internet of Things (IIoT) and Machine to Machine (M2M).

CPS is a cutting-edge technique that can demonstrate the behaviour of tightly coupled, dispersed physical systems that were previously unthinkable, greatly enhancing the effectiveness and productivity of large-scale systems. In the area of computational meditation systems, it aids in the generation of novel theories. It uses a network of actuators and sensors to continuously manage, monitor and improve physical control systems. The integration of embedded systems with the physical environment is what CPS is, in other words (Fig. 2.1 CPS-based technologies raise the standard of living and make advancements possible in sectors like healthcare, medical crises, and other areas.

Extensive implementation of Cyber Physical Systems due to its characteristics refers to the "Industry 4.0", which combines technology and knowledge to achieve autonomy, reliability, systemization, and innovation without the need for human intervention. It represents process control. The technologies like Smart Technology, Cloud Computing, IoT and many more are the key technology trends driving the CPSs.

CPSs working architecture as shown in Fig. 2.2 supports development in extents like Smart Medication, Smart Constructions and infrastructure, Smart Cities, Smart and unmanned Vehicles, Wearable gadgets, Smart Engineering, Mobility systems, Smart and Powerful defence Systems, and Smart technology meteorology. But the fast evolution of applications of cyber physical system raises several security and privacy concerns.

Information security upholds the information's availability, confidentiality, and integrity. The expanding usage of non-wired technologies for data collection, transmission, and reception as well as control orders via wireless sensor networks has increased the need for information security system development in the industry (WSNs). Due to their independence, equipment in CPSs is inaccessible, which raises

Fig. 2.1 Cyber physical system

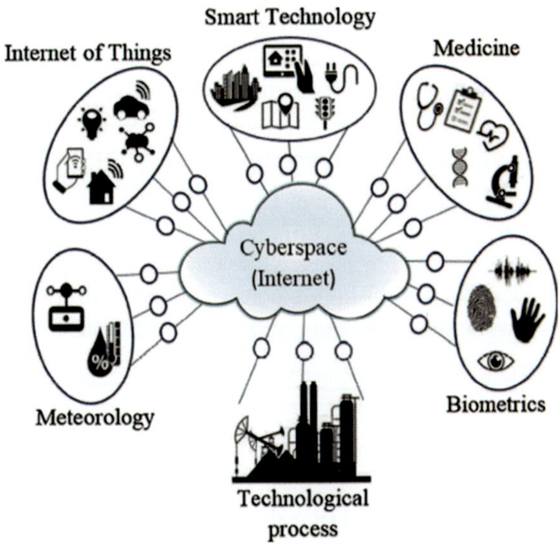

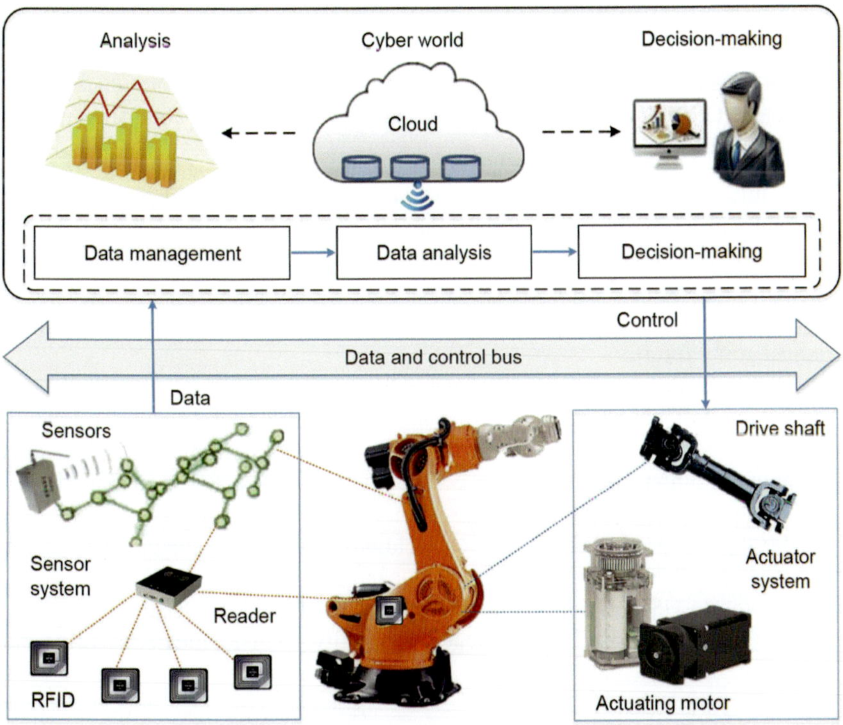

Fig. 2.2 Cyber-physical system architecture and its applications

the possibility of intrusion and attack. When working with many device groups, certain devices can be in danger. New challenges are presented by CPS security. Working with numerous tactics at once can put some of them in jeopardy. The CPS security presents the following new difficulties:

- As the number of IoT devices increases, these systems become more vulnerable to cyberattacks (such as DDoS).
- Modelling of the security intimidations.
- Advancement to assess CPS vulnerabilities approaches.
- Development of highly consistent and fault-tolerant designs to address quickly developing cyber and physical intimidations.

Thus, new techniques are created to satisfy the demands of the cyber physical system for data security, dependability, confidentiality, and specific data. This chapter makes an effort to aggregate and scrutinise the available research on cyber physical system architecture, security, and related topics.

2.2 Cyber Physical System

Helen Gill suggested the term in 2006 at the workshop of US NSF's National Science Foundation. CPSs are now on the US and numerous European countries' priority innovation lists.

- CPS differentiates from existing systems, such as embedded and automated systems, in terms of quality despite having comparable exterior looks. This is made possible by the incorporation of cybernetic, hardware, and software technologies as well as new actuators. Because CPS are a part of their ecosystem, they can recognise changes in it, react to them, note how they were handled, and adjust going future.
- From the standpoint of computer science (Lee 2008). The integration of physical and computational processes is what makes up CPS. These gadgets frequently include feedback and include controllers, network monitors, and embedded computers, among others, where computations are affected both directly and indirectly by physical processes.
- Under the perspective of automation technology, CPSs are customised systems whose functions are governed by computer and communication (Johansson 2014).
- According to US NSF, the future of CPS will perform better than the currently available systems based on efficiency, flexibility, fault tolerance, security, and usability.

2.2.1 History of Cyber Physical Systems

- As embedded systems proliferate, there is a greater requirement for storage space and more memory.
- The complexity and dependability of the CPS algorithm can be influenced by its quality, which raises the computational workload's intensity.
- The response time describes the feedback delay. With longer feedback delays, the accompaniments' quality assurance suffers.
- IoT, smart environments, and other technical trends together in huge systems.
- As information volumes increase, it is vital to outsource some CPS control while maintaining human oversight (Stankovic 2014).

2.2.2 Features of CPS Systems

The main characteristics of CPS as shown in Fig. 2.3 make the system rigid and reliable.

- Mobile and embedded sensing devices.
- Data flows and sensor sources that span domains.
- Cyber and physical components interconnections.
- The capacity for understanding and adaptation.
- Internet of Things
- Employing centralised automatic control to ensure the consistent performance of the systems

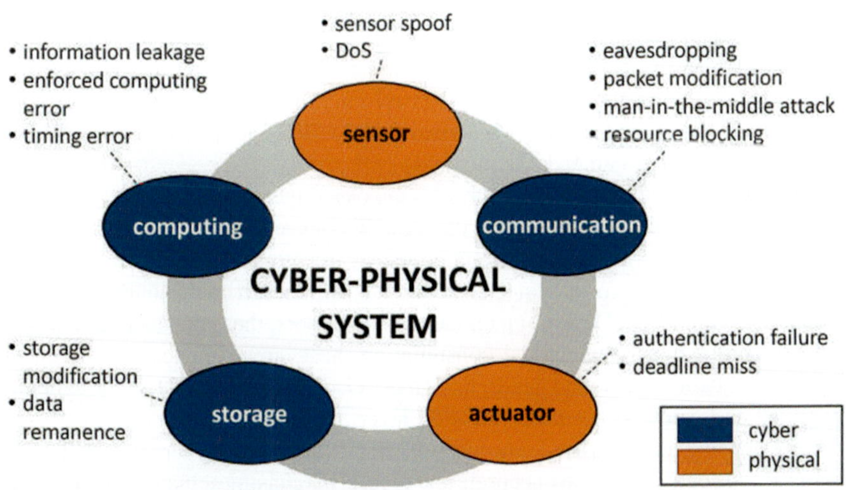

Fig. 2.3 Characteristics of CPS main components

- Communication security via cryptosystems, firewalls, antivirus software, etc., as well as the existence of a shared cyberspace that permits communication between systems and with the outside world.
- In some situations, the operation needs to be dependable and certified.
- Automated intellectual control ensures system robustness.
- Human in/outside the loop.

2.2.3 *Key Attributes of Cyber Physical Systems*

On the web, as shown in Fig. 2.3, physical systems not only act as the bridge between physical and computational approaches, but they also have all physical characteristics that come from the union of two different system types. Some crucial CPS components include (Kumar and Patel 2014):

- Every physical thing has a cyber capacity that is heavily influenced by IT.
- In CPSs, every action is anticipated.
- CPSs use sophisticated sensing.
- All employed software and systems have high levels of confidence and trust.
- There are always one or more feedback loops between a CPS's input and output.
- CPSs self-optimise, self-document, and self-monitor.
- CPSs need to be safely connected to international networks.

2.3 Essential Layers in CPS

Three separate levels and sections make up the game strategy for the CPS structures. These levels and sections communicate with one another through a variety of correspondence advances and shows. The CPS contains three important tiers. Figure 2.4 depicts and describes the Perception, Transmission, and Application Layers. The security breaches at the various CPS divisions are outlined in the study by Ashibani and Mahmoud (Sobhrajan and Nikam 2014).

It is widely termed as the interest layer or the clear layer (Ashibani and Mahmoud 2017). In close proximity to many devices, it connects hardware like sensors, Global Positioning System (GPS), actuators, aggregators and RFID tags. These devices provide clear information to screen, track, and loosen up this ongoing reality (Mahmoud et al. 2015). Depending on the type of sensors, these instances amounted to data coordination for electrical consumption, heat, area, science, and science, giving little attention to sound and light signals (Gaddam et al. 2008). Prior to being integrated and assessed by the application layer, these sensors produce clear data within extensive and nearby connection districts. In order to guarantee that both appraisal and control orders are accurate and secure, purchasing actuators also depend on remaining source awareness (Khan et al. 2012). Overall, it is estimated that young people should terminate the encryption scheme through each degree in accordance

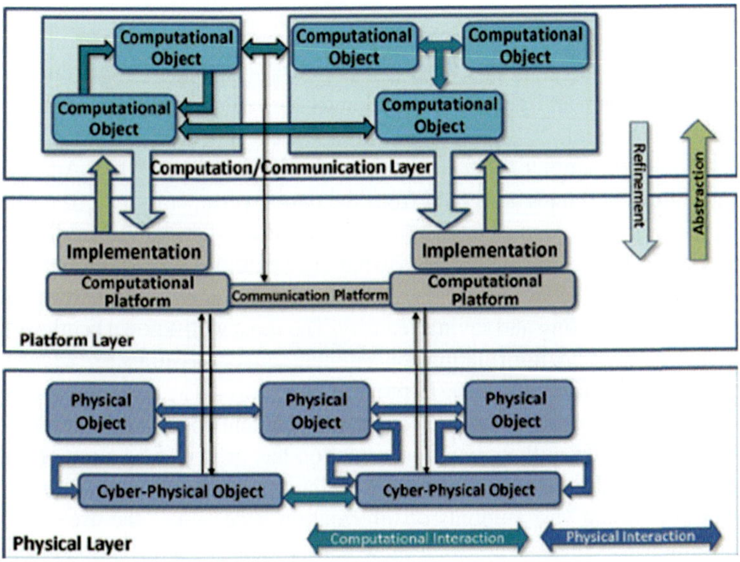

Fig. 2.4 CPS Layers and their interconnections

with the security level (Geng et al. 2006). Along these lines, heavyweight assessments and goliath memory stray pieces would be introduced (Jing et al. 2014). In this situation, there is crucial for a game plan of consistent and lightweight security shows, which contemplate the contraption's abilities and security necessities.

- ***Transmission Layer***:

The vehicle layer, also known as the association layer, is the layer that comes after the CPS layer (Zhao and Ge 2013). Through this layer, data is exchanged and cycled between the data layer and the application layer. Local Area Networks, communication technologies including Bluetooth, Wi-Fi, InfraRed (IR), and ZigBee, as well as various additional advancements, are used to send data and do tasks via the Internet. These are employed to deal with the development of web-related technology, including IPv6 (Internet Protocol Version 6) (Wood and Stankovic 2008). This layer also ensures data sorting and transmission using spread controlling platforms, trade and web Gateways, firewalls, arrangement devices, and intrusion prevention or intrusion detection systems (IDS/IPS) (Wu et al. 2010; Sommestad et al. , 2010). In order to avoid obstacles and damaging attacks like malware, dangerous code injection, Denial of Service (DoS)/Distributed Denial of Service (DDoS), tuning in, and malicious users attacks (Sridharan 2012), it is desperately attempting to obstruct the transport of the data before reclaiming its contents. Given how severely the principal operating and power capabilities are constricted above (Weiss 2010), this is a problem, especially for devices with minimal resources.

- ***Application Layer***:

The third and base layer is this one. It analyses the data acquired from the data communication layer and generates commands for real hardware, such as sensors and actuators (Hu et al. 2013). Strong regions for complicated reasoning about the amount of data are implemented to achieve this (Gao et al. 2013). Additionally, this layer obtains and maintains data from the data layer operating before selecting the appropriately referred motorised rehearsals (Zhao and Ge 2013). Middleware and information mining evaluations are used to handle the information at this tier to ensure proper figuring (Saqib et al. 2015). Protecting confidential information from leakage is necessary for protecting and saving security. The most well-known cautious tactics combine anonymization, information concealment (cover), assurance of security, and mystery sharing (Geng et al. 2006). To prevent unauthorised access and raise honour, this layer also needs strong areas for section endorsement cooperation (Pomroy et al. 2011). The magnitude of the created information has grown to be a major problem because of the development in the number of Internet-related devices (Raza 2013). As a result, obtaining vast amounts of information necessitates the use of valuable security frameworks that can consider these vast amounts of information in a helpful and appropriate manner (Konstantinou et al. 2015).

2.4 Types of Vulnerabilities in Cyber Physical Systems

It is necessary to assess a system's robustness about internal (such as human error) and external (such as power system design failure, software system design faults, and threats (e.g., adversary, environment, and other system threats). Cyber Physical Systems may get affected in three phases' development, maintenance and operation as shown below in Fig. 2.5.

Physiological vulnerabilities in CPS devices (Nagpal et al. 2022) are expanding into the industrial sector due to the provision of an Advanced Metering Infrastructure and Neighbourhood Area Networks along with data metering management devices to ensure the sturdiness of CPS in industrial domains.

In reality, the following three criteria could be used to distinguish physical threats.

Physical Disruption: The electricity grid, power stations, and ground stations are all completely protected since different infrastructure types call for different levels of security. These stations are well-equipped and safeguarded as a result of the implementation of access limitations, authorization, and authentication systems including usernames and access cards, passwords biometrics, and surveillance cameras. However, the main problem is associated with the less secured power-generating sub-stations since transmission lines are vulnerable to sabotage attacks and disruption. There are numerous concerns with smart metres.

In order to tackle this problem, monitoring systems must be challenging to meddle with and may rely on host-based vulnerability scanning or outage monitoring. It

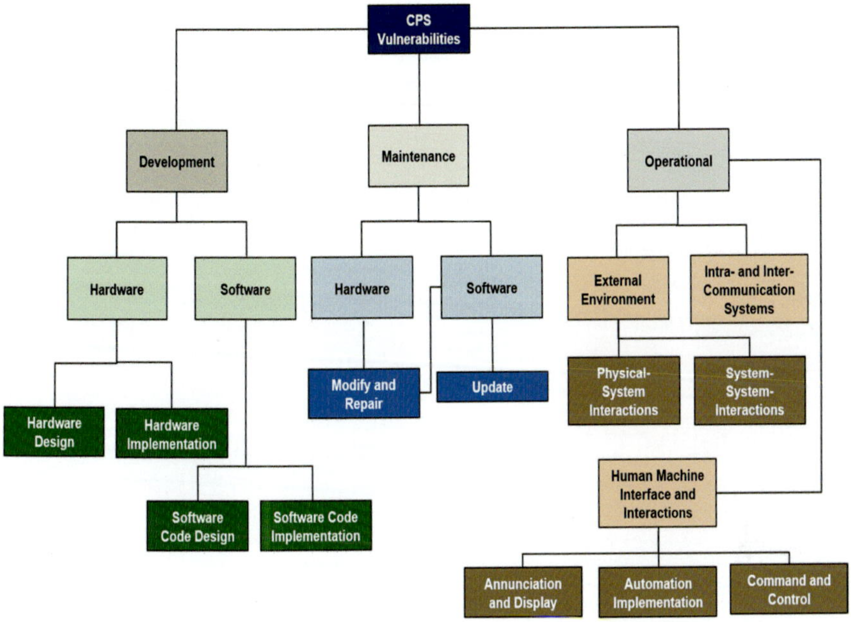

Fig. 2.5 Three levels of vulnerabilities in CPS- development, maintenance and operation

is nearly difficult to avoid physical manipulation or abduction when combating adversaries like Advanced Persistent Threats (APTs).

- *Reduction*: The situation that raises the most alarm is when a malicious attacker repeatedly fails sub-stations. Major urban areas may experience a total shutdown for several hours if the smart grid suffers serious damage. A real-life example is the cascading blackout that the Chinese political structure People Liberation Army (PLA) managed to bring upon the United States.
- *Repair*: It may be built around a self-healing mechanism that examines errors or interruptions, pinpoints the problem, and notifies the connected control system to automatically rebuild the backup resources to meet the demand for the service. The objective is to recover quickly in the lowest amount of time possible. For crucial components, there is, however, either no backup capacity or one that is just partially present. Self-healing can therefore respond to a severe injury more quickly.

2.4.1 Threats Associated with CPS Systems

There are some threats that are associated with CPS systems such as spooling, tracking and many more, as shown in Fig. 2.6.

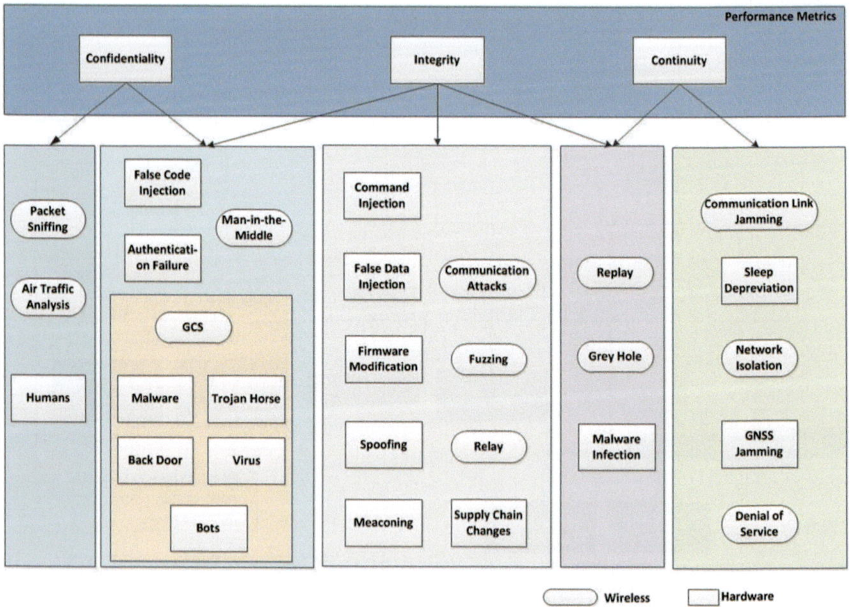

Fig. 2.6 Scope of CPS vulnerabilities and their classification of level phases

1. *Spoofing*: It entails a harmful, unidentified source disguising itself as a reputable entity. Attackers can spoof sensors in this situation, for instance, by providing incorrect or misleading measurements to the control centre.
2. *Sabotage*: Sabotage includes actions like diverting lawful communication traffic and sending it to a malicious party or tampering with the intercultural communication. An attacker might, for example, harm physically vulnerable CPS components dispersed throughout the power system to cause a technical glitch or even a failure of delivery. This can cause a whole or partly blackouts.
3. *Service denial or interruption*: Any device can be physically hacked by an attacker to alter the settings or disrupt a service. This has detrimental implications, particularly when applied to medicinal applications.
4. *Tracking*: Since devices may be physically accessed, an attacker can attach a malicious device, access them, or even track the safe ones. We list the primary CPS weaknesses that the attackers can exploit in the paragraphs that follow.
5. *Tunnelling and encryption (Internet protocol interoperability)*: Ground-anchored communication infrastructure, which are becoming increasingly prevalent, offer measurements of development that require ongoing construction to keep them safe from attack. For these approaches to function effectively, the indications being analysed by avionics systems must have trust in their accuracy, integrity, and availability (continuity). Attacks on networks and software-based firmware are two instances of hazards which could gravely impair upcoming systems.

CPS vulnerabilities are a security flaw that can be exploited for corporate espionage–reconnaissance or we can say active attacks. To discover and analyse the CPS flaws that are currently in place, as well as to determine the best corrective and preventative measures to lessen, alleviate, or even completely remove any vulnerabilities, a vulnerability assessment is necessary. The three major categories of CPS vulnerabilities are as follows:

1. *Network Vulnerabilities*: Unsecured wireless and wireless wired communication and connections are put at risk by man-in-the-middle, espionage, playback, sniffing, masquerading, and connectivity (routing level) attacks. Backdoors, DDoS/DoS, and protocol manipulating assaults are some additional dangers.
2. *Launch Pad (Platform) Flaws*: Vulnerabilities in configuration, System Components (Both hardware and software), and databases are all included (Sztipanovits et al. 2012).
3. *Management Constraints*: Inadequate security measures, protocols, and policies are among them. Numerous factors might lead to vulnerabilities.

2.4.2 Principal Proxies for Vulnerabilities

1. *Confidence and Alienation*: Its foundation is the common "security by obscurity" tendency in CPS architectures. To design a trustworthy and secure system, taking into account the implementation of necessary security services, without assuming that systems are isolated from the outside world, J.A. Yaacoub et al./ Microprocessors and Microsystems 77 (2020) 103,201 7 are focussed here.
2. *Increasing Connectivity*: The attack surfaces grow as connectivity increases. Manufacturers have enhanced CPS through the adoption and use of open networks and open wireless technologies as CPS systems have become more networked in recent years. Up until 2001, most ICS assaults were internal. This was before the use of the internet, which changed attacks to ones from the outside.
3. *Heterogeneity:* CPS applications are created by integrating a variety of third-party components into CPS platforms. Due to this, The CPS system currently has vendor support, and each product is susceptible to distinct security flaws.
4. *USB Utilisation*: Similar to the scenario with the Stuxnet assault that struck Iranian power facilities, the spyware being inside the USB is a significant contributor to CPS risks. When it was plugged in, the malware used replication and exploitation to spread to several devices.
5. *Bad Practise*: It is generally connected to poor coding or insufficient programming skills that caused the code to run indefinitely or become too simple to be altered by a specific attacker.
6. *Spying*: Most spying/surveillance assaults on CPS systems use spyware (malware) types that enter the system covertly and operate for years without being discovered in order to capture delicate or private data.

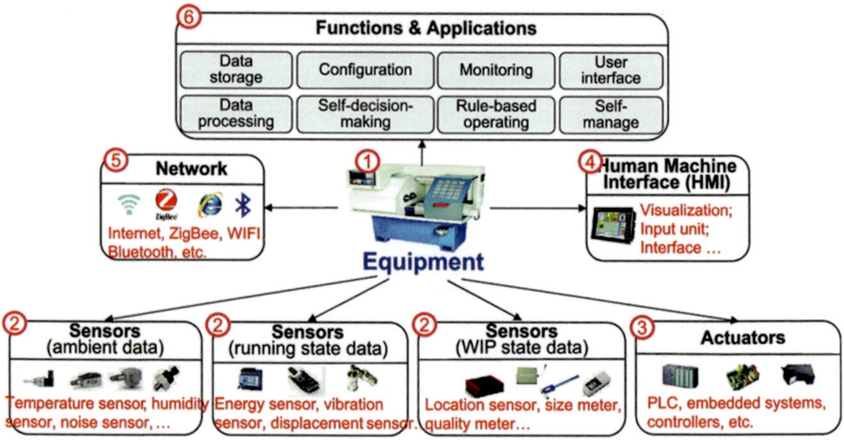

Fig. 2.7 Basic components of CPS prone to threats

7. *Assimilation*: Comparable malware systems have vulnerabilities that, if taken advantage of, might affect the entire surrounding infrastructure. A good example of this is the Stuxnet worm assault targeting Iranian nuclear power installations.
8. *Suspicious Employees*: By undermining and altering the code language, or by providing remote access to hackers by unlocking closed ports or inserting in an infected USB/device, it can purposefully or unintentionally damage or harm CPS equipment. As a result, there are three different kinds of CPS vulnerabilities, including cyber-physical risks (Fig. 2.7). The different activities responsible leading to threats are visualized in the diagram more understanding and analyzing.

2.5 Related Works

The literature work on CPS originates from the integration of physical processes, computational resources, and capabilities of communication; processing units monitor and control physical processes (Ghazani et al. 2012) using sensor and actuator networks. Examples of such systems are transportation networks, water and gas distribution systems, distribution networks, communication systems, control systems and power generation.

The infrastructure based on cyber physical systems (CPS) is one of the important critical structures based on industrial control systems for the last many years and accordingly, there are many cases of computer-based (cyber) attacks (Report: Cyber-Physical Systems Summit. 2008).

A control structure's main purposes are to keep operational goals safe by reducing the likelihood of undesirable behaviour, to meet production demands by maintaining specific process values within established limits, and, finally, to maximise and enhance production profit. Networked agents including sensors, actuators, control processing units like programmable logic controllers (PLCs), and communication devices make up the majority of control systems (Ashibani and Mahmoud 2017). The most significant cyberattack on industrial control systems was Stuxnet, which took place in 2010. It was a sizable piece of malware with numerous features that targeted Siemens industrial control systems and took advantage of four Windows operating system zero-day vulnerabilities (Gaddam et al. 2008). Due to zero-day vulnerabilities, Stuxnet is not only difficult to detect but also has significant implications (Mahmoud et al. 2015). With time, the Iranian nuclear infrastructure began engaging in cyberwarfare. Attacks cannot be stopped by basic antivirus software, but the problems were partially resolved by firms like Kaspersky.

In case of PLC controllers, the victims identify the changes in embedded controllers and code cannot be seen because Stuxnet hides its modifications with sophisticated PLC rootkits and validates the drivers with trusted certificates (Ghazani et al. 2012; Mahmoud et al. 2015).

People use their skillset and mind in cyberattack illegal activities by using vivid ideas to crack a cyber system and are full proof rather can prove advantageous for their nation or for themselves for gaining money. The physical attacks in Cyber physical systems are employed for blackmail or terrorism. Cyberattacks are usually inheritors to physical attacks because of cheap and risky to the attacker (Ashibani and Mahmoud 2017), and additionally, they are easy to replicate and can be coordinated well if at a distance.

2.6 Security Issues in Cyber Physical Systems (CPS)

A combination of societal, specific, and systematic deterrents limits CPS's options. CPS combines a significant number of diverse genuine items and materials with presented and dispersed frameworks that, when combined, should effectively play out the common positions in accordance with the show subtleties (Klesh et al. 2012). The lack of powerful language and expression that must exist to represent computerised genuine affiliation may be the most disturbing problem that such trade-offs face. However, there aren't any crucial first stages for a central affiliation point among structures, real objects, and people, which makes it more difficult for the entire mixture to be interchangeable (Aggarwal et al. 2022).

Human association with CPSs frequently encounters a fundamental barrier while analysing the human–machine collaborative efforts and producing genuine models that consider the present situational measures and natural modifications. These progressions are crucial to the cycle, especially in structures like flying power and military systems (Klesh et al. 2012). Additionally, in complicated CPSs where problematic behaviour should be dealt with promptly employing AI approaches, findings and exercises shouldn't be astonishing or dubious. However, the portions currently prepared for query distribution are currently irrelevant, and the problem is made worse by bad programming strategies, unstable associate connections, and flawed genuine articles (Gaba et al. 2022b).

Additionally, there are difficulties managing the interdependencies between programming and system planning, stresses with compositionality and disengagement for such structures, and difficulties staying aware of a comparable required degree of precision, unwavering quality, and execution of all system components. Security, assurance, and trust are essentially stressed in every cutting-edge development. Politically contentious difficulties include maintaining a CPS's security and constancy and protecting its own data from any usual control. There are security plans in place for a few CPS tiers, including establishment, people, safeguarded development, and items. Since there is a big gap between ensuring that an attack is computerised and real, it is attempting to develop a security system that can swiftly identify both (Sztipanovits et al. 2012).

2.7 Types of Challenges Faced in Cyber Physical Systems

2.7.1 Measures and Challenges in CPS

An examination of the various model kinds, layers, and essential components that make up cyber and physical security. When such attacks are made against any targeted physical or cybernetic system or device, as well as the related vulnerabilities of each such domain, cyber-physical attacks are taken into consideration and analysed. The criteria on which the security is judged are listed below. To estimate the risk and exposure levels for CPS to suggest security countermeasures, a qualitative risk assessment must be conducted (Zhang et al. 2016).

To extract evidence, security measures and their limitations, including the newest cryptographic and non-cryptographic techniques, must be analysed. Cyber forensics techniques are being researched to improve forensics investigations. Numerous life lessons are learned in order to protect authentic data/information communication across CPS devices with limited resources and to achieve CPS security objectives including secrecy, authenticity, reliability, and identification (Gaba et al. 2022a).

For a secure CPS environment, it is advised to minimise and mitigate all threats—cyber, physical, and hybrid–as well as difficulties and problems.

2.8 Conclusion

This chapter paves the way for future advancements in CPS technology. Applications of CPS are better and more versatile because of the improved security parameters. To determine the potential for improvement, the levels of the CPS Architecture were examined. The weaknesses, threats, and attacks related to CPS security are examined.

The serious problems and difficulties encountered are acknowledged. The current security measures are also discussed, and their primary limitations are identified.

References

Aggarwal A, Gaba S, Nagpal S, Arya A (2022) A deep analysis on the role of deep learning models using generative adversarial networks. In: Blockchain and deep learning. Springer, Cham, pp 179–197

Ashibani Y, Mahmoud QH (2017) Cyber physical systems security: analysis, challenges and solutions. Comput Secur 68:81–97

Gaba S, Budhiraja I, Makkar A, Garg D (2022b) Machine learning for detecting security attacks on blockchain using software defined networking. IEEE Int Conf Commun Workshops (ICC Workshops) 2022:260–264. https://doi.org/10.1109/ICCWorkshops53468.2022.9814656

Gaba S, Budhiraja I, Kumar V, Garg S, Kaddoum G, Hassan MM (2022a) A federated calibration scheme for convolutional neural networks: models, applications and challenges. Comp Commun

Gaddam N, Kumar GSA, Somani AK (2008) Securing physical processes againstcyber attacks in cyber-physical systems. In: Proceedings of the national workshop for research on high-confidence transportation cyber-physical systems: automotive. Aviation & Rail, Tyson's Corner, VA, USA, pp 1–3

Gao H, Peng Y, Jia K, Dai Z, Wang T (2013) The design of ICS testbed based onemulation, physical, and simulation (EPS-ICS testbed). In: 2013 Ninth International conference on intelligent information hiding and multimedia signal processing. IEEE, pp 420–423

Geng Y, Rong C-M, Veigner C, Wang J-T, Cheng H-B (2006) Identity-based keyagreement and encryption for wireless sensor networks. J China Univ Poststelecommun 13(4):54–60

Ghazani SHHN, Lotf JJ, Alguliev RM (2012) A study on QoS models for mobile ad-hoc networks. Int J Model Optim 2(5):634–636

Hu W, Oberg J, Barrientos J, Mu D, Kastner R (2013) Expanding gate level information flow tracking for multilevel security. IEEE Embed Syst Lett 5(2):25–28

Jing Q, Vasilakos AV, Wan J, Lu J, Qiu D (2014) Security of the internet of things: perspectives and challenges. Wirel Netw 20(8):2481–2501

Johansson KH (2014) Control of cyber-physical systems: fundamental challenges and applications to transportation networks. In: 27th International conference on architecture of computing systems, Lübeck Germany

Khan R, Khan SU, Zaheer R, Khan S (2012) Future internet: the internet ofthings architecture, possible applications and key challenges. In: 2012 10th International conference on frontiers of information technology. IEEE, pp 257–260

Klesh AT, Cutler JW, Atkins EM (2012) Cyber-physical challenges for space systems. In: 2012 IEEE/ACM third international conference on cyber-physical systems (ICCPS). Beijing, pp 45–52. https://doi.org/10.1109/ICCPS.2012.13

Konstantinou C, Maniatakos M, Saqib F, Hu S, Plusquellic J, Jin Y (2015) Cyber–physical systems: a security perspective. In: 2015 20th IEEE European TestSymposium (ETS). IEEE, pp 1–8

Kumar JS, Patel DR (2014) A survey on internet of things: security and privacyissues. Int J Comput Appl 90(11)

Lee EA (2008) Cyber physical systems: design challenges. In: 11th International symposium on object/component/service-oriented real-time distributed computing, Orlando, Florida, USA

Mahmoud R, Yousuf T, Aloul F, Zualkernan I (2015) Internet of things (IoT) security:current status, challenges and prospective measures. In: 2015 10th International conference for internet technology and secured transactions (ICITST). IEEE, pp 336–341

Nagpal S, Aggarwal A, Gaba S (2022) Privacy and security issues in vehicular Ad hoc networks with preventive mechanisms. In: Proceedings of International conference on intelligent cyber-physical systems. Springer, Singapore, pp 317–329

Pomroy SP, Lake RR, Dunn TA (2011) Data masking system and method. USPatent 7,974,942

Raza S (2013) Lightweight security solutions for the internet of things. Ph.D. thesis, Mälardale-nUniversity, Västerås, Sweden

Report: Cyber-Physical Systems Summit. 2008. [online]. https://iccps2012.cse.wustl.edu/_doc/CPS_Summit_Report.pdf

Saqib A, Anwar RW, Hussain OK, Ahmad M, Ngadi MA, Mohamad MM, Malki Z, Noraini C, Jnr BA, Nor R et al (2015) Cyber security for cyber physical systems: atrust-based approach. J Theor Appl Inf Technol 71(2):144–152

Sobhrajan P, Nikam SY (2014) Comparative study of abstraction in cyber physical system. Int J Comput Sci Inf Technol (IJCSIT) 5(1):466–469

Sommestad T, Ericsson GN, Nordlander J (2010) SCADA system cyber security-acomparison of standards. In: Power and energy society general meeting. IEEE, pp 1–8

Sridharan V (2012) Cyber security in power systems. Ph.D. thesis, Georgia Institute of Technology

Stankovic JA (2014) Research directions for the Internet of Things. IEEE IoT J 1(1):3–9

Sztipanovits J, Ying S, Cohen I, Corman D, Davis J, Khurana H, Mosterman PJ, Prasad V, Stormo L (2012) Strategic R&D opportunities for 21st century cyber-physical systems. Technical report for steering committee for foundation in innovation for cyber-physical systems, Chicago, IL, USA

Weiss J (2010) Protecting industrial control systems from electronic threats. Momentum Press

Wood AD, Stankovic JA (2008) Security of distributed, ubiquitous, and embeddedcomputing platforms. Wiley Handb Sci Technol Homel Secur 1

Wu M, Lu T-J, Ling F-Y, Sun J, Du H-Y (2010) Research on the architecture of internet of things. In: 2010 3rd International conference on advanced computer theory and engineering (ICACTE), vol 5. IEEE, pp V5–484

Yaacoub et al. 2020 Yaacoub JPA, Salman O, Noura HN, Kaaniche N, Chehab A, Malli M (2020) Cyber-physical systems security: limitations, issues and future trends. Microprocess Microsyst 77. https://doi.org/10.1016/j.micpro.2020.103201

Zhang H, Shu YC, Cheng P, Chen JM (2016) Privacy and performance trade-off in cyber-physical systems. IEEE Netw 30:62–66. https://doi.org/10.1109/MNET.2016.7437026

Zhao K, Ge L (2013) A survey on the internet of things security. In: 2013 Ninth international conference on computational intelligence and security. IEEE, pp 663–667

Chapter 3
Sensing and Communication Mechanisms for Advanced Robotics and Complex Cyber-Physical Systems

Kartik Singhal, Pritika Sabharwal, Deepak Kumar Sharma, Chandana Kuntala, Sristi, and Uttam Ghosh

3.1 Introduction

A robotic cyber-physical system (CPS) is a system that combines computational, physical, and communication elements to perform tasks using robots. These systems use advanced technologies like AI and machine learning to sense, learn, and adapt to their environment in real time, and are commonly used in manufacturing, transportation, and healthcare to automate tasks that are too complex or hazardous for humans to perform.

K. Singhal
Department of Manufacturing Process and Automation Engineering, Netaji Subhas University of Technology, (Formerly known as Netaji Subhas Institute of Technology), New Delhi, India
e-mail: kartiks.mp.17@nsit.net.in

P. Sabharwal
Department of Electronics and Communication Engineering, Netaji Subhas University of Technology, (Formerly known as Netaji Subhas Institute of Technology), New Delhi, India

D. K. Sharma (✉) · C. Kuntala
Department of Information Technology, Indira Gandhi Delhi Technical University for Women, New Delhi, India
e-mail: dk.sharma1982@yahoo.com

Sristi
Department of Computer Science and Engineering, Indira Gandhi Delhi Technical University for Women, New Delhi, India

U. Ghosh
Department of Computer Science & Data Science, Meharry Medical College Nashville, Nashville, TN, USA
e-mail: ghosh.uttam@ieee.org

N. Sharma et al. (eds.), *Big Data Analytics in Intelligent IoT and Cyber-Physical Systems*, Transactions on Computer Systems and Networks,
https://doi.org/10.1007/978-981-99-4518-4_3

Sensors in robotics are needed to estimate the local environment variables associated with the physical environment in which the robot is functioning. The signals generated by these sensors are passed onto a controller to enable suitable actions. Sensory units are employed to perform functions similar to sensory organs present in the human body (Svechtarova et al. 2016). Sensing enables robots and complex CPS with the artificial ability to see, hear, touch, and progress as it utilizes algorithms and statistical techniques to model this behavior with the processing part of the system.

In the real world, moving the robot instead of the workpiece, in the case of robotic manipulators, is often more convenient as well as effective when the aim is the compensation of relative positioning errors. The major challenges in such an application include the determination of the reference position and the presence of deformities or irregularities on the workpiece, which may develop in future. Thus, local sensing becomes important for online collision detection and obstacle avoidance. In another domain of CPS, that is wheeled mobile robots, where the base is not stationary and can traverse in a dynamic environment wherein the surroundings are imperfectly modeled virtually, reliable sensors are a necessity.

Considering the above reasons it should be emphasized that. Firstly, sensing is in general concerned with information related to the positioning, relative distancing, or the physical aspects of the mobile robot concerning its target surroundings (Sobh and Elleithy 2015). So, this usually involves measurements on a much lower level as compared to the advanced information processing deployed in visual recognition techniques and similar heavy measurement-driven functionalities. Secondly, to ensure that there are minimum blind spots for the deployed robot, sensory units should be used carefully, although using a large number of sensors may become a hindrance due to the increased utilization of physical space. Thus, the sensors' size itself is an important criterion for selection, especially in the case of manipulators or legged vehicles. Lastly, since the information extraction is on a lower level, the processing, and the response is expected to be more reflexive rather than reflective. As in the case of human response systems, rapid action is taken using nerves in case of sudden environmental changes, response for such sensors also needs to overcome model time constraints. Thus, it is imperative that these sensors themselves have a reasonably quick response time and that all the channels and communication mediums also need to be efficient, with the operating system allocating the required energy and memory to these local sensors and subsystems in CPS.

3.2 Physical Properties and Characteristics of Sensors

Sensor characteristics in CPS can be broadly classified under static and dynamic conditions (National Research Council 1995). Under static characteristics, the following definitions are necessary indicators of performance:

Accuracy of the sensor is defined by the extent sensor signal is correctly measured after it stabilizes, that is, reaches a transient period.

Resolution of a sensor is the smallest amount of unit change it can indicate accurately. For example, a proximity sensor can detect distance in increments of 0.1 cm. The measurement of 0.1 cm will be its resolution. The resolution may or may not be greater than the sensor's accuracy.

Sensitivity is an absolute quantity defined as the measure of the relative change of output to a unit change in its input. It is the smallest amount of change that can be detected by the sensor or causes a change in the sensor's output.

Drift is the deviation from a particular reading of the sensor when the sensor is maintained at the condition for an extended period. It could be also understood as the stability of the sensor to keep a steady level.

Range is simply the allowed extreme limits of the sensor's input or output measurements.

Precision defines how close together the readings obtained are. The values need not be true or accurate but simply how well the sensor can reproduce the measurements.

When the time response of the sensor is considered, it forms dynamic characteristics. Some common dynamic attributes of sensors are as follows.

Rise time is the time taken by the sensor to cover 10%–90% of its steady-state response.

Peak time is the time it requires for the sensor to reach its maximum output for the first time.

Steady-state error is the deviation from the desired value.

Settling Time is a certain period of time taken by a senor to stabilize around the steady state value.

3.3 Sensory Models and Techniques

Humans and animals can perceive their natural state and their environment with the help of either of their basic senses or in combinations of them. These senses have been developed throughout continuing evolution which has provided them with a multitude of highly advanced multi-sensory systems. Similarly, for any effective task, tool actions, and traversal, robotic systems, and complex CPS rely on unique devices called sensors or actuators. Sensors are instruments that respond to a physical stimulus and generate a response signal in the electrical form to be assessed by a computer. In the field of wheeled mobile robotics, sensors are utilized in mapping real-world environments to a digital rendition which is in turn utilized in developing real-time perception and motion of the system (Thrun 2001). This approach is in contrast to a more general methodology in which the environment is partly considered static, that is, provides a generalized view. Such a methodology can be inaccurate in actual implementations as the physical world tends to be more dynamic.

Sensors and sensory models can be classified under various categories. Few of them can be either passive or active sensors or techniques. This type of classification is based on how the sensory system interacts with the environment. If a sensor module emits energy to its environment and measures the response, it is considered active, for example, an ultrasonic sensor emits sound frequencies and then calculates the distance based on the time difference between the receptions of reflected frequency. A passive technique would be which only relies on the reception and does not generate the energy of its own.

As robotic CPS has advanced, so is the requirement for sensing and estimation modules (Christensen and Gregory. 2008). Thus today, these systems employ multitudes of sensor arrays of different kinds which help them in broadly two ways: proprioception, that is, sensing and estimation of its state within an environment, and exteroception, that is, sensing and estimation of environmental variables. For example, a wheel encoder can provide data on how much a vehicle system has traveled since the last known position which can be used to deduce the current state of the system. While a Lidar sensor, based on the illumination of the target, can provide a mapping of its immediate environment.

The following subsection briefs about several existing and in-research sensory models and techniques (Gajjar 2017; Busch-Vishniac 1999).

3.3.1 Proximity Sensors

Proximity sensors are sensors that do not rely on physical contact for the detection or relaying of information. Objects/obstacles are detected using robot sensors when they move and are present in the immediate environment.

The sensors convert the information to an electrical signal which is then relayed to the processing unit. Proximity sensors can be based on many physical properties or phenomena. For example, Hall effect sensors are based on magnetic fields. More specifically, conductors carrying currents perpendicular to magnetic fields develop voltages across them in directions perpendicular to the current and magnetic field, a phenomenon which is termed the "Hall Effect."

Inductive sensors are also based on magnetism-based effects. Induced currents can counter a change in the magnetic field and at frequencies above the self-damping time of the conductor can also reduce the effective inductance. Following this property, inductive sensors can also be termed metal detectors since all metals are mostly good conductors. Several other sensors such as ultrasonic sensors or infrared sensors are also used as proximity sensors. Any sensor that provides distance measurements can be used as a proximity or motion detector. Both infrared and ultrasonic sensors can be defined in a perpetual "on" state. For some sensor models, it is even possible to configure the frequency of emission.

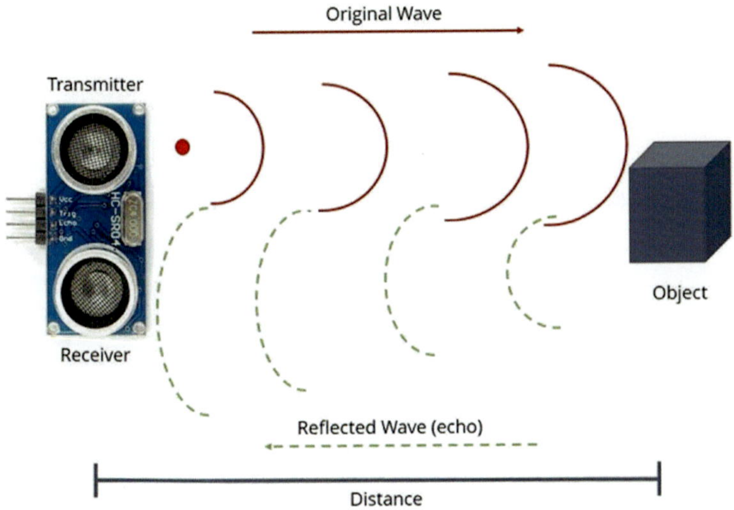

Fig. 3.1 Ultrasonic Sensor

Ultrasonic sensors are based on echolocation that is, utilizing sound waves for object detection. Ultrasonic sensors can be divided into two units, namely, a transmitter that emits waves and a receiver that receives the bounced-back energy. The distance between the object and the system can then be calculated with the knowledge of the time taken between the emission and reception of the signal and the speed of sound in the medium. Since they are based on acoustics, the efficiency of ultrasonic sensors is dependent upon the medium properties and environmental disturbances. For example, beam width can be affected by the sensor diameter and transmitted frequency can be varied. A common example of the application of ultrasonic sensors would be the use of ultrasonic sensors in parking assistance for vehicles. The rear bumper of a vehicle can be fitted with ultrasonic radars to calculate the distance between the vehicle's rear and the nearest object in its field (Fig. 3.1).

Infrared sensors can be used to detect heat as well as the movement of an object. They are either "measurement sensors," which can only detect infrared radiation or they can be dual unit setups like ultrasonic sensors, where they use an infrared (IR) light-emitting diode to produce near-infrared radiation and a photodiode that can detect IR of the same wavelength as the emitted radiation. Infrared radiation emitted is between 820 and 880 nm wavelength. A sensor transmits 100% amplitude-modulated light and measures the transmitted and reflected beams' phase shift (Fig. 3.2).

Further, several proximity sensors can work in underwater conditions or detect approximate fluid levels as well. For this purpose, *capacitive sensors* can be utilized, given the fluid is a dielectric. Fluid flows in between the capacitor plates which causes fluctuation in capacitance. It is possible because the capacitance of a parallel plate capacitor is linearly proportional to the permittivity of material between the two plates.

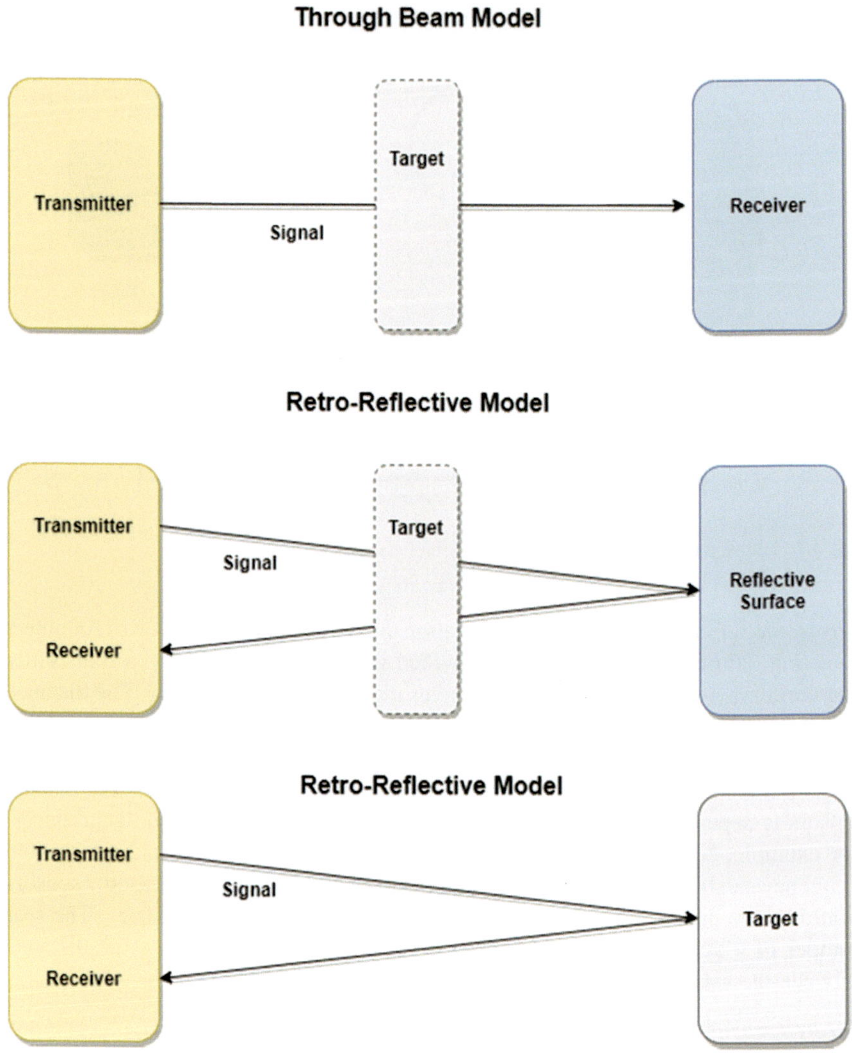

Fig. 3.2 Type of sensing models for detecting the distance between the sensory unit and a target

Proximity sensors or a sensor array can also be utilized in distinguishing the shapes and sizes of obstacles. In the case of robotic manipulators, proximity sensors can be used for collision detection between the robot structure and its environment. Especially in industries and manufacturing units, where a robotic manipulator has to interact with various distinct objects of varying size and shape, they are required to have information added to their programmed trajectories. This is achieved by obstacle detection capabilities that provide alternative trajectories to the predefined path. Additionally, sensors are required to restrict robot links within safe movements

Table 3.1 Different types of technologies for proximity sensors and their applications

Technology	Application	Sensing Range for distance
Inductive sensors	Detection of metal parts, assembly lines	<3 mm–40 mm
Ultrasonic sensors	Distance, depth, level measure, and object detection	<30 mm–3 m
Infrared sensors	Night Vision, infrared tracking, imaging devices	<4 m–5 m
Capacitive sensors	Inspection, measurement of levels of fluid	5 mm–40 mm
Photoelectric sensors	Target detection, distance measure	<1 mm–70 mm

about their joints to not let the links collide with the robot's base. Thus, it can be said that for many tasks, robots require multitudes of sensors working together to achieve their goal. The integration of multi-sensory systems will be discussed in a later part of this chapter. Table 3.1 provides a brief on different proximity sensors and their advantages.

3.3.2 *Contact Sensors*

In the previous section, we discussed sensors that do not rely on a physical contact to function; this subsection will explore the applications of contact-based sensors. These sensors are more reliant on the electro-mechanical functions and changes of the system or its environment for detection. Some common classification of contact sensors is bumper and tactile sensors. Tactile sensors are the more complex of the two, as bumper sensors can only provide information about *"if a collision or touch has happened or not?"* However tactile sensors can be used to gauge the pressure, movement, or even the direction of change.

Tactile sensors respond to physical contact forces such as touch, slip, push or pull, etc. Tactile sensing can be as simple as a bump sensor utilized in hobby kits to detect collision or utilization of a coordinated touch sensor array which provides information about the intensity and direction of the force. Simply, tactile sensors do not exactly measure the forces or torques acting upon them but rather provide information about the touch between the sensor and the contact surface. Sense of touch is as important as a sense of vision in many industrial environments. A sense of touch helps increase accuracy whereas camera systems cannot detect fine clearances very accurately. Thus, sensory feedback is required to enable complex robotic processes such as polishing, welding, or pick & place. These sensors can be accompanied by force-torque sensors which are usually used as end effectors of robotic manipulators to check the application of force from a robot (Fig. 3.3).

Besides the choice of core sensor, the kind of actuator also defines the particular tactile sensations generated and accounted for in a system. To date, a large number of tactile sensing systems have been designed but none of them has been able to fully

Fig. 3.3 Structure of Tactile sensor

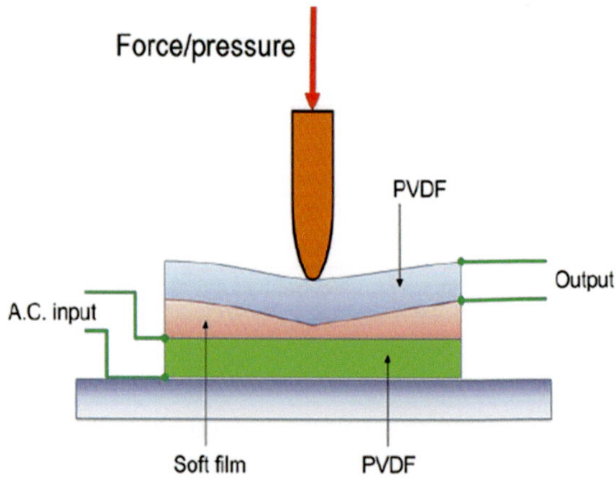

exploit the tactile sensing implications in their entirety. The sensors must have low hysteresis and need not be linear.

In general, these sensors are based on a resistive, piezoelectric, and capacitive or strain gauge mechanism. A strain gauge consists of a thin conductor and metal foil arranged in a specific pattern. On the application of mechanical stress, the conductor deforms and by calculating the change in electrical resistance the deformation can be quantified into a measurable force. Such a feature makes these sensors highly sensitive to touch, but due to the presence of such an intricate mechanical structure, these sensors are also susceptible to overloading and damage. In piezoelectric sensors, the application of force is used to produce electrical energy proportional to the deformation. Piezoelectric sensors are frequently used in dynamic sensing owing to their good high-frequency response. These sensors are based on various materials such as crystals, polyvinylidene fluoride, ceramics, etc.

Further, a tactile sensory system can be sensitive to static or dynamic forces and can be used for both proprioceptive as well as exteroceptive sensors. As proprioceptive sensors, they can be used in manipulators to compute joint torque and limb positions. As exteroceptive sensors, they can be used to determine the contact surface and pressure when in contact with an object of interest.

3.3.3 *Soft Sensors*

Soft sensors are virtual sensors or software which is utilized for estimating processes that are hard to measure otherwise. Soft sensors explore multiple new challenges, which are often highly complex to model accurately. For instance, an ideal soft sensor must be capable of providing information with high omni-directional compliance. Just like our skin, which works as a continuous network of a multifaceted sensory system, artificial sensors are required to perceive multiple physical parameters.

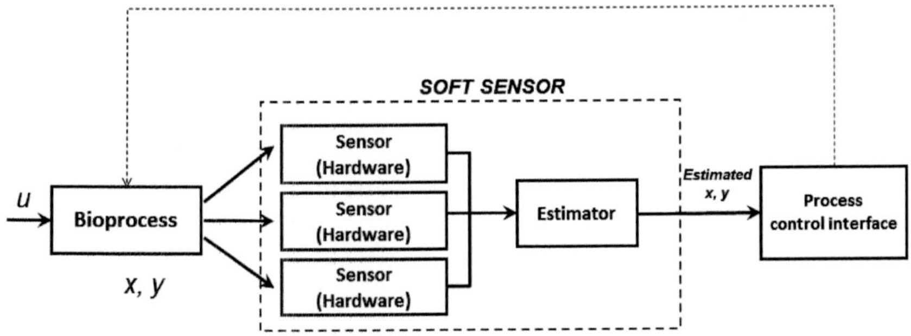

Fig. 3.4 Basic principle of soft sensor

An impending obstacle in modeling soft sensors is their accompanying dynamic system, which exhibits a highly non-linear time-variant behavior. Further, the sensors' own influence should not interfere with the dynamics of the system. Due to their continuous and omnidirectional behavior, the information collected by these sensors could provide singularities, i.e., a non-unique solution set. Thus it can be said that their mathematical modeling and analysis is a highly complicated field, but the potential of soft sensors in revolutionizing biotic and industrial automation cannot be overlooked (Fig. 3.4).

One major area of their application is explored in robotic manipulators. For fault detection purposes in the mechanism, several redundant sensor subsystems are incorporated. The outputs from each subsystem are compared for consistency; this process is termed hardware redundancy. The other approach is of analytical redundancy which utilizes sensory data for estimation along with a quantitative or a qualitative representation of the physical system. The sensory value estimates are then compared with actual measurements to process a residual value. If the residual exceeds a predefined threshold value, a fault is registered. Overall, the process of fault recognition can be divided into three parts (Vemuri et al. 1998):

1. Detection of a malfunction,
2. Diagnosis of the problem and isolating the fault,
3. Reconfiguration to self-correct the system or the subsystem underdiagnosis.

Although there has been advancement in the evolution of soft sensors, there remains a considerable adeptness gap in soft robotic systems to work in hostile environments and provide accurately interpretable data, which restricts increased exploration into their application in control, manipulation, prosthetics, etc. Nevertheless, the integration of soft computing techniques such as fuzzy logic modeling of soft sensors and neural networks becomes an improved task. With the use of such techniques, it has become possible to develop a learning architecture that may work as online approximates for analyzing the system behavior, thus providing a diagnosis scheme to layout system failures and accommodate reconfiguration control.

In (Lunni et al. 2018), Dario Lunni et al. introduced an approach based on Kalman filters to estimate the shape of a soft robotic arm. Soft robots have flexible links or joints, thus they require sensors that are stretchable and work omni-directionally along their body. This poses a challenge to accurately measure their configurations without interfering with their dynamics. Several methodologies including computer vision and image processing have been studied in past but these approaches require external sensors and mathematical models to properly work. In their study, Dario et al. utilized an Adaptive Extended Kalman Filter (AKEF) with an embedded plastic optical fiber curvature sensor. The sensory feedback is collected through accelerometers and is quantified against AKEF which provides a better accurate estimation model than using model prediction alone.

3.3.4 Holographic Techniques

To handle objects in a workspace, a robot needs to first recognize the object in its three-dimensional space. This can be achieved by imparting a projection of the object in the "brain" of the robot, which details the location of the object relative to the robot itself. For this purpose, TV cameras, Microsoft kinetic, and similar types of equipment can be used as detectors. These devices either provide a 2-D projection of the 3-D objects or the 3-D data points can be converted to 2-D representations, as implemented in several SLAM applications (Jüptner 1988). Figure 3.2 shows the difference between image projection for a human and a robotic operator (Fig. 3.5).

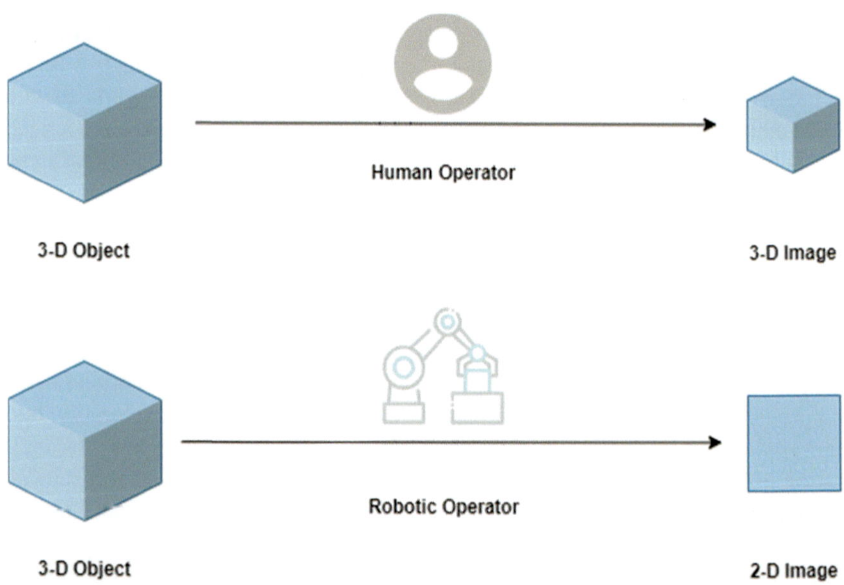

Fig. 3.5 The difference between human imaging and robotics imaging based on holography

Methodologies of this type involve collecting data points. A few optical procedures, such as multidirectional viewing or light intersection, can be used to evaluate the third dimension. However, for even one object, dozens of frames will have to be processed. Thus, the information stored for pre-processing may very soon exceed some million bytes. This can be a lot of memory used for a system with continuous applications. In order to overcome these problems, holographic techniques can be applied. Holography involves the two-dimensional coding of three-dimensional information. Through holography, the image of a three-dimensional object is stored with all its depth information in a plane of two dimensions. For this reason, holography is one of the best methods for converting 3D to 2D.

Holography, whether optical or acoustic, revolves around reconstructing wavefronts based on their phase and amplitude distribution across a defined surface. Two steps are involved in the process: recording and reconstruction. As a result of the collection of data, holograms are formed. Recognizing objects is done after the reconstruction of their surfaces. To achieve this, the reconstructed wave field takes on higher magnitudes at all locations where a sound source or reflector is present, whereas, at other points, it has a lower magnitude. So, a suitable thresholding method isolates contours within the reconstruction region from the rest of the image.

A number of studies have used holography to achieve localization and control of mobile robotic CPS. However, an issue with holographic interferometry involves the interpretation and recording of fringe patterns caused when a wavefront reconstructed from a hologram is caused to interfere with a wavefront generated earlier.

3.3.5 *Photoacoustic Imaging*

In many fields, especially medical robotics, optical or acoustic techniques alone cannot sustain in hard-to-reach applications. For example, in the case of medical investigations of tissues, conventional optical imaging is obstructed by scattering which restricts the resolution beyond an optical diffusion limit, which is usually around 1–2 mm. On a similar note, in surgical settings, the clutter and acoustic reflection from the presence of metals complicate the pose estimation of surgical tools. In both of these cases, augmenting optical technique with acoustics and vice versa increases their productivity.

Photoacoustic imaging offers an improved alternative to standalone visualizing techniques.

In the context of CPS, photoacoustic imaging can be a useful tool for providing real-time, high-resolution visual information about the internal structure and function. These can be utilized in guiding robotic tools in surgery or in the navigation and localization of nanobots in invasive operations. The working of optical imaging is based on the scattering and absorption of photons. In biomedical applications, a short pulse laser is used to illuminate a tissue, which is then allowed a thermoelastic expansion due to induced heat. This causes a shift in pressure inside the tissues and the tissue emits ultrasound waves. These wave signals are amplified and processed for data acquisition. As mentioned earlier, the penetration power of optical imaging is limited, thus utilizing the acoustic signal is used to compensate for the same. Further, this provides a method to check for system mismatches on the bases of data from both sources.

Photoacoustic imaging does not suffer from the drawbacks seen in both acoustic and optic-based techniques. This allows photoacoustic imaging techniques to produce high-quality images with increased depth (Attia et al. 2019). Effects of disturbances from an acoustic cluster are present in the usage of this technique. It can be observed that disturbances cause serious degradation of raw signals. A photoacoustic imaging system has two nodes, a signal generator, and a receiver. The generator emits pulse radiation which is designed to provide a good sensor-to-noise ratio in contrast to a continuous wave. The detector receives the signal that was generated by the sample under observation. There are several factors that decide the sensitivity of photoacoustic imaging.

Controlling and guiding micro-robots is a difficult task, especially in a translucent environment such as the human body. Moreover, medical practices require real-time visualization for tracking and detecting the motion of robots with sizes on the scale of nanometers. Thus, in a lot of biomedical practices, traditional detection methods can fail to control nanobots. By utilizing photoacoustic imagery, this limitation is overcome. And since the photoacoustic imaging technique is based on absorption, metallic objects such as nanobots are suitable candidates to be observed with high-resolution imagery. Yan Yan et al. (Yan et al. 2020) presented a noninvasive approach for tracking microbots in a nontransparent environment using an integrated ultrasound and photoacoustic imaging system.

3.4 Sensor Data Fusion Techniques

For simple tasks, a single sensor or a single type of sensory feedback might be enough to successfully execute a task. For example, for a robot gripper designed to sort objects based on color, a single RGB color detector unit is required, but to localize the gripper around the object will require multiple sensors to work in coordination to update several parameters, including but not limited to position from object, gripper orientation, joint localization, etc. As can be observed, these sensors are recording varying parameters working under different constraints. Thus, cooperation between multi-sensory modules becomes imperative for improved or advanced applications (Kam et al. 1997).

Thus, in such cases, a multisensory system is significantly more useful as it provides increased reliable information and is rich in data. Since robots operate in highly uncertain environments substantial coherent information is required for effective task executions. The goal of fusion techniques is to draw consistency within the streams of data (which may or may not be homogeneous in nature) generated and provide the necessary information with utmost accuracy and precision (Yager 2004). Further, these techniques should also be able to address disturbances and noises present with sensors. In this section, we highlight a few techniques that are commonly used in data fusion in many areas of robotics.

3.4.1 Kalman Filter

The Kalman Filter is an estimation technique that provides estimates of variables under observation on the basis of measurements processed within a time frame (Pei et al. 2017). In application, they are used to define the state of a model varying with time. This is done in three basic steps:

1. Utilizing the knowledge of kinematics and dynamics of the model and information about the previously known state, say position, to predict its evolved state over time,
2. Measuring/Processing the sensor readings to get the data's actual state at the given time and then comparing it against the predicted state,
3. Updating the state knowledge on the basis of prediction and actual measurements.

In robotics, Kalman Filter is utilized in localization, tracking, and navigation problems. The algorithm allows the incorporation of various sensor models and state variables like location, velocity, or acceleration. Further, the Kalman filter approach has been proven to be one of the most powerful approaches to solving estimation and sensor fusion problems even with low computational costs. The estimation of state x can be defined by the following state-space model:

$$x(t+1) = \Phi(t).X(t) + G(t).u(t) + w(t) \tag{3.1}$$

$$z(t) = H(t).x(t) + v(t) \tag{3.2}$$

Here, $\Phi(t)$ is the state transition matrix, $G(t)$ is the input transition matrix and $u(t)$ is an input vector. While $z(t)$ is the measurement at time t of the parameter of interest X. $H(t)$ is a measurement matrix, and w *and* v *are* Gaussian variables of covariance matrices $Q(t)$ and $R(t)$**.** The estimation $x(t)$ is is provided by

$$x(t) = x(t|t+1) + K(t).(Z(t) - H(t).x(t|t-1)) \tag{3.3}$$

$$x(t+1|t) = \Phi(t).x(t|t) + G(t).u(t) \tag{3.4}$$

$$K(t) = P(t|t-1).H^T(t).\left[H(t).P(t|t-1).H^T(t) + R(t)\right]^{-1} \tag{3.5}$$

where K is the filter gain.

$$P(t+1|t) = \Phi(t).P(t).\Phi^T(t) + Q(t) \tag{3.6}$$

$$P(t) = P(t|t-1) - K(t).H(t), P(t|t-1) \tag{3.7}$$

In Kalman filter-based estimation, the error is measured as Gaussian noise, and the system is described as a linear model varying with time. Since it is an iterative process, with each new predicted state the covariance vector concerning errors is predicted with each new sensor reading. Kalman Filters are very commonly used in autonomous robotics and vehicular systems as their features make them reliable for estimation and sensor data fusion in cases where the intended process and noise can be modeled as Gaussian or symmetric sequences (Tsihrintzis et al. 2016). For their application to non-linear processes, a modified version, known as the extended Kalman filter, is used (Lefebvre et al. 2001).

3.4.2 Bayesian Networks

Bayesian theory is based on the works of Thomas Bayes. Since then it has been implemented both independently and in conjunction in various fields (Bayes' 2003). In robotics, it is widely used for target recognition, occupancy grids, mapping, etc. (Thrun 2002). Bayes Theorem deals with conditional probability, that is it is related to the conditional probability of events A and B. Though Bayes theorem can easily be utilized for a more complex number of events, the Bayes theorem for two events A and B can be stated as

$$P(A|B) = \frac{P(B|A).P(A)}{P(B)} \tag{3.8}$$

where $P(A|B)$ is the conditional probability of event A given B, $P(B|A)$ is the conditional probability of B given A, and finally $P(A)$ indicates the prior probability of A and $P(B)$ is the prior probability of B.

In data fusion techniques, Bayes theorem forms the basis of statistical inferences that is deriving information about the subject state based on the observation. The relation between the state X and the observation Z can be modeled as events A and B described in the above equation. If the state information is independently available before observation, the probability density function $P(Z|X)$ can be improved to accurately model results.

$$P(X|Z) = \frac{P(Z|X).P(X)}{P(Z)} \tag{3.9}$$

Maximum a posteriori is a method by which the approximation can be improved by maximizing the posterior distribution or the numerator part of Eq. 1.9 and can be formulated as:

$$X_{MAP} = argmax(P(X|Z) \propto P(Z|X)P(X)) \tag{3.10}$$

In a minimum mean square error estimator, the sum of squared errors is minimized, i.e., the distance between the true state and estimated state is minimized. For 2 sensors it can be written as

$$P(X|Z1, Z2) = \frac{P(Z1|X)P(Z2|X)P(X)}{P(Z1)P(Z2)} \tag{3.11}$$

In (Vechet and Jiri 2010), Bayesian network is used to correctly position a robot relative to its previous orientation. Three readings including a compass, Odometry reading, and orientation of the robot using a steering angle are used to describe the position. Each mechanism's accuracy is calculated experimentally. It is observed that the compass provides measures with an equal probability of being accurate in all directions. P (Compass = 0) = 0.95. But in the case of Odometry and steering, the probability varied with direction as P (Odometry = 0) = 0.95, P (Odometry = 90) = 0.5, and P (Odometry = 180) = 0.3. Thus, the influence of Odometry values on decision-making is comparatively less than that of readings from the compass.

Despite being one of the oldest standing theories, Bayes's theorem in application suffers from a few drawbacks. Firstly, it is based on *priori* information for each occurrence as seen above. Secondly, it falters in case of gaps in data or missing data. Most importantly, as it is based on probabilistic measures the method is susceptible to flaws incurred due to uncertainties and difficult-to-model environments.

3.4.3 Dempster–Shafer Theory

Dempster–Shafer theory is most frequently accepted as one of the most famous methods of data fusion (Castanedo 2013). This is attributed to the fact that it has an advantage over other methods in terms of that it can represent ignorance and does not require prior information about the mechanism. Both drawbacks of the Bayesian networks. Dempster–Shafer theory is based on the works of Dempster with Bayesian probabilities with subjective evidence and its extension by Shafer in the theory of everything.

This method can integrate with any type of numerical and multidimensional data. In the presence of rich information, methods like Bayesian networks are comprehensible but in cases where model uncertainties and sensor defects are present conflicting results may be obtained. The advantage of Dempster–Shafer theory lies in the fact that not only does it work in models with information defects but also it does not assume data. The method is based solely on basis obtained data. Evidence of a mechanism is represented as a Shafer belief function. Before gathering information the total belief of the world is taken as ignorance. As more evidence is collected, it gradually replaces ignorance. Using Dempster–Shafer theory, evidence from sensor observations and implicit domain knowledge can be used directly in the decision-making stage.

In this approach, weight is assigned to each information source. Then the standard deviation of data for a set of data and data validation from each sensor is determined.

If the standard deviation is smaller than a specified threshold, say alpha, the sensor is termed as reliable and if the standard deviation is more than the reliance on the sensor is reduced.

$$\begin{aligned} \sigma^2 &= \tfrac{1}{N}\sum_{i-1}^{N}(x_i - \mu)^2 \\ \sigma^2 &\leq alpha - High\ Reliability \\ \sigma^2 &\geq alpha - Low\ Reliability \end{aligned} \tag{3.12}$$

Here, N is the number of the data source. With each new dataset, the variance is updated. Dempster–Shafer theory is also referred to as evidential reasoning as in contrast to probabilistic methods it not only places belief on elements and sets but also on sets of sets. So while the discipline of probabilistic methods for a universal set X is all possible subsets X, the discipline of evidential reasoning is power set 2^X

For a detailed study, refer to Kim et al. (2002), in which Young-Chula Kim et al. have explored the Dempster–Shafer approach in map building for an autonomous vehicle's environment. Zou et al. in Yi et al. (2000) have also provided a comprehensive report on the utilization of Dempster–Shafer theory in robotics. It is to be noted that, while Dempster–Shafer theory does not need us to state prior probabilities, it does need some initialization of preliminary masses that reflect our understanding of the concerned system.

3.5 Communication in Robots and Cyber-Physical Systems

Robots communicate utilizing different modules and units which allows them to transfer and receive video, audio, or data signals (Gupta et al. 2008). The communication system deployed can be either wired or wireless and often makes use of a common domain language to transmit and receive commands or orders. With advancements in radio modules and wireless protocols, wired communication is being increasingly replaced although is not entirely absolute. The usage of set protocols has become important for successful communication and the performance of any task.

A communication protocol describes a set of rules to establish information exchange between the devices in any CPS. A variety of protocols are used to achieve communication with robotic CPS. Some of the most popular ones are listed below:

1. Ethernet/IP: It is based on the standard TCP/IP and communication is carried forth using an existing network setup. TCP and UDP ports are used along with an Ethernet physical layer. It ensures high-speed Internet and connectivity for remote control.
2. Control Net: The system uses an RG-6 coaxial cable and a single bus. In addition to 5Mbit/s speed, it supports data upload/download, peer-to-peer communication, and up to 99 nodes.

3. Device Net: The data link layer is based on the CAN (Controller Area Network) bus framework. Two twisted pair cables link a host processor, a controller, and a transceiver on a CAN bus. The speed goes from 1 Mbit/s at 40 m to 20 Kbit/s at 1200 m.
4. Profibus: In a bit-serial field, master–slave, slave–slave, and master–master communication are used. Twisted pairs or optical fiber cables can provide 9.6 Kbits/s–12 Mbits/s (maximum 1200 m) in the physical layer.

3.5.1 Human–Robot Communication

The usefulness of any form of communication system deployed between humans and robots depends on the ability to communicate with parties to understand each other. This is particularly important as the robot is expected to take actions upon the instructions of a human and these instructions need to be conveyed adequately so that the robot can rightly interpret these and take actions accordingly. In HRC (human–robot communication), the basic problems which arise include (Mavridis 2015):

1. The purpose of Communication: What is the type of information which needs to be transmitted? And what is the need to transfer it? That is, what is the purpose being served? It is equally important to consider the level of abstraction for the transmission of this information from human to robot.
2. The medium of Communication: How are we aiming to transfer this information? There could be many possible ways depending on the subsystems deployed and the need for the transfer of information. For example, verbally, using gestures, or by explicitly stating.
3. The direction of Communication: Whether we need to only establish a one-way flow of information from humans to robots or the flow must occur both ways between the concerned parties.

To establish collaborative communication, a peer-to-peer interaction style is adopted. Typically, robots maintain two models: a decision model and a perception model, which are implemented probabilistically to derive a two-way information exchange. *Probabilistic data determines what to communicate, and internal inference mechanisms determine when to communicate.*

3.5.2 Application of Augmented Reality

Manual teleoperation has been the major methodology used for conveying information or commands for the longest time, however, due to technological constraints, this method presents many limitations (Green et al. 2008; Milgram et al. 1993). For example, in the master–slave mode, manual controllers are often kept occupied in the control loop, which can be a hindrance when the operations are complex or rather exceedingly long. In contrast to these, we have the concept of autonomous robots, which are particularly useful in manufacturing environments where the environments are well known and often stable and the tasks are repetitive. Considering the present scenario use of autonomous robots in fields that require flexible manipulation is still not explored in detail. It becomes essential to feed in a lot of prior detailed information and task-related instructions into the autonomous robots for them to be able to function effectively and smoothly.

University of Colorado Boulder roboticists investigated how utilizing augmented reality to assist robots connect with people might make the bots feel safer, more productive, and more like part of a collaborative team. The major aim of Augmented Reality (AR) is to convey "motion intent" about what the robot is going to do shortly. AR lets the human operator see in real life the motion of the robot and their surrounding environment without requiring any translation from the virtual world or vice versa (Coworkers xxxx). The researchers in this study carried out a series of experiments to check the quality of communication:

1. NavPoints: It is used to represent a robot's trajectory using a sequence of x lines and waypoints. The lines join the path from the current location to the future destination. The robot's location in 3D space is visualized using spheres each destination sphere also casts a drop shadow on the ground underneath it. Two radial timers are used, the inner one keeps track of arrival time and the outer one denotes the departure time of the robot
2. Arrow: A blue arrow travels across 3D space to reach a spot that the robot will require x seconds to achieve. The path to the robot is guided by the line at the tail of the arrow. Users may observe the path the arrow has followed, which the robot will likely follow, by using this line.
3. Gaze: It generates virtual images by superimposing an x-meter-diameter white sphere directly on the aerial robot, converting it from a multirotor to a flying eye. The eye model looks at its current goal until it reaches a predefined distance limit of y meters between itself and the existing destination, at which time it begins pointing to the robot's next target place.

AR is a technology that allows computer imagery to be integrated with real-world applications or challenges. It differs from virtual reality in the sense that, in a virtual environment, we create the entire local environment using computer graphics whereas AR enhances instead of exactly replacing. In the context of CPS, AR can be used to provide real-time, context-specific information to users, enabling them to interact with and understand their environment more effectively. AR can be a powerful tool for enhancing the functionality and usability of CPSs, and it has the potential to revolutionize a wide range of industries and applications. AR-based systems possess three main characteristics:

1. They are involved in the combination of virtual and real objects.
2. The virtual items appear as real-world ones.
3. It is possible to interact with virtual objects in real time.

The usage of AR to facilitate human–robot interaction is highly efficient because of the following reasons:

1. AR systems provide the ability to enhance reality.
2. They are enabled to provide a seamless interaction between the real and virtual environments.
3. It provides a method to share remote reviews.
4. It helps to visualize robots in real-world task environments and program them accordingly.
5. It provides spatial cues for both local and distant cooperation.
6. It provides support for transitional interfaces, allowing for a seamless shift from reality to virtuality.
7. It allows for the use of a tangible interface metaphor.
8. It offers resources for improved teamwork, particularly when several people are working with a robot.

The above-mentioned features let AR facilitate natural dialogue by using visual cues essential for communication as well as the maintenance of situational awareness. AR provides the use of spatial dialogues; dielectric gestures along with adaptable autonomy by providing assistance to numerous users and allows humans to communicate virtually using graph overlays in the real world. The application of AR enables a tangible UI, where material objects are controlled to influence modifications in the 3D scene (Billinghurst et al. 2010).

3.5.3 AR in Collaborative Applications

AR may be used actively to improve face-to-face recognition. A shared space project, for example, coupled the usage of AR with physical and spatial UI inside a face-to-face collaboration setting. In this application, users were given an HDM, head-mounted display, alongside a camera fixed to the top. The output was given to a computer and fed back into the HDM, enabling users to view the world through this video. It is called the video see-through AR interface. Several labeled cards with square fiducial patterns were placed in the real world, which were identified using computer vision techniques. The exact orientation and camera position were also calculated and 3D virtual images were then displayed. Thus physical markers were exploited for interaction with virtual content.

AR toolkit software allowed the users to communicate in a 3D AR environment, by enabling them to interact by tracking physical markers. Combining the results of the physical markers in animation. The shared space system was tested at the SIGGRAPH99 Emerging Technologies exhibit, with all participants, over 3000, using the application collaboratively. Thus, face-to-face communication is facilitated through the shared space interface enabling the participants to view the body responses, reactions, and facial expressions of the others, enabling them to work together.

The MagicBook is another example of how AR may be used, which allows the transition to virtual environments from the physical world. This consists of an actual book which also allows AR content popping out of it to be viewed by the user. AR placement and setup are achieved using a computer vision library for the AR toolkit.

3.5.4 Verbal and Non-verbal Interactive Communication

The earliest robots equipped with verbal interactive properties date back to the 1990s. Conversational robots like MAIA (Antoniol et al. 1993), RHINO (Burgard et al. 1998), and AESOP (Versweyveld 1998) appeared around that time and were designed for multiple purposes ranging from delivering objects to acting as museum guides. These robots could perceive and act upon simple signals like hand waving and were pre-programmed with fixed verbal descriptions. Despite providing basic verbal communications and tools for replying to standard commands, these subsystems could not act flexibly and presented many limitations.

Limitations raised in Verbal Interactive Communication:

1. Accepted and acted on a set of fixed pre-programmed commands with pre-decided answers.
2. Only one-way communication was supported, the robots could only act upon commands.
3. Inability to handle effective speech, not responsive to emotional speech, and cannot produce emotion-carrying prosody.
4. Hand gestures, stride, and facial expressions are not observed, hence non-verbal communication is essentially non-existent.
5. There is no genuine speech planning and purposeful dialogue generation.

What do we desire from a conversational Robot?

To achieve effective communication, either verbally or non-verbally, we begin by stating the basic subtasks we must achieve and explaining them in detail.

1. Breaking the simple commands.
2. Carrying forward mixed speech acts.
3. Effective usage of mixed-initiative dialogues.
4. Affective interaction.

Breaking the simple commands: The traditional human–robot interaction has been based on the master–slave model which continues to restrict the robot's conversational abilities to merely motor command request acceptance and execution. What this means is that the robot is merely following the instructions and the onus of carrying out communication is dependent only on the human. The robot produces motor responses to simple instructions given to it. In this paradigm, even the designer pre-decides the word mapping, and changes in the orientation of sentences can lead to confused reactions. As an example:

Carrying forward mixed speech acts: Austin's Simple Act Theory (Austin 1962) defines a speech act as an utterance with a performative purpose in language and communication. The emphasis is on the purpose and function of content rather than its form. Apart from giving simple motor requests, one might also need the robot to respond to other detailed queries. Assertive commands, as well as directive commands, need to be understood and we also need to contemplate the kind of response we want our system to generate for such types of commands. Such statements need to be framed keeping in mind the "mental" or "situational" model that the robot has been programmed for. Instigating mental or internal tokens is the aim of this communication. Another issue with the speech acts is their classification by the robot after hearing them. Conventional methodologies rely on linguistic information primarily; however, we also need to work on prosodic affect identification for speech acts. As an example:

Human: "How large is the blue one?"

Human: "There is a green object at the right"

Human: "Let us call the small dog Dexter"

Effective usage of mixed-initiative dialogues

The human initiative command is not the only way to carry forward conversations; another axis is the one dependent on robot initiative conversation. As an example:

Robot: "Hello! Are you Nimit Nielson"

Human: "Yes".

Robot: "Hello Nimit! How are you?"

Human: "I am good!"

Robot: "Awesome! Do you know that our friend Ajit Kahnwald has posted a new photo?"

Human: "No!"

Robot: "Well, I am sending you a message for this, check it out!"

On the other hand, a dialogue by robot BIRON at Bielefeld is an example of a dialogue with limited mixed initiative:

Robot: "Hey there!"

Human: "Hi! Who are you?"

Robot: "My name is BIRON if you show me something I can remember it"

Human: "OK, look at this mug"

Robot: "Sorry?"

Human: "This is a mug here"

Robot: "Sorry, can you use some other words?"

Human: "This is a mug!"

Robot: "Sorry, I cannot find it. Are you pointing to it with your finger?"

Human: "Yes!"

Robot: "I know it is sometimes difficult with me, but don't feel discouraged!"

Robot: "OK, I have found it, it is nice!"

In the example given above, the onus of carrying forward the dialogue is not on any single entity and the initiative changes multiple times within the conversation. Karlsruhe Humanoid, which is one of the Biron and Barthoc systems located at Bielefeld is another example of a mixed-initiative framework.

Affective Interaction: The affective dimension in robot-human interaction is necessary to ensure important functions like persuasion, learning, and empathy. The effect is signified by both semantic contents as well as the prosody of speech. Gait, facial expressions, body posture, heart rate while speaking, breathing rate, and galvanic skin response are some other examples of affect markers. Several pieces of research in this field of human–robot effective interaction include virtual avatars like Rea, Steve, and Greta. Designed to handle a variety of gestures and facial expressions, Cynthia Breazeal comprises an interactive emotion subsystem and a driving system for the Kismet robot.

Multi-linguistic emotional speech corpus with the Sony AIBO robot (Batliner et al. 2004) is another example of an effective interaction-enabled robot. Affective storytelling by robots along with the help of facial expressions is a recent area of research. Recognition of facial expressions is achieved using SHORE and Seeing Machines product FaceApi. For producing facial expressions, dynamics is an important factor to ensure believability. If we consider semantic content as well, fields like sentiment analysis and shallow affect identification also get associated. As an example, using products like Affectivas, Q sensors, and other methods such as galvanic skin response, heart rate monitoring, and breathing rate measurement, etc., could potentially enhance human–robot effective interactions, with certain caveats.

3.5.5 *Sensing and Communication in Robotic Manipulators*

Robotic manipulators are commonly used in a variety of applications, from manufacturing industries, nuclear plants, and underwater exploration to hospitals. In many of the applications, the main operator is distant from the workstation and could even be at a very different location than the robotic CPS itself. In such scenarios, communication architecture is utilized to implement a network between the joint controlling unit of the manipulator and the main control system. Depending upon the application, remote operation of manipulator robotics can fall under several categories. A few such applications are detailed below (Saravanan and Sivaramakrishnan 2019):

Command-Controlled Manipulator: In a command-controlled manipulator, joint movements of the robotic parts are controlled by a human operator using a control panel. Common examples include research vessels used in ocean beds, where the movements are monitored using a viewport.

Master–Slave Manipulator: These manipulators are also operated by humans who need not be in the close vicinity of the machines. They use a master–slave mechanism, sharing common functions. A joint is provided between the master and slave and this joint transmits motion from master to slave. Such manipulators are commonly used; exposure to harmful gases or radiation is needed.

Semi-Automatic Manipulator: These are controlled using a joystick, which could be at a remote place or connected to the main control panel. These manipulators allow a higher degree of freedom and allow movement in all directions. A signal is produced by the control system on the motion of the joystick and this is further converted into a control signal which is then supplied to the actuator in the manipulator.

Supervisory Controlled Robots: These are fully functional robots that have well-defined programs written in their subtasks in their control systems. Human involvement during real-time operations is minimum. The human controller simply needs to assign the tasks or goals to the robot and the robot performs its operations independently.

Interactive Robots: These robots are more developed versions of the supervisory controlled robots as they can determine actions on their own using environment sensing techniques. Hybrid robots are a combination of two or more types of robots listed above.

Sensing in robotic manipulators is affected by various conditions, especially their intended application. Simple structured tasks in a known environment may not require as expensive sensory arrays as robotic manipulators working in an industrial environment requiring the manipulation of a variety of objects. A sensing strategy in the case of robotic manipulators can be divided based on the parts of the manipulator, as different actuators and links are tasked with different functions. A brief description of sensing in robotic CPS based on a similar concept is presented below:

1. Force pedestal sensing: By using a platform fitted with appropriate sensors, it is possible to measure the force present between the robot and the local environment obstructions. We aim to monitor the surrounding environment and not the robot in this case. The platform may or may not be static depending on the requirements. The sensors fitted on the platform are used for measuring the force components along with the mutually perpendicular pair of axes, which can be used to find the application point of the force. The system proposed is similar to the case of cooperating manipulators in robots. In this light, it seems highly likely that the possibility of integration of all sensory subsystems into a single "advanced"

robot exists. The positioning functions can then be taken care of by a secondary robot. The advantages of this arrangement include high stiffness and sensitivity and the major drawback is the need for a complex object positioning system.
2. Joint torque-sensing: This involves sensing torque generated typically in the joints of the manipulator. It is commonly used in master–slave manipulators. It is capable of detecting the force applied at other points of the manipulator and not just the hands. The joint sensor is also protected from the robot link-associated inertia and arm stiffness. The major disadvantage is that forces are not directly measured but inferred, which may lead to problems in heavyweight and large-sized robots. Joint torque can be measured simply using the motor currents. For example, Wu used strain gauge-based joint torque sensors to provide a complete compliance control. This sensor has many advantages like simple linear relations required for obtaining torque applied at each joint.
3. Wrist sensing: It usually consists of a mechanical structure that is subject to changes or deformities in the external environment. Suitable sensing mechanisms are deployed to detect the deflection in elastic components and the transducer measures the three to six components of torque or force operating in the desired direction. Several different force-sensing wires also were created using diverse methods for sensing like strain gauge, potentiometric, and electro-optic. These are small in size, easy to use, and reliable; therefore many force-sensing wrists make use of strain gauges. To retrieve useful information like force vector values, we employ a decoupling matrix, the shape of which provides a qualitative analysis of the sensor. For example, researchers at SRI developed a force-sensing wrist that tries to mechanically decouple the effects of different force components and detects forces and moments independently.

3.6 Case Study: Communication Between Two Robots—Detection and Tracking

Sfax, Tunisia University students created a system to track moving targets with mobile robots, with one robot acting as a moving target and the other acting as a tracker (Bousnina 2011). The idea involves two sub-steps: robot tracking and robot detection. To detect robots, Gabor filters are used to extract robot features, and Support Vector Machines (SVMs) are used to classify them based on their features. Once the Kalman filter is applied, the concept of tracking robots can be realized. The ZigBee technology is also employed in the communication of the two robots.

Tracking and detecting robots is crucial in multi-robot systems. In collaborative settings, robots often exchange information and communicate. This is more critical in cases where robots do not share the same communication protocols. This project uses SVM and Gabor filters to detect autonomous mobile robots, the Kalman filter to track them, as well as ZigBee communication for inter-robot communication.

Overview: For the setup, two robots, the Bioloid Biped robot, and the IRobot Create robot were used as trackers and targets, respectively. These robots are mainly

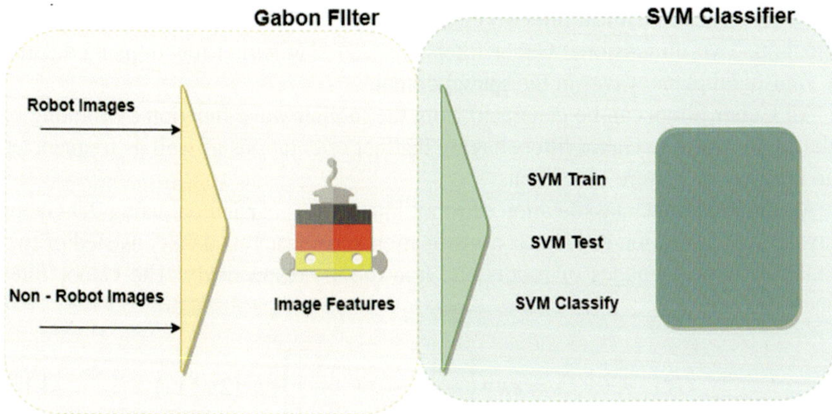

Fig. 3.6 Application of SVM classifier

used for educational research. These have programmable behaviors, sounds, and movements and allow additional electronic components to be attached. These robots were programmed to navigate the indoor environment securely. The IRobot Create robot followed a trajectory and obstacles were avoided using light-intensity actuators and infrared sensors (Fig. 3.6).

With ZigBee, robots can communicate with each other in a multitude of fields. ZigBee networks use low data rates in medical, scientific, and industrial bands based on IEEE 802.15.4 specifications. It permits communication at very low power consumption and has high durability.

"Bioloid Biped robot," a tracker robot using these modules for wireless communication. A ZIG-100 wireless communication module and ZIG2Serial are employed. When the ZIG-100 is linked to the ZIG2Serial, it allows communication between robots (remote control communication) and control of robots through a PC where the ZIG-100 is embedded in the CM-5. A wireless camera is installed on the tracker for capturing the local environment. An intelligent supervised learning method is set up to detect target images. ZigBee makes it easier to deploy and has many other benefits like low cost, low power requirement, and long battery life.

Robot Detection: Digital image processing techniques were used to detect objects in frames. Background subtraction, point detectors, supervised classifiers, and segmentation was used to detect objects. Object detection and recognition focused mainly on:

1. Representation of object.
2. Learning: employing machine learning algorithms to learn properties common to objects.
3. Recognition: identifying objects using learning models.

Gabor filter is used for representing robot features and SVM (supervised vector machine) is used for the classification and recognition of robots.

Gabor filter: It is a linear filter developed by Dennis Gabor and is utilized for object detection. Two-dimensional Gabor filters are Gaussian kernel functions modulated by sinusoidal plane waves in the spatial domain.

All Gabor filters can be generated from the mother wave function by rotating or dilating it. Multiple Gabor filters having distinct orientations as well as frequencies can be used for feature extraction.

Automatic image classification of robot images was carried out using 2D Gabor wavelet representation and linear classification schemes. Test data consisted of two databases having images of robots and non-robots, respectively. The Gabor filter function used was

$$G(x, y; \theta, f) = exp\left\{-\frac{1}{2}\left[\frac{x_\theta^2}{\sigma_x^2} + \frac{y_\theta^2}{\sigma_y^2}\right]\right\}cos(2\pi f x_\theta) \quad (3.13)$$

$$\begin{aligned} x_\theta &= x \cos\theta + x \sin\theta \\ y_\theta &= -x \cos\theta + x \sin\theta \end{aligned} \quad (3.14)$$

Here, θ denotes the orientation of the Gabor filter, f stands for the frequency of the cosine wave, σ_x and σ_y are the standard deviations of the Gaussian envelope along the x and y axes, respectively, x_θ and y_θ define the x and y axes of the filter coordinate frame.

Support Vector Machine: It is a collection of supervised learning approaches that are mostly used for data pattern recognition. It is mainly useful in regression analysis for classification. It takes input data and performs prediction to select one of the two possible scenarios. It is a non-probabilistic and linear classifier. Based on the SVM classifier principle, IRobot created non-robot images and the robot images were used for feature extraction. Images of the environment were used to track the robot's presence in the surrounding. Gabor filter was applied to the feature and the image vector was further tested utilizing learning models, to get precise position coordinates of the robot.

Robot Tracking: Kalman filter was used to estimate the process state and minimize the mean square error as it provides an efficient and recursive computational method. Kalman filters are majorly used for controlling and predicting the behavior of dynamic systems. It is an optimal and effective estimator for large-scale systems as well.

Kalman filter is used for two main purposes: Prediction and update, we need to evaluate each constraint individually on output. The process is done multiple times and the data gathered is used to find the state. Using Kalman filters for robot tracking is based on the assumption that robot motion is constant across frames. State variables, measurement matrix, and dynamic matrix are used for predicting the location of robots. The tracking algorithm involves target tracking in a sequence video using frame-by-frame extraction.

To use Kalman Filter, the study assumed robot motion as a constant throughout the frames following which the location of the robot could be estimated. As Kalman Filter is an iterative process, the procedure was repeated for all the frames.

The study concluded that the combination of SVM classification and Gabor texture features provided impressive results along with the Kalman Filter for robot tracking. This provided a low-budget and efficient technique for tracking a wheeled mobile robot.

3.7 Conclusion and Future Scope

Robotics and complex CPSs are rapidly growing to be one of the most impactful fields of interest in the twenty-first century. Each age of progress is triggered by innovation and breakthroughs in science and technology. It is very evident from the world around us that robotics and artificial intelligence will indeed be the major driver of the transition to a more prosperous digital age. In future, CPSs are expected to have a significant impact on a variety of industries and applications, including manufacturing, transportation, health care, agriculture, and energy. The scope of CPSs is likely to continue to grow as they become more sophisticated and widely adopted. Consequently, robotics is a broad spectrum of research and application of which sensors and communication technologies are an integral part. Today, sensing technology has developed beyond what could have been only imagined in science fiction a few decades ago and the development is only in its infancy. Thus, sensing and communication in itself is a very interesting domain of robotics and cannot be possibly presented in its entirety within a single chapter.

This chapter in its capacity first introduces the basic sensing technologies which continue to shape our experiences with robotic CPSs. A brief overview of sensing techniques that are commonly used has been presented here. Details on physical phenomenon-based sensors broadly classified under proximity sensors, force, and tactile sensors were provided for the reader to provide an introduction to simpler sensing techniques. This is further expanded as the domain of multi-sensor units is explored and statistical methods are employed for deriving knowledge from the data captured. The reader should note that different methodologies have their usage in different scenarios and thus it is important to consider the trade-offs that are a part of these techniques. As seen in the case of traditional Bayesian techniques and their advanced versions such as the Dempster–Shafer Theory, the role of the user's prior knowledge impacts the application of these methods used. Thus, a designer should always consider the benefits of the application and the complexity associated with the method. This complexity can be either one or a combination of the following: accuracy/trust in the sensory unit, computation cost, physical space occupied by the unit, data processing, etc.

The chapter also presents data fusion techniques used in more advanced applications using popular Kalman Filters and their extensions. Such methods of estimation are widely used in both linear and nonlinear systems for gathering information on the environment's state and robotic system. Moving forward, the chapter discusses more recent advancements in robotics including collaborative robotics, AR, and human-like verbal communication methods. For more information, consider the references and numerous papers that deal with this field which are by no means an exhaustive list.

References

Antoniol G, Cattoni R, Cettolo M, Federico M (1993) Robust speech understanding for robot telecontrol. In: Proceedings of the 6th International conference on advanced robotics. pp 205–209

Attia ABE, Balasundaram G, Moothanchery M, Dinish US, Bi R, Ntziachristos V, Olivo M (2019) A review of clinical photoacoustic imaging: Current and future trends. Photoacoustics 16:100144. https://doi.org/10.1016/j.pacs.2019.100144. ISSN 2213–5979

Austin J (1962) How to do things with words. Oxford

Batliner A, Hacker C, Steidl S, Nöth E, D'Arcy S, Russell MJ, Wong M (2004) "You stupid tin box"-children interacting with the AIBO robot: a cross-linguistic emotional speech corpus. In: LREC

Billinghurst M, Grasset R, Seichter H (2010) Tangible interfaces for ambient augmented reality applications. In: Human-centric interfaces for ambient intelligence. pp 281–302. https://doi.org/10.1016/B978-0-12-374708-2.00011-5

Bousnina S (2011) Detection, tracking and communication between two robots student: Sonda

Burgard W, Cremers AB, Fox D, Hnel D, Lakemeyer G, Schulz D, Steiner W, Thrun S (1998) The Interactive museum tourguide robot. In: Proc. of the Fifteenth national conference on artificial intelligence (AAAI-98)

Busch-Vishniac I (1999) Electromechanical sensors, and actuators. Springer New York, New York

Castanedo F (2013) A review of data fusion techniques. Sci World J 2013:1–19. https://doi.org/10.1155/2013/704504

Christensen H, Gregory H (2008) Sensing and estimation. https://doi.org/10.1007/978-3-540-30301-5_5

Augmented reality makes robots better coworkers. IEEE Spectrum. https://spectrum.ieee.org/augmented-reality-makes-robots-better-coworkers

Gajjar MJ (ed) (2017) Chapter 10–Usability. In: Mobile sensors, and context-aware computing. Morgan Kaufmann, pp 267–302. ISBN 9780128016602

Green SA, Billinghurst M, Chen X, Chase JG (2008) Human-robot collaboration: a literature review and augmented reality approach in design. Int J Adv Rob Syst. https://doi.org/10.5772/5664

Gupta GS, Mukhopadhyay SC, French JR (2008) Wireless communications and control module of a web-enabled robot for distributed sensing applications. In: 2008 IEEE instrumentation and measurement technology conference. Victoria, BC, pp 393–398. https://doi.org/10.1109/IMTC.2008.4547067

Joyce J (2003) Bayes' theorem (Stanford encyclopedia of philosophy)

Jüptner W (1988) Holographic techniques. In: Dario P. (ed) Sensors and sensory systems for advanced robots. NATO ASI Series (Series F: Computer and Systems Sciences), vol 43. Springer, Berlin, Heidelberg

Kam M, Zhu X, Kalata P (1997) Sensor fusion for mobile robot navigation. Proc IEEE 85(1):108–119. https://doi.org/10.1109/JPROC.1997.554212

Kim YC, Cho SB, Oh SR (2002) The Dempster-Shafer approach to map-building for an autonomous mobile robot with fuzzy controller. In: Pal NR, Sugeno M (eds) Advances in soft computing—AFSS 2002. AFSS 2002. Lecture notes in computer science, vol 2275. Springer, Berlin, Heidelberg

Lefebvre T, Bruyninckx H. De Schutter J (2001) Kalman filters for nonlinear systems: a comparison of performance

Lunni, D., Giordano G, Sinibaldi E, Cianchetti M, Mazzolai B (2018) Shape estimation based on Kalman filtering: towards fully soft proprioception. 541–546. https://doi.org/10.1109/ROBOSOFT.2018.8405382

Mavridis N (2015) A review of verbal and non-verbal human–robot interactive communication. Robot Auton Syst 63(1):22–35. https://doi.org/10.1016/j.robot.2014.09.031. ISSN 0921–8890

Milgram P, Zhai S, Drascic D, Grodski J (1993) Applications of augmented reality for human-robot communication. In: Proceedings of 1993 IEEE/RSJ international conference on intelligent robots and systems (IROS '93). https://doi.org/10.1109/iros.1993.583833

National Research Council (1995) Expanding the vision of sensor materials. National Academies Press

Pei Y, Biswas S, Fussell D, Pingali K (2017) An elementary introduction to Kalman filtering. Commun ACM 62. arXiv:1710.04055

Saravanan N, Sivaramakrishnan R (2019) Command and control of industrial manipulator through speech-based interfaces in indic languages. J Supercomput 75:5106–5117. https://doi.org/10.1007/S11227-019-02790-0

Sobh T, Elleithy K (20015) Innovations and advances in computing, informatics, systems sciences, networking, and engineering. Springer International Publishing, Cham

Svechtarova M, Buzzacchera I, Toebes B, Lauko J, Anton N, Wilson C (2016) Sensor devices inspired by the five senses: a review. Electroanalysis 28(6):1201–1241. https://doi.org/10.1002/elan.201600047

Thrun S (2001) A probabilistic on-line mapping algorithm for teams of mobile robots. Int J Robot Res 20(5):335–363. https://doi.org/10.1177/02783640122067435

Thrun S (2002) Learning occupancy grids with forward sensor models

Tsihrintzis G, Virvou M, Jain L (2016) Intelligent computing systems. Springer, Berlin Heidelberg

Vechet S, Jiri K (2010) Sensors data fusion via Bayesian network. https://doi.org/10.1007/978-3-642-05022-0_38

Vemuri BC, Cao Y, Chen L (1998) Fast collision detection algorithms with applications to particle flow. Computer Graphics Forum 17:121–134. https://doi.org/10.1111/1467-8659.00233

Versweyveld L (1998) Voice-controlled surgical robot ready to assist in minimally invasive heart surgery. In: Virtual medicine world monthly

Yager R (2004) A framework for multi-source data fusion. Inf Sci 163:175–200. https://doi.org/10.1016/j.ins.2003.03.018

Yan Y, Jing W, Mehrmohammadi M (2020) Photoacoustic imaging to track magnetic-manipulated micro-robots in deep tissue. Sensors 20:2816

Yi Z, Khing HY, Seng CC, Wei ZX (2000) Multi-ultrasonic sensor fusion for autonomous mobile robots, sensor fusion: architectures, algorithms and applications IV. Proc SPIE 4051:314–321

Chapter 4
Deep Learning-Based Anomaly Detection in Cyber-Physical System

Sangeeta Oswal, Subhash K. Shinde, and M. Vijayalakshmi

4.1 Introduction

Cyber-physical systems are characterized by their tight integration of hardware and software resources for computing, communication, and control task. The worldwide Cyber Physical System market is projected to rise from its 2019 level of USD 6596.1 million to a projected USD 9563.9 million by 2025, representing a CAGR of 9.7% over the forecast period (https: globalmarketvision.com, reports, global-cyber-physical-system-market, 4270 2023).

In Fig. 4.1, we present a conceptual representation of CPS in which the Physical System is linked to the Computational System through a network (IoT). We are currently in the midst of the fourth Industrial Revolution, which extends far beyond the industry. Smart, linked technologies are altering the design, production and maintenance of parts and products. Moreover, by introducing a digital world, they are altering organizations. A Cyber-Physical System (CPS) is a platform consisting of a computer-controlled mechanical system that is closely connected to the Internet and its networked users. Here, the software components, represented by computers and networking devices, are tightly linked to the physical–mechanical components, represented by intelligent sensors and actuators. Multiple subsystem

S. Oswal (✉)
Research Scholar, Computer Engineering LTCE, Navi Mumbai, India
e-mail: sangeeta.oswal@ves.ac.in

Subhash K. Shinde
Professor and Vice Principal, Lokmanya Tilak College of Engineering, Navi Mumbai, India
e-mail: skshinde@ltce.in

M. Vijayalakshmi
AI and Data Science VESIT, Mumbai, India
e-mail: m.vijayalakshmi@ves.ac.in

N. Sharma et al. (eds.), *Big Data Analytics in Intelligent IoT and Cyber-Physical Systems*, Transactions on Computer Systems and Networks,
https://doi.org/10.1007/978-981-99-4518-4_4

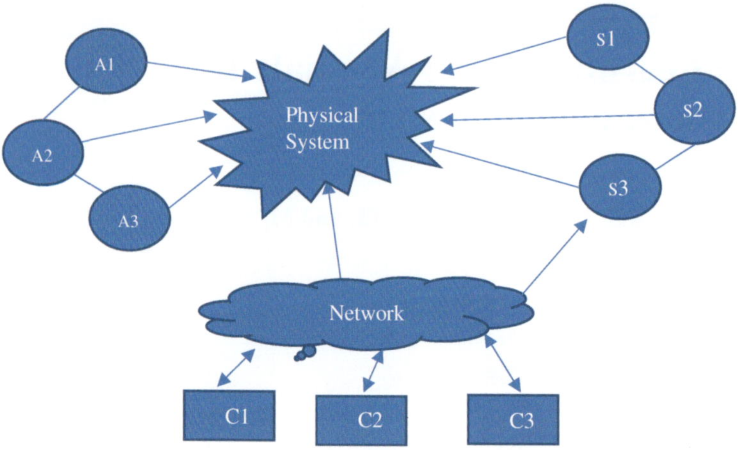

Fig. 4.1 A conceptual view of cyber physical system. (A−Actuator S−Sensor and C−Computational system)

components, including sensors, controllers, and actuators, are coupled via a communication network to create a closed control loop in a CPS. Industrial Internet of Things (IIoT) and cyber-physical systems fuel Industry 4.0. Cyber-physical systems are the brains of Industry 4.0, enabling computers to monitor and control equipment, robots, and vehicles.

4.2 Block Diagram of CPS

CPSs consist of two interdependent components (see Fig. 4.2). First,

- 'Physical' component, in which sensors and actuators monitor and control various physical processes. Second,
- 'Cyber' component made up of software parts like a computer system that keeps tabs on the physical world.

Connected to sensors and actuators through a network, (PLCs) (Alur 2015) that implement control logic take sensor data as input and provide commands to actuators. To top it all off, the PLCs are linked to a SCADA system, which can keep an eye on the control procedure.

The counter measures against CPS attack can be broadly classified as:

- Normal rule-based checking of sensors and actuators state to check its normal working which is a standard rule checker for industrial cps.
- Anomaly detector at SCADA.

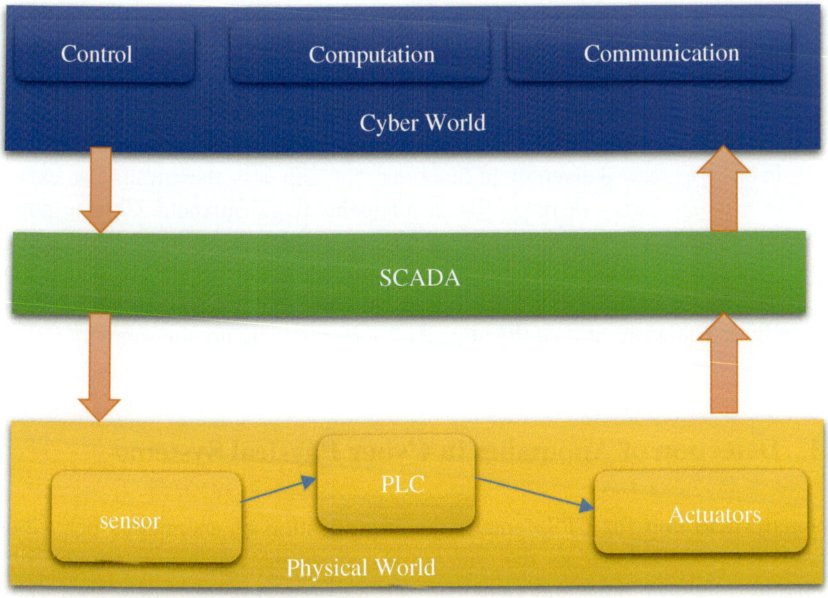

Fig. 4.2 Integration of cyber and physical world

4.3 Applications of CPS

Application of CPS includes.

- Embedded computing device-based process control systems Examples include the pharmaceutical, chemical, and textile industries.
- Autonomous vehicles with intelligent sensors, cameras, actuators, and intelligent data analysis and motion control
- Electrical power management utilizing smart grids entails the balanced and coordinated distribution of power to all users based on a particular set of parameters.
- Motion-manipulating robots with applications in healthcare, hazardous areas, and precise manufacturing, among others.
- Supply–chain with edge computing at several touchpoints to provide effective customer services.
- End-to-end automated production line; for example, in the automotive industry, with monitoring of assembly processes and alarming prior to the occurrence of a defect.
- The sectoral applications area includes Aerospace Engineering, Pharmaceutical Industry, Medical imaging and diagnostics, Farming and Agriculture, Water distribution management, Intelligent Manufacturing, Transportation systems, Military and Defence systems, Energy Systems. Etc… to name a few.

Since technologies are rapidly progressing, we can anticipate an improvement in the usefulness, scalability, security, and usability of CPS. Consequently, CPS anticipated to transform how people engage with engineering systems, similar to how the Internet has altered how people interact with information to serve a diversity of functions. CPSs have evolved to be sophisticated, diversified, and integrated platforms in order to serve a diversity of functions. Nevertheless, these traits also expose CPSs to a greater array of risks. Recent incidents (e.g., Stuxnet,[1] Ukraine power grid outage,[2] auto-driving accidents,[3] and robot malfunction[4]) have demonstrated that advanced and complex systems are susceptible to failure. Insidious attacks (and errors) can have catastrophic effects on the economy, ecology, and society including human lives Therefore, it is of the utmost importance to assure the safety of CPSs.

4.4 Detection of Anomalies in Cyber Physical Systems

Anomaly detection for CPSs entails the identification of anomalous behaviours (anomalies), i.e., behaviours that are not shown during regular operation. These anomalies might be the result of assaults on the control components, network, or physical environment; however, they could also be the consequence of failures, operator errors, or even common software bugs or misconfigurations. Thus, the capacity to notice anomalies serves as a protective mechanism and facilitates the growth, maintenance, and repair of CPSs.

Anomaly detection approaches are offered to detect attacks and unanticipated faults in CPSs and mitigate these dangers. The normal state of CPSs is evaluated using statistical models (such as the Gaussian model and histogram-based techniques) (Lun et al. 2016). However, these technique requires a domain expert and a normal underlying distribution. Methods of machine learning may not require domain experts (Chandola et al. 2009), but they fail to capture the spatial–temporal correlation between the properties of CPSs. There is a need to monitor the CPS system to secure them against the attacks in critical applications like transport, water treatment plant, power grid etc. CPS become more complicated by capturing the physical properties of the sensors and semantics of the control system and the attacks become more stealthy. Anomaly detection needs to adapt to these changes and this chapter discuss deep learning-based anomaly detection techniques for CPSs. current research has investigated several neural network designs (e.g., ConvLSTM) for mitigating threats in various CPS domains (e.g., smart grid). However, as this research are rarely presented in a coherent manner, a thorough examination of existing approaches and guidance for future solutions is required.

---

[1] https://en.wikipedia.org/wiki/stuxnet.

[2] https://en.wikipedia.org/wiki/ukraine_power_grid_hack.

[3] https://en.wikipedia.org/wiki/Self-driving_car.

[4] https://edition.cnn.com/2015/07/02/europe/germany-volkswagen-robot-kills-worker/.

4.5 Taxonomy for AD Based on Deep Learning in CPS

As the complexity and size of sensor data increase, the ability of individuals to manually monitor them declines. This necessitates automated anomaly detection strategies that can spot anomaly in high-dimensional data quickly and provide context for their significance to human analysts, allowing them to diagnose and respond to the anomaly as fast as feasible. Due of the nonlinearity of CPS data, deep learning techniques are appropriate.

In this part, we show the CPS system's taxonomy for detecting anomalies. It is described along four axes. Anomaly type, Input data, the Deep learning network utilised, and Dataset used (Fig. 4.3).

- **Types of Anomaly** Anomalies can be categorized as attacks and defects (fault), with the latter referring to a CPS device malfunction. Examples include mechanical failure, sensor damage, environmental changes that affect actuators, incorrect sensor value reporting, etc. CPS systems are at risk of being targeted by attacks such as denial of service, man in the middle, packet injection attacks at the network layer, and malware or false signals transmitted to actuators that compromise the system.
- **Input Data** To detect specific anomalies, DLAD methods must first determine what type of input data to accept. On the basis of the layer and the origin of the data, we classify the input into three categories:

  (1) Data from sensors and actuators.
  (2) Network traffic data, including system calls and
  (3) Time series data.

  To compensate for the dearth of labeled data, DLAD methods combine semi-supervised and unsupervised learning (especially for anomalous data).

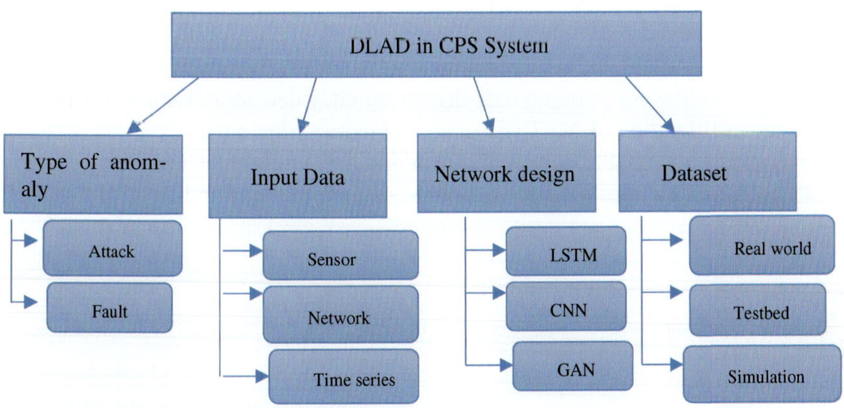

Fig. 4.3 Taxonomy for DLAD in CPS system

- **Network Design** For detecting anomalies, neural network architectures based on LSTM, CNN, GNN, and GAN are utilized. The deep network may be hybrid (LSTM and CNN) or stacked LSTM or GRU/LSTM with Autoencoders. Deep learning models may be centered on prediction or reconstruction error as an anomaly measure. Prediction-based approaches forecast future values and find anomalies based on the discrepancy between the expected and actual values. Reconstruction-based, the input data value is compressed and reconstructed, and the error is calculated based on the reconstruction error.
- **Dataset** The environment for data collecting consists of
  - Data from real-world systems.
  - Testbeds
  - Simulations.

Although the data is difficult to obtain and the number of systems is restricted, the data from real-world systems represent the essential principles of actual systems. Simulation is easy to implement, but it cannot reproduce difficulties that occur exclusively in real systems. A tiny testing platform testbed is more popular.

4.6 Anomaly Detection in CPS's ICS Using Deep Learning

The use of cyber-physical systems has become a commonplace in the monitoring and management of industrial processes at key infrastructure assets such as power plants, gas pipelines, and water treatment facilities. Active monitoring of sensor readings and actuator states in such systems for early identification of unexpected system behavior is vital for preventing possible economic and environmental damage due to system breakdowns. That way, we can take preventative measures like checking for faults, doing preventative maintenance, and shutting down the system before any serious damage is done. Nevertheless, as the complexity of contemporary CPS increases, it becomes impossible to rely solely on manually specified evolution rules for anomaly identification. As a result, it is now common practice to employ Deep learning algorithms for building data-driven anomaly detection frameworks in CPS due to the proliferation of AIoT (AI+IoT) techniques (Fig. 4.4).

In this section we talk about application of deep learning in CPS. Deep learning advantage of extracting features is capable of yielding outstanding result in CPS

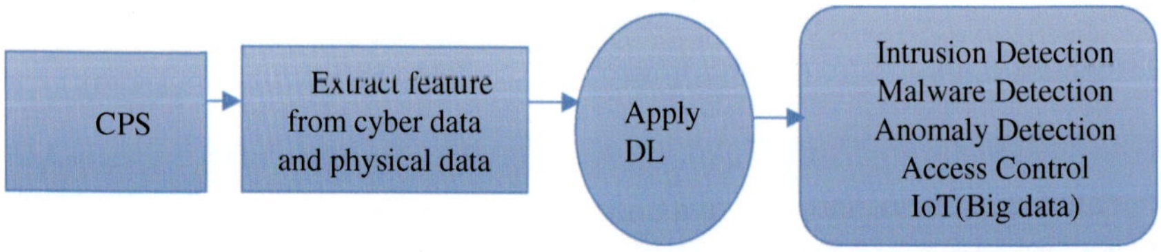

Fig. 4.4 Application od DL in CPS

Table 4.1 Industrial CPS datasets

Dataset	Domain	Entity	Channels	% anomalies	Link
SWaT	Water treatment	1	51	4.65	https://itrust.sutd.edu.sg/
WADI	Water treatment	1	123	5.76	https://itrust.sutd.edu.sg/
MSL	Space craft	27	55	12.02	https://github.com/khundman/telemanom
SMAP	Space craft	55	25	12.04	https://github.com/khundman/telemanom
SMD	Server monitoring	28	38	4.21	https://github.com/NetManAIOps/OmniAnomaly

security applications, which includes industrial anomaly detection (wind turbines, power plant, water plant etc.), log based anomaly detection, fraudulent activities in massive interconnected IoT network etc.

Developments in technology may increase the incidence of network abnormalities and expose ICS to more sophisticated cyber–physical assaults by exposing the system to a broader range of vulnerabilities. There has been a dramatic rise in the use of smart devices in ICS as a weapon in cyberattack incidents, posing serious operational issues. Industrial cyber physical system attacks, unlike traditional cyber-attacks, result in lasting physical harm to the manufacturing process, generating economic loss and disastrous effects.

Different perspectives of anomaly detection based using deep learning in CPS includes.

- Input Data: The data to CPS we consider for this survey is sequential time series
- Availability of Labels
- Training objective.

To demonstrate how to use DL to spot abnormalities in an industrial control system, we give examples based on already-existing models. Table 4.1 displays a sample of the publicly accessible MVTS datasets.

4.6.1 Testbed and Dataset

SWaT is a six-stage water treatment facility that records 51 variables (sensors and actuators) for 11 days. Raw water is ingested, essential chemicals are added, it is filtered by an Ultrafiltration (UF) system, de-chlorinated using UV lamps, and then fed to a Reverse Osmosis (RO) system. The SWaT system has been the subject of numerous experiments designed to probe the nature of cyberattacks and the ways in which the system reacts to them. For a comprehensive account of the assaults, see the

SWaT website.[5] While gathering information for SWaT 2016, a total of 36 attacks were carried out.

WADI testbed is a natural extension to SWaT which record data from 123 sensors and actuators for 16 with 2 days in attack scenarios.

MSL, SMAP, and SMD are multi-entity datasets in which each entity represents a distinct physical unit with identical dimensions. By physically counter-promising the performance of a steady-state system, anomalies in datasets other than SMAP and MSL were induced on purpose. Anomalies reported by SMAP and MSL are manually identified by professionals using operational data from prior mission reports (Hundman et al. 2018).

4.6.2 Baseline Anomaly Detection Methods

Extensive research has been conducted on simulation- and model-based anomaly identification for CPSs. In addition to its applicability to CPSs, anomaly detection is a well-studied area (Chandola et al. 2009). In industrial anomaly detection, the DLAD approach identifies both attack and malfunction. The vast bulk of prior research identifies abnormalities using widely accessible sensor and actuator readings. Very few focus on system logs and network traffic.

Deep learning methods utilizing LSTM, CNN, AE, and, more recently, GAN are common. LSTM is used to capture the temporal dependency between sensors and actuators, whereas AE is implemented in an unsupervised configuration. Typically, anomaly scores are determined using prediction-based algorithms. We focus on time series data as input to an industrial CPS application in this paper. The employed neural network models are.

- LSTM: The LSTM model is used to simulate temporal context. However, it remains challenging to capture sensor temporal behavior in the context of spatial and logical properties in multidimensional (time-series of numerous sensors and actuators) data.
- CNN: is able to jointly extract features from multidimensional data using convolution operations (Munir et al. 2018). The hyper parameter and window size for modelling data must be selected with care.
- GAN: (Lee et al. 2019) presented a GAN-based framework to capture the multidimensional data's spatial–temporal correlation. Both the generator and discriminator are used to identify abnormalities caused by reconstruction and discrimination errors when both the generator and discriminator are employed to find anomalies resulting from reconstruction error and discrimination loss. LSTM models are used to build the generator and discriminator as well.
- Autoencoders: AE are used in the LSTM/RNN (Su et al. 2019) setup to compress the input data and use the reconstruction error as anomaly measure. Variations in

[5] https://itrust.sutd.edu.sg/itrust-labs_datasets/dataset_info/.

AE like VAE and GRU along with GAN or LSTM are also used in some studies (Yin et al. 2021).

Here we present some of the DLAD models used in Timeseries data as input for anomaly detection. Most solutions evaluate an anomaly's deviation using the prediction error. Typically, testbeds are employed to evaluate proposed approaches, and the SWaT testbed is a typical evaluation platform.

1. LSTM-PRED (Goh et al. 2017) is a time series anomaly detection technique that makes predictions using a regressor constructed from a recurrent neural network.
2. DAGMM: DAGMM (Zong, et al. 2018) is a density-based anomaly detection approach. The model employs a deep autoencoder to generate a low-dimensional representation and reconstruction error for each input data point, which is then fed into a Gaussian Mixture Model (GMM). To alleviate the curse of dimensionality, density estimation is conducted in a low-dimensional space that is latent.
3. OmniAnomaly (Su et al. 2019) Characterizes non-Gaussian distributions of latent space using reconstruction probabilities of input samples as anomaly scores and a deep generative model with stochastic variable connection and planar normalizing flow.
4. USAD (Audibert et al. 2020) employs a reconstruction model based on autoencoders to find anomalies in time series using residual errors. To be more precise, anomalies in reconstruction errors are purposefully boosted through adversarial training of the autoencoder.
5. EncDec-AD (Malhotra et al. 2016) Normal time series reconstruction can be trained using an LSTM Encoder Decoder model. The anomaly scores a (i) for each point x (i) in the test time-series is then calculated using the reconstruction errors to determine the statistical significance of the anomaly at that point. The more likely it is that a given point is anomalous, the higher its anomaly score.
6. FID-GAN (Freitas De Araujo-Filho et al. 2021), an unique fog-based, unsupervised intrusion detection system (IDS) for CPSs employing GANs, is employed to detect cyberattacks on CPSs. As the assaults must be halted before they compromise the system, the author discusses the stringent latency requirements. The author utilized GANs to detect network anomalies in the security area.
7. DAEMON (Chen et al. 2021) is an Adversarial Autoencoder Anomaly Detection Interpretation that addresses the mostly overlooked anomaly interpretation method of recent work. To figure out what the root cause of an anomaly is, the reconstruction error of each dimension of the observation is calculated. Then, the top k dimensions with the largest reconstruction error are returned as the root cause of the anomaly.
8. (GAN-AD) (Malhotra et al 2016), using a generative adversarial network researchers explore principled techniques for selecting the latent dimension and PC dimension, and apply feature selection for multivariate anomaly detection.
9. LSTM-GAN-XGBOOST (H. Z. 2, H. L. H. S. Xin Xu1 2021) uses Ball Bearing time series data in a hybrid approach that utilizes long short-term memory

network (LSTM) to extract time-dimensional aspects of time series data, generative adversarial networks (GAN) to efficaciously extract deep features of normal data, and extreme gradient boosting (XGBOOST) to categorize the features extracted and export anomaly scores.

10. TadGAN (Geiger et al. 2020) using spacecraft telemetry signals comprising two datasets: Mars Science Laboratory (MSL) and Soil Moisture Active Passive (SMAP) provided by NASA. It incorporates an Encoder, which maps the time series sequences into the latent space, and G, which turns the latent space into the reconstructed time series data.
11. MADGAN operates in a GAN and LSTM/RNN setup; the author leveraged the spatial–temporal correlation and other dependencies among the system's various variables (sensors/actuators) to discover abnormalities. It employs several window sizes to capture the system's state at various resolutions.
12. Robot-Assisted Feeding: A Multimodal Anomaly detector addresses the difficult problem of fusing high-dimensional and heterogeneous modalities that can be tackled with the help of an LSTM-based variational autoencoder (Park et al. 2017). Five sensors provide data for the robot. The author introduces a dynamic threshold that varies based on the estimated status of a task's execution.
13. Hundman et al. (2018) uses LSTM for telemetry data uses dynamic error thresholding to mitigate false positive in spacecraft anomaly detection using the SMAP and MSL dataset.
14. Using Spatial–Temporal Adversarial Networks to Identify Abnormalities in Spacecraft Telemetry data with the help of a convolutional neural network (CNN) and a long short-term memory (LSTM), the authors of Yu et al. (2021) create a GAN-based model for extracting spatial and temporal information from the telemetry data.

4.7 Evaluation

Proposed metrics for measuring the efficacy of DLAD approaches. We conclude that precision, recall, and F1 score are the most often employed metrics. Precision is defined as

$$\text{Precision} = \text{TP}/(\text{TP} + \text{FP}) \tag{4.1}$$

where TP and FP are respectively True Positives and False Positives. The recall is defined as

$$\text{Recall} = \text{FN}/(\text{FN} + \text{TN}) \tag{4.2}$$

where FN signify False Negatives and TN is True Negative.

$$\text{F1} = 2 * \text{Precision} * \text{Recall}/(\text{Precision} + \text{Recall}) \tag{4.3}$$

Also in addition, the Receiver Operating Characteristic (ROC) curve is employed to manage trade-offs between FP and TP. Error-based measures, such as Mean Absolute Error (MAE) and Relative Errors, are also used to evaluate the prediction and reconstruction performance records.

4.8 Current Challenges

Here we briefly describe some of the current difficulties encountered while using anomaly detection techniques based on deep learning in CPS.

- **Understandability**: The majority of researchers demonstrate outstanding performance of DL-based algorithms, yet existing approaches are opaque to the user. The opaque nature of DL and its lack of theoretical context reduces human confidence. Explainable AI is the latest frontier of study for explicating anomalies.
- **Regularization**: Modification in the algorithm to reduce the generalization error helps to improve the algorithm efficiency. Implicit regularization consists of stochastic GD, Batch Normalization, and the number of layers employed, among other things. Selection of proper DL methods and regularization techniques are the key for success (Fig. 4.5).
- **Runtime performance**: In order to make DLAD approaches more applicable, we argue that running performance is also essential. Rather just modelling offline previous data, the dynamic change in real-time data should also be accounted for, or else the accuracy of the actual environment is diminished.
- **Distinct neural network designs**: CPS kinds and anomalies have different data type and since they use different neural network design for anomaly detection. For e.g. in ICS time series data is more common and typically LSTM network are used for its modeling. The prevalence of FDI assaults within the smart grid uses AE, hence we recommend that researchers modify their models in light of these findings.
- **Incremental learning**: Updating model parameters on the fly and integrating new data at the same time is a new research path.

Technique	Regularization methods
LSTM-RNN	Weight sharing, L2 regularizer
CNN	Drop Out, Pooling, Data augmentation
GAN	Loss function, Gradients, normalization
AE	Sparsity regularization, weight decay, drop out

Fig. 4.5 Regularization methods for neural network models

4.9 Conclusion

The paper discusses DL based Security application for the Cyber Physical system. Attackers are becoming smarter day by day by using new techniques and exposing systems' vulnerability. Therefore, generalization of the system in CPS is essential. In this research work, we first suggested a taxonomy to classify the salient features of DLAD approaches. In addition, we provided an overview of the state-of-the-art approaches to DLAD as well as the most relevant research outcomes using our taxonomy. The main challenge of gaining users trust can be addressed using explainable AI which opens the black box off deep learning methods. This study, by systematizing cutting-edge deep learning-based CPS security solutions, is meant to aid the community in determining where to focus their research efforts in order to best solve urgent deployment challenges in CPS anomaly detection.

References

Alur R (2015) Principles of cyber-physical systems. MIT Press

Audibert J, Michiardi P, Guyard F, Marti S, Zuluaga MA (2020) USAD: unsupervised anomaly detection on multivariate time series. In: Proceedings of the ACM SIGKDD International Conference on Knowledge Discovery and Data Mining. pp 3395–3404. https://doi.org/10.1145/3394486.3403392.

Chandola V, Banerjee A, Kumar V (2009) Anomaly detection: a survey 41(3): 1–58. https://doi.org/10.1145/1541880.1541882.

Chen X et al (2021) DAEMON: unsupervised anomaly detection and interpretation for multivariate time series. In: 2021 IEEE 37th International Conference on Data Engineering (ICDE), vol 2021. pp. 2225–2230. https://doi.org/10.1109/ICDE51399.2021.00228

Freitas De Araujo-Filho P, Kaddoum G, Campelo DR, Gondim Santos A, Macedo D, Zanchettin C (2021) Intrusion detection for cyber-physical systems using generative adversarial networks in fog environment. IEEE Internet Things J 8(8): 6247–6256. https://doi.org/10.1109/JIOT.2020.3024800.

Geiger A, Liu D, Alnegheimish S, Cuesta-Infante A, Veeramachaneni K (2020) TadGAN: time series anomaly detection using generative adversarial networks. In: Proceedings of 2020 IEEE International Conference on Big Data (Big Data) 2020. pp 33–43. [Online]. https://arxiv.org/abs/2009.07769v3.

Goh J, Adepu S, Tan M, Lee ZS (2017) Anomaly detection in cyber physical systems using recurrent neural networks. In: Proceedings of IEEE International Symposim on High Assurance Systems Engineering. pp 140–145. https://doi.org/10.1109/HASE.2017.36.

H Z 2, H L H S Xin Xu1 (2021) LSTM-GAN-XGBOOST based anomaly detection algorithm for time series data. In: IEEE conference. [Online]. https://doi.org/10.1109/PHM-Jinan48558.2020.00066.

(2023) https://globalmarketvision.com/reports/global-cyber-physical-system-market/4270

Hundman K, Constantinou V, Laporte C, Colwell I, Soderstrom T (2018) Detecting spacecraft anomalies using LSTMs and nonparametric dynamic thresholding. In: Proceedings of the 24th ACM SIGKDD international conference on knowledge discovery & data mining, vol 18. pp 387–395. https://doi.org/10.1145/3219819.3219845

Lee CK, Cheon YJ, Hwang WY (2019) Studies on the GAN-based anomaly detection methods for the time series Data. https://doi.org/10.1109/ACCESS.2021.3078553.

Li D, Chen D, Goh J, Ng SK (2017) Anomaly detection with generative adversarial networks for multivariate time series. [Online]. https://github.com/LiDan456/GAN-AD.

Lun YZ, D'Innocenzo A, Malavolta I, Di Benedetto MD (2016) Cyber-physical systems security: a systematic mapping study. J Syst Softw 149:174–216. https://doi.org/10.1016/j.jss.2018.12.006

Malhotra P, Ramakrishnan A, Anand G, Vig L, Agarwal P, Shroff G (2022) LSTM-based encoder-decoder for multi-sensor anomaly detection. [Online]. https://arxiv.org/abs/1607.00148v2.

Munir M, Siddiqui SA, Dengel A, Ahmed S (2018) DeepAnT: A deep learning approach for unsupervised anomaly detection in time series. https://doi.org/10.1109/ACCESS.2018.2886457.

Park D, Hoshi Y, Kemp CC (2017) A multimodal anomaly detector for robot-assisted feeding using an LSTM-based variational autoencoder.

Su Y, Zhao Y, Niu C, Liu R, Sun W, Pei D (2019) Robust anomaly detection for multivariate time series through stochastic recurrent neural network. https://doi.org/10.1145/3292500.3330672.

Yin X, Han Y, Xu Z, Liu J (2021) VAECGAN: a generating framework for long-term prediction in multivariate time series. Cybersecurity 4(1):1–12. https://doi.org/10.1186/S42400-021-00090-W/FIGURES/6

Yu J, Song Y, Tang D, Han D, Dai J (2021) Telemetry data-based spacecraft anomaly detection with spatial-temporal generative adversarial networks. IEEE Trans Instrum Meas 70:3515209. https://doi.org/10.1109/TIM.2021.3073442

Zong B et al (2018) Deep autoencoding gaussian mixture model for unsupervised anomaly detection.

Chapter 5
Security Issues and Privacy Challenges of Cyber-Physical System in Smart Healthcare Applications

Soumya Samarpita, Ritunsa Mishra, Rabinarayan Satpathy, and Bibudhendu Pati

5.1 Introduction

One of the most significant research fields for technology developers and designers is emerging as the healthcare sector. More security is being included into networks by researchers for data and communication. Any one alteration in a patient's data could have drastic effects on the patient's life. Researchers are motivated to examine different security solutions including multi-layered data cryptography, cryptosystem, and other techniques because of cyber-attacks on medical data. Medical device development has changed quickly as a result of advances in embedded network and software connectivity. The use of standalone devices to independently track and manage patients is being phased out in the healthcare sector. Therapeutic healthcare appliance systems, also known as Healthcare Cyber-Physical Systems (HCPS), are created by the combination of embedded technology monitoring devices, networking capabilities of healthcare equipment, and the complicated physical features shown by patients' bodies (Priyadarshini et al. 2021). IoT also has many uses in the healthcare sector, including remote healthcare monitoring, hardware accessibility and availability, patient inventory tracking, and utilization of health services.

S. Samarpita
FOS, Sri Sri University, Cuttack, Odisha, India

R. Mishra (✉)
FET, Sri Sri University, Cuttack, Odisha, India
e-mail: ritunsa.m@srisriuniversity.edu.in

R. Satpathy
FET, FOS, Sri Sri Univeristy, Cuttack, Odisha, India

B. Pati
Department of CS, Rama Devi Women's University, Bhubaneswar, Odisha, India

N. Sharma et al. (eds.), *Big Data Analytics in Intelligent IoT and Cyber-Physical Systems*, Transactions on Computer Systems and Networks,
https://doi.org/10.1007/978-981-99-4518-4_5

The idea of a smart factory is based on artificial intelligence (AI) and the IoT, and it results from industrial automation. Examples of industrial automation and CPSs include adaptive supply chain management, human–robot interaction, production quality, and predictive maintenance (Latif et al. 2022). A creative span known as Industry 4.0 that introduces CPSs has been brought about by the necessity for consistent communication, management, and coordination for streamlined and efficient systems for quality of service. Helen Gill of the NSF in the US first used the term CPS in 2006 (Verma, 2022).The latest technology providing cyber security for physical systems and processes is known as the CPS, or cyber system merged with physical phenomena (Kurde et al. 2019). A cyber physical system is a system that combines different physical components with computational components and operations. CPS is also referred to as the Industry Internet, Embedded Systems, and the Internet of Everything. Although cyber physical systems have long existed, a new intellectual system has emerged in recent years. As we cannot simply link physical processes and cyber computer components directly, CPS design and incorporation must be restricted.

The terms healthcare and medicine refer to the problems defining various physiologic elements of the patient. Since they present important research opportunities for the CPS community, medical applications in CPS research are given particular attention (Gunes et al. 2014). Systems known as "CPS" link the real environment with the digital world of cognitive processing. It's possible that sensors and actuators make up the physical universe. The HCPS is the network of healthcare systems brought together to achieve high-quality healthcare. An overview of HCPS, which includes physical systems and devices, is shown by Fig. 5.1. An overarching system of various physical systems is created by combining the physical systems.

Fig. 5.1 The mapping procedure

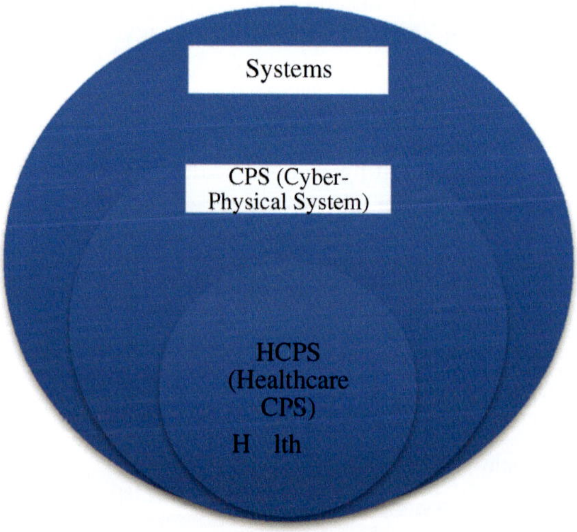

CPS is a novel method for examining digital systems. Actually, CPS is the latest era of digital systems, focusing primarily on complex interdependencies and interconnections between cyberspace and physical reality (Jamal et al. 2021). With regard to communications, computational resources, sensors, and physical aspects, CPS takes an integrated perspective. In such a scenario, cyberattacks might put underlying systems in danger. Conventional approaches to CPS security design looked at the cyber and physical systems separately and were unable to define network-related risks. Big Data analytics can aid in the government's ability to provide its residents with improved services. Big Data can help governments enhance vital industries like healthcare and public transportation, thereby assisting in the creation of a more effective modern society. Big Data analytics and techniques of machine learning, for instance, can offer inventive solutions for complex issues like health stress prediction or improvement of the transportation services provided by the government to the general public (Iqbal et al. 2020).

5.1.1 Cyber-Physical System (CPS)

Primarily CPS is an intelligent system that is related to computer intelligence which may accommodate some enhanced applications than human intelligence. The basic structure of CPS may include design, modelling, and simulation of information. CPSs' are physical assets or manual systems with a computing and interaction centre that manage, synchronize, merge, and give commands to the operations of the said system (Cabello et al. 2020). CPS implemented in different sectors i.e. Economical, Industrial, and in healthcare units as well to control and monitor networks, clients, patients, and system. CPS can be characterized as systems that involve computational things that are in close contact with the physical environment and its ongoing activities while simultaneously providing and utilizing services for data access and processing (Liu & Wang, 2020).

CPS involves different emerging technologies for the future trends such as: big data architecture, IoT, Machine Learning, human machine communication as well as machine to machine communication (Mishra et al. 2022). Water, smart grid systems, transport, gas, energy, and healthcare, which are critical infrastructure, depend on CPS. These systems contain IoT devices, which produce enormous amounts of data and transfer that data to a centralized server. The qualities of blockchain, including as decentralization, data integrity, decentralized trust, enhanced security, cryptographic protocols, cryptocurrency, quicker settlements, and minting, can all be used to address various CPS concerns (Khalil et al. 2022).

Now a days our day to day life is transformed over internet revolutionization and semi conduction for the interaction and which may lead to the growth of information technology. HCPSs' are used in various fields to enable process optimization and some enhanced functionality (Rho et al. 2016). Neither cyber security nor physical security can save CPS because the vulnerability can never be introduced by removing patient variation and allow the responses with subject to one treatment (Fink et al.

2017). CPSs' are named as the cyber network system that include communication, computation, and cyber physical system itself considered the actuators and sensors (Ashibani & Mahmoud, 2017).

This chapter has the following structure: By using an illustration, Sect. 2 explores the concepts of SHCPS. The characteristics of HCPS are illustrated in Sect. 3. Modern HCPS technologies are provided in Sect. 4. In Sect. 5, we discuss the challenges of HCPS. Security mechanisms are elaborated in Sect. 6, and in Sect. 7, we draw a conclusion.

5.2 Smart Healthcare Cyber-Physical System (SHCPS)

CPSs are created as integrated systems that incorporate physical, networking, and computing processes. This provides convenient interaction between physical objects and cyber services. A CPS is a tool that uses computer-based algorithms to monitor and manage a system (AlZubi et al. 2021). A contemporary context-aware network for healthcare Cloud storage, IoT technology, and integrated devices are all included in CPS. The healthcare system was changed into a class of CPSs known as HCPS thanks to the integration of integrated medical equipment, networking capabilities, and complicated physical dynamics.

The health sector has advanced significantly recently as a result of the rise of modern technologies. HCPS is an expeditious emerging field, which may reach every characteristics of life from now on. A crucial part in this is played by the SHCPS. It is made up of numerous healthcare gadgets that are connected via a network to ensure flawless operation. Doctors have easy access to the patient's electronic health record (EHR) after it is collected and stored on the cloud. Cyberattack awareness has always been high in the healthcare sector. In the healthcare field, it is essential for systems to be dependable, secure, and cost-effectively store and share patient and institution-specific data. In CPS, the combination of passive and active user input, such as sensing devices and/or smart devices in healthcare facilities, can enable the data collection for effective decision-making (Haque et al. 2014). This combination of data collection and a decision-making system has not yet been thoroughly investigated in healthcare applications, hence it is a topic of great research interest. The emergence of synchronized interoperation of self-governing and adaptive devices, set trends for directing and functioning healthcare systems using computation and regulation, miniature implantable smart devices, body area networks, programmable equipment, and novel fabrication techniques are, for instance, opportunities for using CPS in healthcare.

A SHCPS is a special CPS that combines the complicated physical dynamics of patients in the modern medical field with networking capabilities and application security control devices. HCPS's data are created digitally, maintained electronically, and used remotely by healthcare professionals or patients as part of the communication, gadget, and communications system interaction of the HCPS, which is shown in Fig. 5.2.

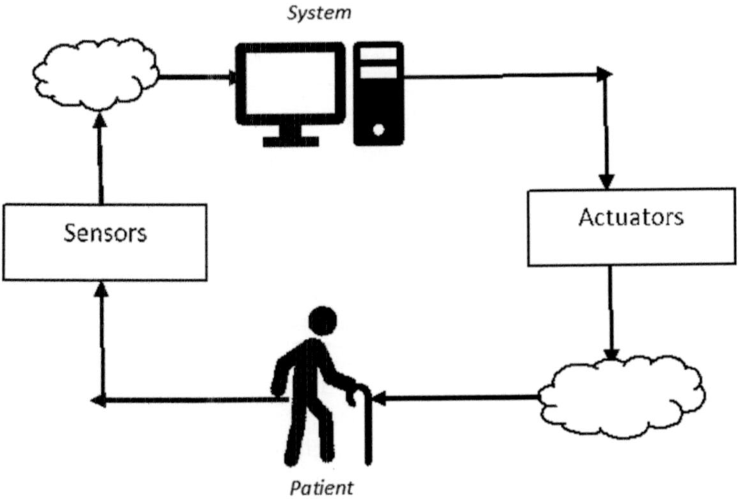

Fig. 5.2 CPS in smart healthcare

With the use of visualization techniques, the CPS, which is made up of sensors and actuators, assists humans in the majority of their daily activities. Pacemakers for humans and robot-controlled medical procedures are among the functions (Krishna et al. 2021).

As a result, HCPS emerges as an unstoppable force that have a significant impact on the medical industry, particularly in light of the development of the market for medical devices that uses integrated software and networking interface. With the development of medical technology, it became intriguing to integrate devices using cloud applications that simultaneously analyse the patient's various physiological features in place of equipment that were created to treat patients independently of one another (Dey et al. 2018).

5.3 Characteristics of Healthcare Cyber-Physical System (HCPS)

Future systems called SHCPS will help the medical community effectively manage pandemic situations. The integration of technologies, organizational domains, life cycles, and "smartness" are characteristics of CPSs. The following characteristics—technical specialization, cross-cutting elements, degree of automation, and life-cycle integration—can be used to describe CPS. CPS in healthcare offers a variety of applications, including care for the elderly, assisted living, and hospitals. The particular application has a big impact on the system complexity. Depending on the relevant area, certain organization of architectural elements may be required. Architecture may have managed aspects in a controlled environment, like an intensive care unit in a medical. Contrarily, it can be necessary to incorporate several automated features in the architecture of an assisted living facility. The two categories of CPS in healthcare applications are controlled and aided (Haque et al. 2014).

HCPS can be categorized at different levels (Verma, 2022). These are as follows.

(a) Unit Level: –Patient monitoring and control are provided by fundamental cyber-physical healthcare systems at the hospital or unit level for patients who are hospitalized. It continuously keeps track of the patients' physiological parameters, including their pulse, hypertension, respiratory rate, etc.
(b) Integration Level: –It is the second level of HCPS. At this level, hospitals connect with smart homes to offer patients remote controlling and medical services.
(c) System Level: –Third level of HCPS is the system level. A Smart City HCPS is formed at this level by a variety of automated systems that assist the HCPS.
(d) Acceptance Level: –Researchers, technicians, engineers, and health professionals collaborate at this level to execute the health sector and make the healthcare system effective.
(e) Evolutionary Level: –Future HCPS systems that have the characteristics of autonomy and personality are evolutionary level HCPS systems.

5.4 Modern HCPS Technologies

HCPS integrates a network of medical equipment, which is essential to healthcare. Hospitals employ these systems more and more to provide consistent, high-quality healthcare. The healthcare network's verification and validation of various devices, followed by encrypted transmission, is an essential component since it validates the credentials of the users (Adil et al. 2022).

The involvement of numerous technologies, including IOT, robotics, artificial intelligence, machine learning, blockchain, cloud computing, and big data in the field of HCPSs is presented in this section. This is illustrated in Fig. 5.3.

5.4.1 Internet of Things (IoT)

The IOT ushered in a novel age of machine-to-machine (M2M) communication, that can be done via Bluetooth, wireless networks, Near Field Transmission, radio communication, and other technologies (Aceto et al. 2019). The IoT enables smart, cyber-physical healthcare systems at the integration level where sensor networks produce large volumes of data that are then sent to distant computers for control and monitoring functions. There are some gadgets linked to mobile devices, which transmit data to cloud servers for healthcare (Verma, 2022).

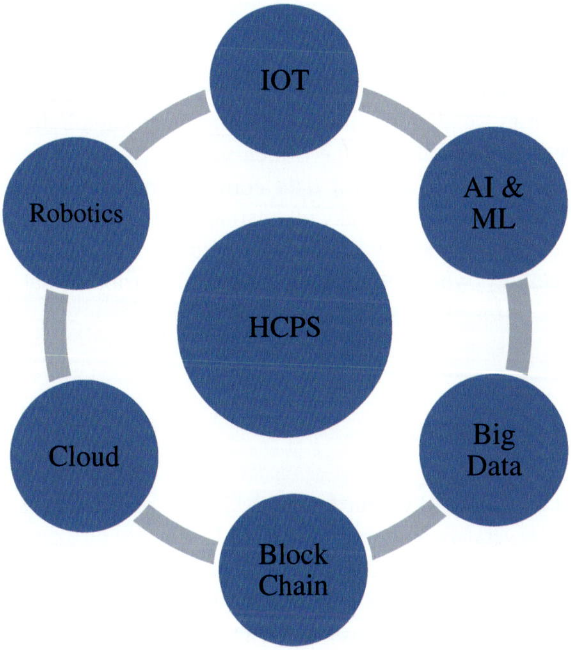

Fig. 5.3 Technologies of HCPS

5.4.2 *Artificial Intelligence & Machine Learning (AI & ML)*

A variety of areas of healthcare, such as medications, healthcare implants, patients, and diseases, can be improved by AI & ML (Verma 2022). ML-based solutions are used to offer different treatment options, personalize therapies, increase the general effectiveness of hospitals and healthcare systems, and reduce overall healthcare costs. (Pourhomayoun & Shakibi, 2021). Symptoms, pre-health condition, and demographic elements that crucially affect the disease activity of different patients.

5.4.3 *Big Data Analysis*

Healthcare CPSs can now discover hidden samples and linkages in massive amounts of information that have been obtained from multiple sources using latest computing paradigms with excessive processing power and big data technology (Verma 2022). High-end computer tools and deep learning models make it possible to analyze huge datasets and find these obscure patterns. Deep learning, machine learning, and statistical methods can be used to reveal these underlying patterns.

5.4.4 *Blockchain*

Blockchain technology has been investigated for use in a range of applications, including smart cities, smart agriculture, and smart healthcare. Different approaches have been suggested to deal with the difficulties and issues, whether they relate to contract tracing records, medical data coming from the Internet of Medical Things (IoMT), mobile gadgets, or any other smart device. Electronic health records (EHR) kept on cloud servers may be accessed securely and authentically, thanks to blockchain technology, which offers a decentralized distributed database (Xu et al. 2019).

5.4.5 *Cloud Computing*

Combining and storing IoMT data for analysis and processing is its main goal. Users and cloud servers are authenticated to IoMT devices utilizing OTP validation techniques and user-id or password generation (Vangipuram et al. 2021). The difficult problem is ensuring the confidentiality depending on the security environment of patient healthcare records used to protect the key generation center, which keeps the secret keys of all genuine users (Xu et al. 2019).

5.4.6 *Robotics*

CPSs use robots extensively, including robotic surgery to aid with operations and patient care, industrial robots to carry out production activities, and monitoring robots to ensure safety and protection. Robots like Moxi carry out a variety of tasks during pandemics (Verma 2022), including distributing PPE kits, carrying out COVID 19 tests, and picking up and dropping off patients. By taking patients' temperatures and facilitating video conferences between patients and their loved ones, the robot Mitra supports medical staff.

5.5 Challenges in HCPS

Potential issues are linked to various systems, including CPS connected to healthcare and IoT, embedded software and green crypto, smart building safety, privacy concerns in CPS, confidence and protection in CPS, emerging security schemes for embedded security, sensitive information handling in CPS, and other unresolved issues brought on by the diversity of technologies used in IoT integrated CPS (Jamal et al. 2021). Healthcare cyber physical system (HCPS) is the combination of a network of critical

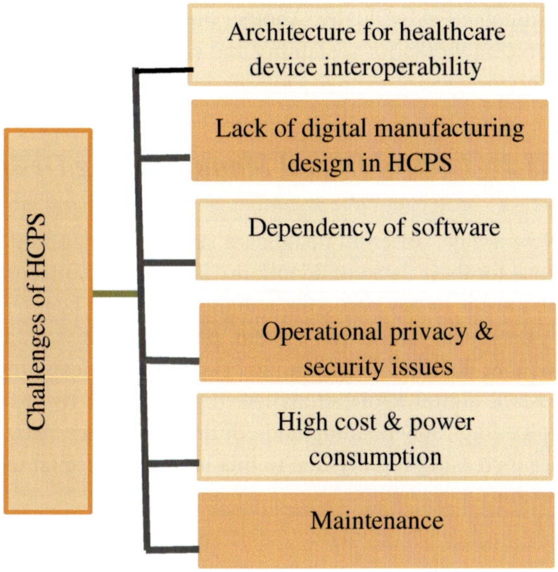

Fig. 5.4 Challenges of HCPS

medical gadgets. These systems are progressively used in different healthcare sectors to achieve very high quality outcomes by adding more intelligence and providing an ability to process major factors for CPS application. Here some issues and challenges described below which may provide some better opportunities to the healthcare world.

Designing a cost-effective healthcare system is a challenge. The goal of CPS research in this field is to create a smart sensing system that can control health, anticipate a patient's future condition, provide treatments depending on that patient's condition, provide virtual healthcare, use semi-automated robots to help patients with physical activities, and more (Kurde et al. 2019). Different challenges of HCPS are represented in Fig. 5.4.

5.5.1 Architecture for Healthcare Device Interoperability

Recently this distributed HCPS are built by integrating different inter-device communication protocols along with healthcare professionals. CPS faces so many limitations of the computational complexity and physical dynamics such as time and cost management, system and process integration, correctness of data, and structural standards (Giansanti 2021). The architecture of HCPS is highly dominant for the performance and feature of the system. The basic model of HCPS needs to be developed based on the concern of the application and integration of the system, and required information for the user. The HCPS based architecture for healthcare can be derived from the point of infrastructures such as server based architecture which is small and maintained individually. Another one is about the cloud based model which may use in recent works for scalability and easy accessibility.

5.5.2 Lack of Digital Manufacturing Design in HCPS

Now-a-days different healthcare sectors focus on different cyber security applications for their data confidentiality and manage information. The healthcare application needs to pay particular attention for some healthcare networks, communication system, user interaction system, pictorial archiving, and some wearable healthcare tools & devices (Giansanti 2021). In digital designing model this may go through several digital applications like digital health records, digital work station, image uploading, downloading and editing etc. As we know in today's world cyber risks are increasing rapidly due to this increasing digital technologies. We need to focus on this.

5.5.3 Dependency of Software

Dependency consists of two other factors namely safety and reliability which ensures the rights of Quality of Services (QoS). Dependency of a software mainly involves digital affliction, cyber vulnerabilities, consequences' & professional repercussion, digital security and digital everything. To understand the dependency of software regarding data collection and assessment, the network security may go through several steps like: IT software and hardware, level of deployment, risk assessment and establish many strategies (Tong 2022).

5.5.4 Operational Privacy and Security Issues

In healthcare, security and privacy is an essential discussion for patient-data & patient-doctor confidentiality from legal and ethical perspectives. As a matter of fact it is an important task to make sure the confidentiality of collected healthcare information along with patient data. Operational security was taken into consideration for an analytical process, rival collection capabilities, and accentuate the value of sensitive and critical information.

Different levels of security are co-related with higher computational costs, which are not only involved some training programs, up-gradations, and several operational phases, but also restricted to spending for improvement of any properties. So to secure, the security of information is an important matter of question from information and application level that can be considered by encoding the data and privacy setting.

5.5.5 High Cost and Power Consumption

The assumption of security measures in healthcare has many advantages when it associated with the protection of CPS components, levels and areas. But apart from that there are some limitations regarding the performance of the system due to time complexity which was associated with higher power consumption along with high cost as well. Power consumption is an important factor, especially for battery backup and resource constrains, which may affect the system performance and reduce it. The more power consumption means a shorter lifespan of battery life and a higher cost to maintain their availability. More power consumption and higher cost are corresponded to each other. When it consumes more power, then it may increase the cost for access and maintenance. Security measures are associated with more estimation costs, which may never be confined with initial phase of security operations, but also include restore, prepare and upgrade for operations (Yaacoub et al. 2020).

5.5.6 Maintenance

In Computer network management, security and maintenance, systems are still facing many inconsistencies' and safety contradictions. The maintenance of the HCPS basically focuses on risk assessment of medical information, establishment of enhanced approaches and also management of hardware equipment's. The healthcare units incorporate with network security and maintenance into the information construction system and need to focus on the possible risk factors to innovate and renovate the computer network management system and manage it.

5.6 HCPS Security Mechanisms

The security and privacy issue has facinated a lot of awareness and has been creating a lot of issues and controversies now often. Here some of the security mechanisms that were identified in the smart and digital healthcare system (Nasiri et al, 2019). In today's healthcare systems EHR is an enhanced and a necessary model of security mechanism for the patients' data storage, access, and retrival. The basic structure of security in EHR is based on the cloud storage system. All kinds of informations regarding patients particulars, physicians, and healthcare professionals are stored in the cloud based cyber physical system. That can be accessed by some specific users who were having the cyber security access. All these database services are managed by the EHR manager.

Security of the healthcare devices and medical data is analyzed by the "Integrated Fuzzy AHP-TOPSIS" technique (Alzahrani et al. 2022). Security mechanism in healthcare is functioning a design and process it to provide several security services

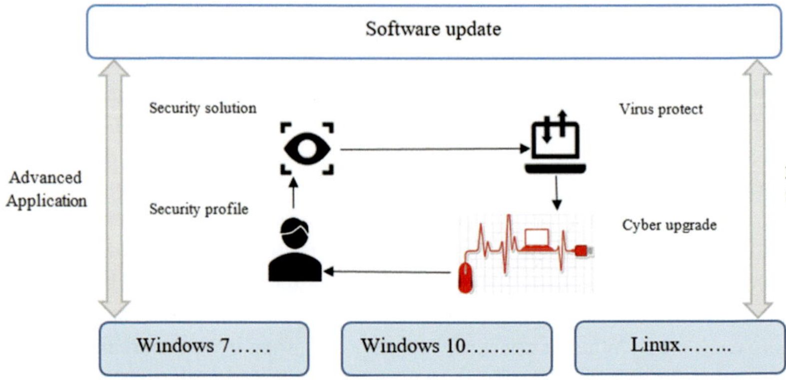

Fig. 5.5 Mechanism of healthcare security

for a particular result in terms of accuracy, assurance, and strength of a system. The Fig. 5.5 represents the basic workflow structure of security mechanism. This may include different types of operating system which can be accessed by the user. For the porpose of data privacy we need to focus on some advanced applications that are based on the security profile of the user and it's solution if any theft arises. The solution of cyber theft can be reduced by upgradation of cyber security applications and protection from any unethical accessses.

5.6.1 Security Mechanism in Block Chain Scheme

There are some security mechanisms corresponding to the block chain mechanism which is based on different assumptions such as: cryptography, fingerprint optimization, data encryption, management system in healthcare database. Block chain technology has emerged rapidly associated with the cloud domain and internet of healthcare security (IoHS) (Pelekoudas-Oikonomou et al. 2022). The healthcare units and tools are correlated in such a way that anyone can access the information whenever they need from any location. IoHS based healthcare management system provides customized and costumer focused healthcare duties by reducing the limitations as: Time, Power, Cost, Maintenance, and Locality. The computational work for digital and smart healthcare system can be reduced in terms of block chain technology. In the field of healthcare the block chain based application needs to be deployed. Edge networking is a frequent type of block chain based security mechanism specially designed for the server based security application (Fig. 5.6).

Block chain system may be considered in two types. First one is the Public block chain mechanism and the other one is the Private block chain mechanism. For todays' healthcare sectors and other organizations as well private block chain is primarily used for the data security and the unauthorized access from the EHR is blocked.

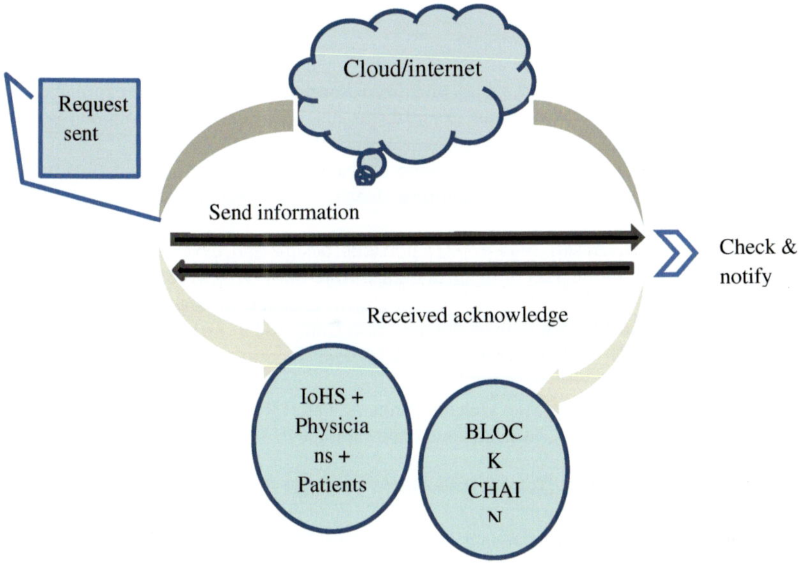

Fig. 5.6 Block chain based mechanism

5.7 Conclusion

Hospital facilities include the use of healthcare equipment and tools. Healthcare gadgets are the main resources for both patients and medical professionals, from patient health scanning and disease diagnosis to report and treatment. The healthcare facility must integrate computer network security into the creation of the system for storing medical and patient information as part of a specified implementation process. IoT and CPS devices are targets for fraudsters, attackers, and other unethical users that are captivated by the vast amount of information disseminated across medical devices. This information might be extremely harmful to anyone involved if it ends up in the wrong hands. IoT-CPS security considered AI-enabled algorithm is therefore required for the future. The security and privacy concerns of CPS in healthcare applications are examined in this chapter. We assume that these issues and challenges will provide enough inspiration for future discussions and an interest in research on security issues for CPS in healthcare applications.

References

Aceto G, Persico V, Pescapé A (2019) A survey on information and communication technologies for industry 4.0: state-of-the-art, taxonomies, perspectives, and challenges. IEEE Commun Surv & Tutor 21(4):3467–3501.

Adil M, Khan MK, Jadoon MM, Attique M, Song H, Farouk A (2022) An AI-enabled hybrid lightweight authentication scheme for intelligent IoMT based cyber-physical systems. IEEE Trans Netw Sci Eng.

Alzahrani FA, Ahmad M, Ansari MTJ (2022) Towards design and development of security assessment framework for internet of medical things. Appl Sci 12(16):8148

AlZubi AA, Al-Maitah M, Alarifi A (2021) Cyber-attack detection in healthcare using cyber-physical system and machine learning techniques. Soft Comput 25(18):12319–12332

Ashibani Y, Mahmoud QH (2017) Cyber physical systems security: analysis, challenges and solutions. Comput Secur 68:81–97

Cabello JC, Karimipour H, Jahromi AN, Dehghantanha A, Parizi RM (2020) Big-data and cyber-physical systems in healthcare: Challenges and opportunities. In: Handbook of Big Data Privacy. pp 255–283

Dey N, Ashour AS, Shi F, Fong SJ, Tavares JMR (2018) Medical cyber-physical systems: a survey. J Med Syst 42(4):1–13

Fink GA, Edgar TW, Rice TR, MacDonald DG, Crawford CE (2017) Security and privacy in cyber-physical systems. In Cyber-physical systems. Academic Press, pp 129–141.

Giansanti D (2021) Cybersecurity and the digital-health: the challenge of this millennium. Healthcare 2021(9):62

Gunes V, Peter S, Givargis T, Vahid F (2014) A survey on concepts, applications, and challenges in cyber-physical systems. KSII Trans Internet Inf Syst (TIIS) 8(12):4242–4268

Haque SA, Aziz SM, Rahman M (2014) Review of cyber-physical system in healthcare. Int J Distrib Sens Netw 10(4):217415

Iqbal R, Doctor F, More B, Mahmud S, Yousuf U (2020) Big data analytics and computational intelligence for cyber-physical systems: recent trends and state of the art applications. Futur Gener Comput Syst 105:766–778

Jamal AA, Majid AAM, Konev A, Kosachenko T, Shelupanov A (2021) A review on security analysis of cyber physical systems using Machine learning. Mater Today: Proc.

Khalil AA, Franco J, Parvez I, Uluagac S, Shahriar H, Rahman MA (2022) A literature review on blockchain-enabled security and operation of cyber-physical systems. In: 2022 IEEE 46th Annual Computers, Software, and Applications Conference (COMPSAC). IEEE, pp 1774–1779.

Krishna M, Chowdary SMB, Nancy P, Arulkumar V (2021) A survey on multimedia analytics in security systems of cyber physical systems and IoT. In: 2021 2nd International Conference on Smart Electronics and Communication (ICOSEC). IEEE, pp 1–7.

Kurde S, Shimpi J, Pawar R, Tingare B (2019) Cyber physical systems (CPS) and design automation for healthcare system: a new era of cyber computation for healthcare system. Structure 6(12).

Latif SA, Wen FBX, Iwendi C, Li-li FW, Mohsin SM, Han Z, Band SS (2022) AI-empowered, blockchain and SDN integrated security architecture for IoT network of cyber physical systems. Comput Commun 181:274–283

Liu H, Wang L (2020) Remote human–robot collaboration: a cyber–physical system application for hazard manufacturing environment. J Manuf Syst 54:24–34

Mishra A, Jha AV, Appasani B, Ray AK, Gupta DK, Ghazali AN (2022) Emerging technologies and design aspects of next generation cyber physical system with a smart city application perspective. Int J Syst Assur Eng Manag 1–23.

Nasiri S, Sadoughi F, Tadayon MH, Dehnad A (2019) Security and privacy mechanisms of internet of things in healthcare and non-healthcare industry. J Health Adm 22(4):86–105

Pelekoudas-Oikonomou F, Zachos G, Papaioannou M, de Ree M, Ribeiro JC, Mantas G, Rodriguez J (2022) Blockchain-based security mechanisms for IoMT Edge networks in IoMT-based healthcare monitoring systems. Sensors 22(7):2449

Pourhomayoun M, Shakibi M (2021) Predicting mortality risk in patients with COVID-19 using machine learning to help medical decision-making. Smart Health 20:100178
Priyadarshini I, Kumar R, Tuan LM, Son LH, Long HV, Sharma R, Rai S (2021) A new enhanced cyber security framework for medical cyber physical systems. SICS Softw-Intensiv Cyber-Phys Syst 35(3):159–183
Rho S, Vasilakos AV, Chen W (2016) Cyber physical systems technologies and applications. Futur Gener Comput Syst 56:436–437
Tong H (2022) Maintenance of network security in hospital information construction based on the internet of things. Int Trans Electr Energy Syst.
Vangipuram SL, Mohanty SP, Kougianos E (2021) CoviChain: a blockchain based framework for nonrepudiable contact tracing in healthcare cyber-physical systems during pandemic outbreaks. SN Comput Sci 2(5):1–16
Verma R (2022) Smart city healthcare cyber physical system: characteristics, technologies and challenges. Wireless Pers Commun 122(2):1413–1433
Xu J, Xue K, Li S, Tian H, Hong J, Hong P, Yu N (2019) Healthchain: a blockchain-based privacy preserving scheme for large-scale health data. IEEE Internet Things J 6(5):8770–8781
Yaacoub JPA, Noura M, Noura HN, Salman O, Yaacoub E, Couturier R, Chehab A (2020) Securing internet of medical things systems: limitations, issues and recommendations. Futur Gener Comput Syst 105:581–606

Part II
Internet of Things

Chapter 6
Application of Machine Leaning for Intrusion Detection in Internet of Things

Ravi Sharma, Nonita Sharma, and Aditi Sharma

6.1 Introduction

The Internet of Things (IoT) is essentially the concept of connecting devices, computers, or sensors through a wired or wireless communication medium (Khan et al. 2012). Every device that connects to the internet is a part of the IoT system. Figure 6.1 shows the different areas that use IoT devices. Health services, transportation, and smart grids are some domains that are heavily dependent on IoT, and the demand for IoT devices is increasing daily. That is why the IoT has become a part of our lives. These days, every device is connected to other devices to facilitate communication and share information.

Electronic devices have changed our lives since the inception of the computer. With the help of the internet and technological advancement, every device we use has some automated features that make these devices smart. These features not only make devices easy to use, but they also help us in multiple ways. For example, smartwatches are IoT devices. They are different from normal analog and digital watches. With the help of these smart watches, we can track our sports activities. We can measure heart rate and SpO2. They also show the phone's notifications. There are many such areas where IoT devices are used and provide reliable and accurate services. But some major concerns make these devices a threat to users and organizations that use them.

R. Sharma (✉)
Department of Computer Science and Engineering, NIT Jalandhar, Jalandhar, India
e-mail: ravis.cs.19@nitj.ac.in

N. Sharma
Department of Information Technology, IGDTUW, New Delhi, India
e-mail: nonitasharma@igdtuw.ac.in

A. Sharma
Department of Computer Science, Nazarbayev University, Astana, Kazakhstan
e-mail: aditi.sharma@ieee.org

N. Sharma et al. (eds.), *Big Data Analytics in Intelligent IoT and Cyber-Physical Systems*, Transactions on Computer Systems and Networks,
https://doi.org/10.1007/978-981-99-4518-4_6

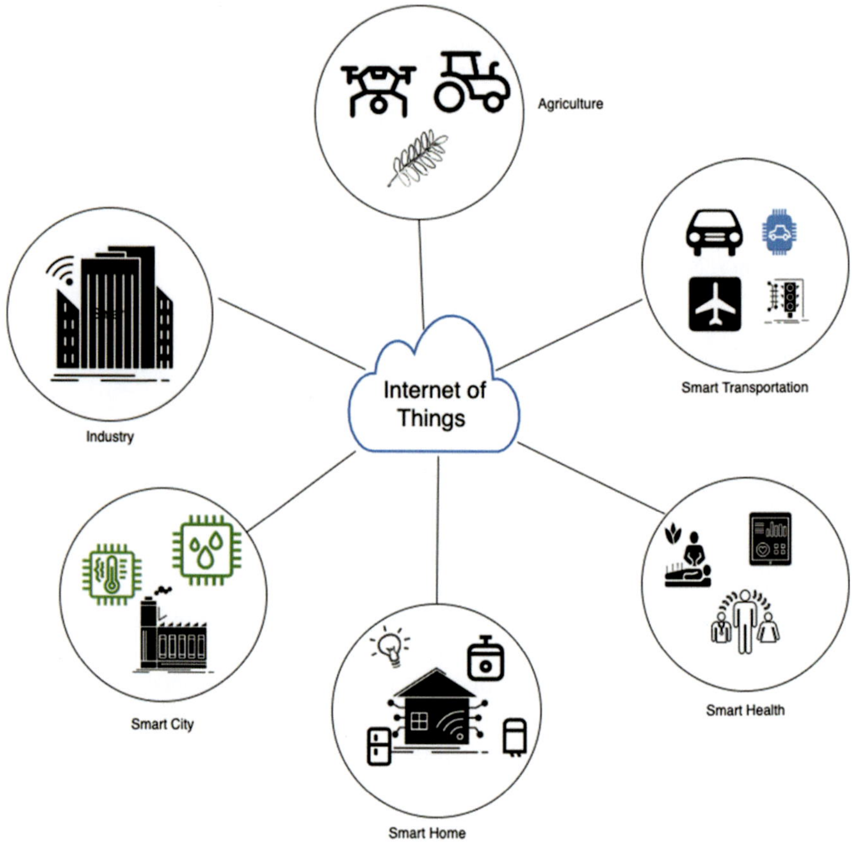

Fig. 6.1 IoT applications

The major issues with IoT devices are privacy and security (Das 2015; Sharma and Sharma 2022).

IoT devices are made in such a way that they should consume fewer resources so that they can be used in remote locations. IoT devices are traditional devices with extra functionalities. For example, a smart bulb is an IoT device. The main function of the bulb is to illuminate the area, and now we can control it with our smartphones. To control a smart bulb with a phone, we need to add some sensors and integrated chips into the bulb, but we must be aware that to add extra functionality, we should not forget the device's main function and size. To make the best combination, we must make the devices in such a way that they will consume fewer resources. Optimum resource utilization (ORU) is the main aim of any IoT device. ORU has made IoT devices more vulnerable. Data is encrypted when we transfer it from any electronic device, laptop, or mobile. There should be ample resources available to encrypt it; to protect these devices, these systems use firewalls. But if we want to achieve high security in IoT devices, they will become costlier.

In this study, we use machine learning (ML) models that are used these days for intrusion detection in the IoT. The remaining sections of the chapter are Sect. 2 is about related work, which describes recent work in this field; Sect. 3 is about how to use ML for intrusion detection, which explains how to analyze data and what different tools and techniques were used for intrusion detection. Section 4 is a result section where we analyze our ML model results. The conclusion is the final section, provides a final summary and what future work can be done in this field.

6.2 Related Work

In recent years, the use of machine learning algorithms for intrusion detection in the IoT has received significant attention. IoT comprises many interconnected devices and sensors that collect and process data. The data that these devices and sensors collect can be used to train machine learning models that can be used to find intrusions and attacks.

Several studies have applied machine learning algorithms to intrusion detection in cyber physical systems. In (Vijayanand et al. 2017), the authors use a support vector machine (SVM) to detect anomalies in network traffic data collected from a power grid system. They demonstrate that their approach can accurately detect a wide range of known and novel attacks. In (Sawadogo et al. 2021), the authors use a deep learning approach to detect intrusions in industrial control systems. They show that their system can accurately detect various types of attacks.

In (Rathore and Park 2018) ELM classifier and Fuzzy C-mean-based attack detection framework for IoT. They've used fog computing. ELM was trained on labeled data. The fuzzy C-mean cluster algorithm was trained on labeled and unlabeled data to handle new data and attacks. For testing, they used the NSL-KDD dataset. The dataset's attack detection accuracy was 86.53% and 11 ms. This framework can detect IoT attacks in smart homes and cities.

Authors in Punithavathi et al. (2019) proposed cancellable biometric authentication for IoT devices. The traditional biometric authentication uses a single biometric template. Without the template, it's impossible to revoke service. Cancellable biometric authentication is canceled if the template is tampered. This study aims to develop a lightweight cloud-based CBS for IoT user authentication. This cloud-based cancellable biometric system uses random projection transformation. Performance is checked using four databases. Each database has 100 fingers and eight images. EER was used for accuracy. This model is accurate in all four databases compared to the previous study. All operations are performed on the cloud, and data offloading from IoT devices to the cloud is not discussed.

In (Outchakoucht et al. 2020) the authors modeled IoT access control. Authors say all Internet-connected devices are IoT. Some devices, like irrigation sensors, are resource-constrained, while others are powerful. They've categorized these devices as C1, C2, and C3. C1 has least powerful devices, C3 the most. They proposed an access control model and did a smart city case study. All three device categories can be covered in a smart city case study. In this case study, an access control policy is set for a car rental agency with normal, VIP, and blacklisted customers.

Authors in Verma and Ranga (2019) proposed an anomaly-based DoS intrusion detection system. An anomaly-based IDS detects system attacks. Single classifiers (MLP, CART) and ensemble classifiers (RF, AB, GBM, XGB, and ETC). These classifiers were tested on CIDDS-001, UNSW-NB15, and NSL-KDD datasets. Different classifiers and datasets should eliminate bias. Two statistical methods were used to compare classifiers. They've used FPR, AUC, etc., to measure classifier performance. According to this study, CART and XGB classifiers can be used to build anomaly-based IDS for IoT.

In this study, the authors (Xiao et al. 2018) did a literature review of IoT attacks and ML-based security solutions in IoT. They discussed DoS, jamming, spoofing, man in the middle, software, and privacy leakage attacks. For handling the above attacks, they have discussed four security solutions: authentication, malware detection, IoT offloading, and access control. Each security solution can handle different attacks. An authentication-based security solution is used to handle eavesdropping and sybil attacks. Similarly, access control is used to handle privacy leakage, malware detection, and DoS attacks. They also reviewed ML techniques used to achieve these security solutions in IoT.

Authors in Cui et al. (2018) surveyed machine learning applications used on the Internet of Things. In this study, the authors try to include all domains of IoT where ML is used. They have divided this survey into six sections, i.e., traffic profiling, IoT device authentication, security, edge computing, software-defined network, and IoT applications. Each section is further divided into subsections discussing ML techniques used in different areas of that domain. In each section, challenges and open issues were discussed. That shows challenges and hurdles come in that domain to implement Machine Learning. In this study, the authors try to include all domains, but in-depth details of the domains are not provided.

The literature shows that machine learning can effectively detect intrusion detection in cyber-physical systems. The specific type of machine learning algorithm used will likely depend on the data collected by the system. Additionally, future work should focus on developing more robust methods for detecting novel attacks.

6.3 Experimental Setup

ML and Deep Learning (DL) are the techniques that make machines intelligent. These techniques have more accuracy than humans in some fields (image classification). ML and DL work on data. If we have more data, they could provide better results. On the other hand, IoT devices are generating huge amounts of data, so we can analyze the data and get insights from data. With the help of machine learning, we can detect different anomalies in the network. As we know, IoT devices are resource constrained, and they have no encryption and security features that make them an easy target for intruders.

IoT devices generate huge amounts of data, which is used to detect intrusions in the network. With machine learning, there are different steps for attack and intrusion detection in the IoT. In this part, we'll discuss tools so you can detect attacks and analyze any data.

6.3.1 Wireshark

Wireshark is an open-source network packet analyzer. This will give insights into the network packets transmitted in a network. There is a different information that Wireshark can provide. With Wireshark's help, we can also monitor our system network. It will be helpful to check the source port, destination port, what message is in transfer over the network, what protocols are used for transmission, and many more.

In this study, Wireshark is used to analyze the raw data of an IoT network. There are lots of datasets that are available online. This data is mostly available in pcap (packet capture) format. A pcap file contains the whole network data, and these are very large files. So, it's a good practice first to check this data on Wireshark and analyze it. Wireshark is not only used to analyze the data, but it can also filter the data, and we can also export this data in CSV format. The CSV format is one of the most common python dataset extensions.

As we already mentioned, Wireshark is open-source software available for all operating systems. It can be downloaded from Wireshark Download (2022). After installing, we can import any pcap file and analyze the packet. Figure 6.2 shows the Wireshark home windows. In the bottom half of the window, we can check the data transferred over the current network. Figure 6.3 shows the different features of a dataset imported into Wireshark.

6.3.2 Anaconda Navigator

Anaconda navigator is a very popular python distribution platform. This platform has many preinstalled applications. Some of the applications that come with this navigator are jupyter notebook and Rstudio, which run the Python and R languages, respectively. In this study, we used a jupyter notebook to run the python code. Another advantage of the anaconda navigator is that it comes preinstalled with python libraries, making it very simple to write python code without waiting for library installation. An anaconda navigator can be downloaded from reference (Anaconda | Anaconda Distribution 2022). This study uses python programming to implement the ML model, and a jupyter notebook is used to write the code. Figure 6.4 shows the home window of the navigator. Then we must select the jupyter notebook. The welcome window of the notebook is shown in Fig. 6.5.

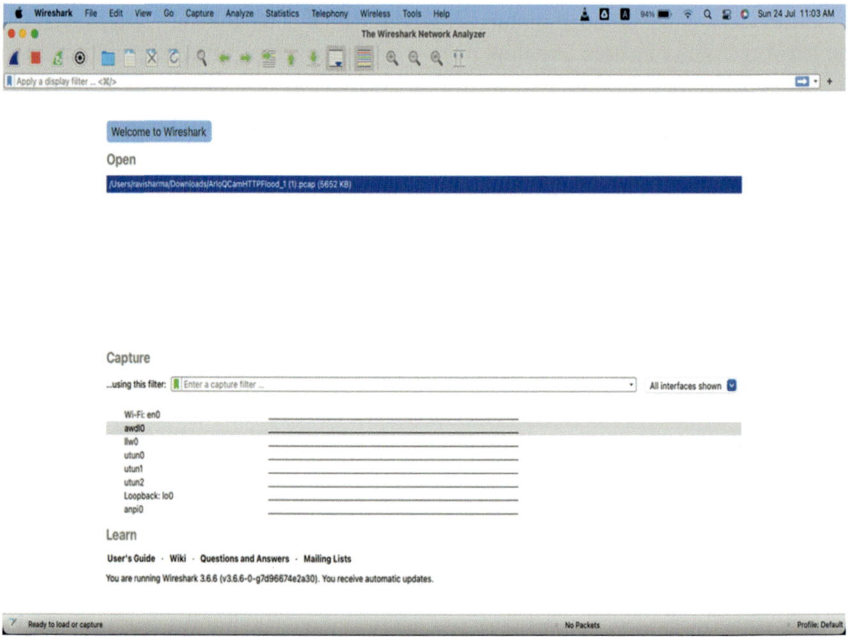

Fig. 6.2 Wireshark window

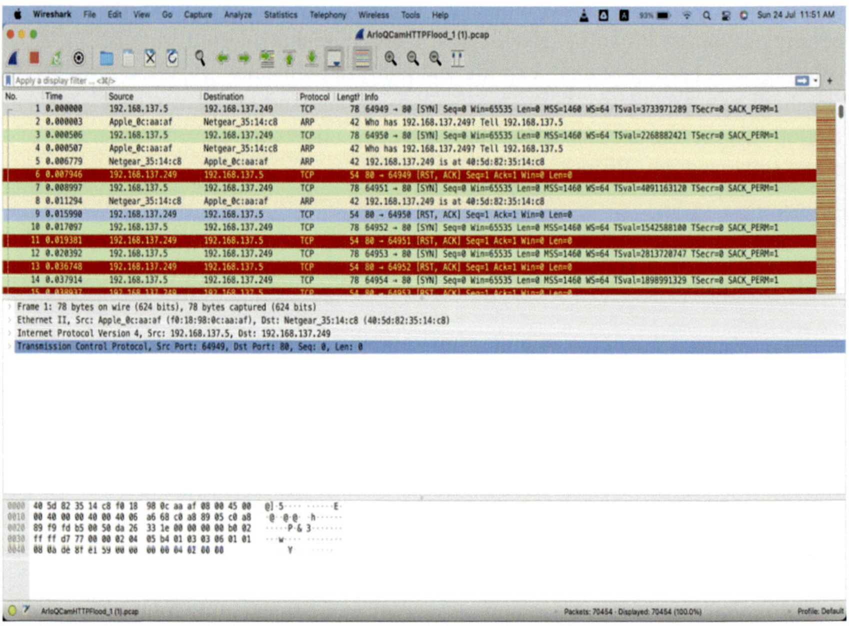

Fig. 6.3 Packet analysis with wireshark

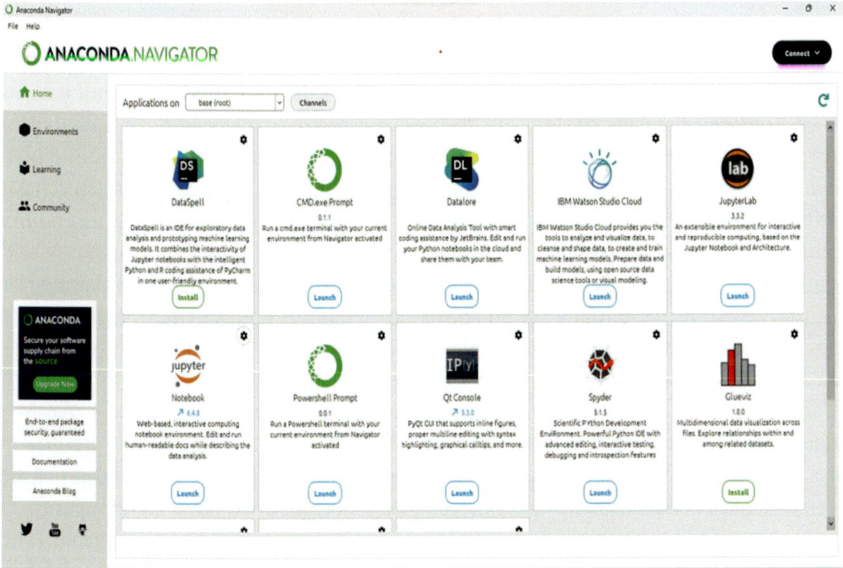

Fig. 6.4 Anaconda navigator

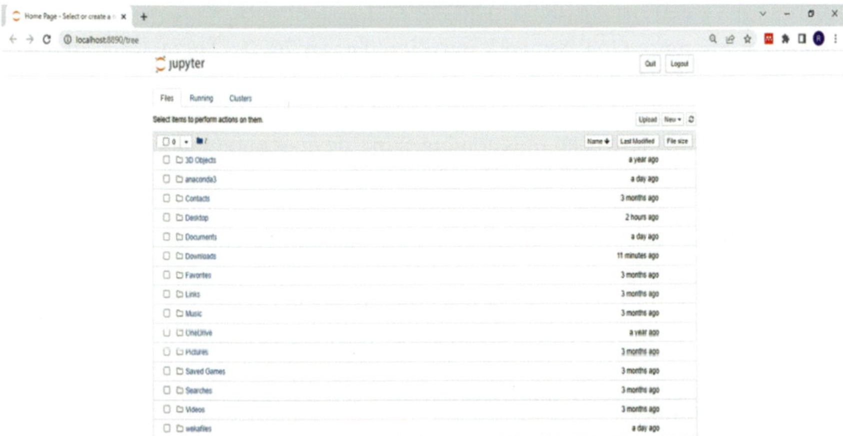

Fig. 6.5 Jupyter notebook

6.3.3 Dataset

A dataset is the backbone of any ML and DL model. If these models are not trained on well-defined datasets or datasets do not have features that can have meaningful information, then these models are just a line of code and provide less accurate results. Model accuracy is dependent on the quality of the dataset. The dataset should have features related to the predicted class, and the amount of data should be large enough so that we can train our model. Obtaining a massive amount of data is difficult and requires lots of resources and different phases. Before we get the final dataset for machine learning techniques on an intrusion detection dataset, there are multiple steps. The first step is to make a testbed to collect the data. This data is raw network data and is stored in the form of pcap files. Then this data is analyzed and divided into different features. These features will be provided as input to ML models. An ML model can be supervised or unsupervised. There is one label called the prediction class. If this class is given in the dataset, then the dataset has a label, and it can be given to a supervised model, but if there is no label, then we have an unsupervised model. After finalizing the feature and label, this data is stored in CSV format. The ML model now uses this dataset for training and testing. Before training the ML model, we normally analyze the data to provide good results, which is called the data preprocessing phase. These steps are covered in the result section. For this study, we have used the IoTID20 dataset (Ullah and Mahmoud 2020). This dataset is freely available in reference (IoT Network intrusion dataset 2020). There are 80 features in this dataset and 600 thousand instances.

The IoTID20 dataset was generated from the pcap file (Kang et al. 2019). The testbed to generate these pcap files combines IoT devices and interconnecting structures. They have generated 42 pcap files. With the help of smart home devices EZVIZ Wi-Fi camera and SKT NGU, a smart home environment was developed. These two devices are linked to a Wi-Fi router in a smart home. Laptops, tablets, and smartphones are some of the other electronic devices that can be connected to the smart home router. IoT victim devices include the SKT NGU and the EZVIZ Wi-Fi cameras, while all the other devices in the testbed are attacking devices (Omar and George 2021).

Dataset Description

IoTID20 the dataset has 83 features, three target label features, and 625,783 network packet instances. In the first output class features, these instances are divided into normal and anomaly instances. So, this label can be used for binary classification. In the 2nd label, they have divided instances into five categories. The number of instances with category names is shown in Table 6.1. The third label categorized the previous five categories into a subcategory. The name of these three columns in the dataset is a label, Cat, and Sub_Cat. For this study, we have used Cat as a predicted class.

Table 6.1 Details of dataset instances

Category	Number of instances	Percentage
Mirai	415,677	66.42
Scan	75,265	12.03
DoS	59,391	9.49
Normal	40,073	6.41
MITM	35,377	5.65
Total	625,783	100

6.3.4 Data Pre-processing

First, data is collected from various sensors and systems within the IoT environment. This data is then processed and fed into an ML algorithm. The algorithm is trained to recognize patterns that are indicative of an intrusion. Once the algorithm is trained, it can be deployed within the IoT environment to monitor intrusions in real-time.

Data preprocessing is an important step in machine learning. This step is crucial in ensuring that the data is ready for modeling and that any potential biases are removed. One of the most important aspects of data preprocessing is normalization. This involves rescaling the data so that it is within a certain range. This ensures that no single value has too much of an effect on the model so that the machine learning algorithm can learn from the data more accurately.

6.3.5 Machine Learning

ML is a learning process that is used to make a machine intelligent. Different learning techniques are used, and they are broadly classified into three categories: unsupervised, supervised, and reinforcement learning. Supervised learning is a learning where the model is provided with label data, which means the machine knows for a specific input the output during the training phase. In the testing phase, the machine is given data where the label is provided only on the input data and tries to predict the output label.

On the other hand, in unsupervised learning, there is no target label during any phase. The machine tries to find relations between features in the data. Reinforcement Learning works on the concept of an agent and environment, where agents learn from the surrounding environment and try to improve their performance. ML techniques include, but are not limited to, decision trees, K Nearest Neighbor (KNN), Random Forest, Support Vector Machine (SVM), Logistics Regression, and Artificial Neural Network (ANN). ANN is a technique that is inspired by the human neural system. In the human neural system, millions of neurons work together to make decisions. Similarly, when we use several artificial neural networks to make intelligent machines, it is called a Deep Neural Network, and a new term that comes into the picture is called Deep Learning (DL).

In this study, we have used different supervised ML models on labeled datasets. Different decision trees, SVM, and neural network models were used to detect different attacks in the dataset. The following sections provide details of these models.

Decision Tree

A decision tree is a machine learning model used to predict the probability of an event occurring. Decision trees split data into smaller sets until only one outcome is left. Each branch of a decision tree represents a test on some aspect of an object (for example, its age, the number of siblings, and so on). The leaves are the outcomes of the branch and can be either "success" or "failure," depending on the objective of your prediction model. Decision trees are often confused with other types of modeling, such as neural networks. However, they are very different. Neural networks are connected graphs with many layers. Their connections are often referred to as "weights." Decision trees have no hidden layers, weights, or parameters. Decision trees are one of the simplest machine learning models out there. Figure 6.6 shows the graphical representation of the decision tree. In this model root node and non-leaf node are the input features. Leaf node are the output class.

SVM

A Support Vector Machine is an ML model that can be used to solve both classification and regression problems. This model is mostly used to solve classification problems. The main function of this model is to find the optimal hyperplane that can divide the data into different classes. In binary classification, a hyperplane is a line that separates data into two classes, but many such hyperplanes may exist. To find out the optimal hyperplane, we first find out the border data points of these classes, and these points are called support vectors. With the help of these support vectors, we find the hyperplane.

Fig. 6.6 Decision tree

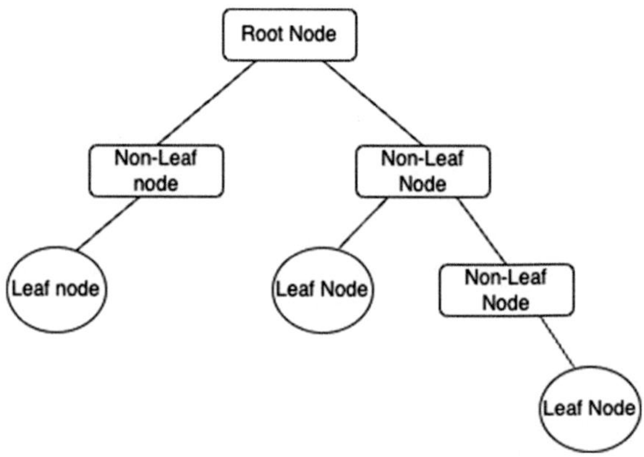

Neural Network

A neural network is a model inspired by the human neural system. In this model, different neural networks take the input, process it, and then give an output. A neural network is a layered model; there are three layers: input, hidden, and output. In machine learning, we just use one hidden layer. If we use multiple hidden layers, the model will become deep, and this type of model comes under Deep Learning.

Performance Metrics

Every machine learning model is used to train on data, and after training, the model's performance is checked on testing data. When testing, the model can give correct or wrong output. So, there are exactly four outputs that we can get. These outputs have standard names that are: True Positive (TP), True Negative (TN), False Positive (FP), and False Negative (FN). These outputs are used in performance metrics. The most popular performance metrics are as follows:

Accuracy: Total accurately predicted data divided by total data is used to calculate accuracy.

$$\frac{TP + TN}{TP + TN + FP + FN}$$

Precision: Precision is used when we want to find the predicted positive value out of the total positive value predicted via the model.

$$\frac{TP}{TP + FP}$$

Recall: Recall is also called sensitivity. This matric is also predicted to be positive but out of the actual total positive value.

$$\frac{TP}{TP + FN}$$

F1 Score: This metric is used to find out the harmonic mean of precision and recall, and this is the most widely used performance metric in machine learning.

$$2 * \frac{Precision * Recall}{Precision + Recall}$$

Confusion Matrix: This is a tabular or graphical representation of the output. This will give an output in n*n matrix form. Where n is the number of output classes, for binary classification, it will generate a 2*2 matrix.

6.4 Results

In this section, we will discuss the results of our study on intrusion detection in IoT systems using machine learning. We will also share some ideas and suggestions on making machine learning models for this task work better.

We have used decision trees, SVMs, and neural network models to detect attacks in datasets. To implement these models in Python, we have used the Scikit learn library (Scikit-learn: Machine Learning in Python 2022). This library supports all machine learning models, so you must import it into your Python code and then run a specific model as needed. Other than Scikit, there are some other standard libraries that we have imported for different tasks. Figure 6.7 shows the different libraries that we used in this study.

In python, the pandas' library is used to read and import the dataset. After reading the dataset, we can display the dataset by typing df in the python console. It will show the first five and last five instances of the dataset. At the end of the output, it will also show the total instances and features. In this dataset, there are 625,783 instances and 86 features. Figure 6.8 shows the code to read and display the dataset.

Data preprocessing is one of the main tasks before applying any machine learning model to a dataset. The data set can be imbalanced or can have outliers and null values. All these factors will affect the performance of machine learning models. So, we will check for these constraints. As mentioned in the previous section, we have analyzed our machine learning model on the cat label of this dataset, so we also checked for data imbalance on the same label. Figure 6.9 shows the code and output. This dataset is very large and imbalanced. We have used the sampling method to make a dataset in the range and balance. This will randomly sample the dataset. Figure 6.10 shows the code for sampling.

After sampling, we combine all the new data frames into one so that all instances should be in one data frame. Machine learning models need the data in numerical form. But in this dataset, some values are in the nominal form. So we drop these features and Label and Sub_cat features because we can analyze the performance of a machine learning model on one output feature at a time. After dropping these

```
import numpy as np
import pandas as pd
import seaborn as sns
import matplotlib.pyplot as plt
from sklearn import preprocessing
from sklearn.impute import SimpleImputer
from sklearn.model_selection import train_test_split
from sklearn.metrics import classification_report,confusion_matrix,accuracy_score,precision_recall_fscore_support
from sklearn.metrics import f1_score
from sklearn.ensemble import RandomForestClassifier,ExtraTreesClassifier
from sklearn.tree import DecisionTreeClassifier
```

Fig. 6.7 Imported libraries

```
df = pd.read_csv('C:/Users/NITJ/Desktop/Springer Book Chapter/IoT Network Intrusion Dataset.csv')
```

```
In [4]: df
```

Out[4]:

	Flow_ID	Src_IP	Src_Port	Dst_IP	Dst_Port	Protocol	Timestamp	Flow_Duration	Tot_Fwd_Pkts	Tot_Bwd_Pkts	...	Active_Std	Active_Max	Active_Min	Idle_Mean
0	192.168.0.13-192.168.0.16-10000-10101-17	192.168.0.13	10000	192.168.0.16	10101	17	25/07/2019 03:25:53 AM	75	1	1	...	0.0	0.0	0.0	75.0
1	192.168.0.13-222.160.179.132-554-2179-6	222.160.179.132	2179	192.168.0.13	554	6	26/05/2019 10:11:06 PM	5310	1	2	...	0.0	0.0	0.0	2655.0
2	192.168.0.13-192.168.0.16-9020-52727-6	192.168.0.16	52727	192.168.0.13	9020	6	11/07/2019 01:24:48 AM	141	0	3	...	0.0	0.0	0.0	70.5
3	192.168.0.13-192.168.0.16-9020-52964-6	192.168.0.16	52964	192.168.0.13	9020	6	04/09/2019 03:58:17 AM	151	0	2	...	0.0	0.0	0.0	151.0
4	192.168.0.1-239.255.255.250-36763-1900-17	192.168.0.1	36763	239.255.255.250	1900	17	10/09/2019 01:41:18 AM	153	2	1	...	0.0	0.0	0.0	76.5
...	...	...	...	...	...	...	...	...	...	...	...	...	...	...	...
625778	192.168.0.24-210.89.164.90-56112-8043-17	192.168.0.24	56112	210.89.164.90	8043	17	25/07/2019 03:25:00 AM	277	1	1	...	0.0	0.0	0.0	277.0
625779	192.168.0.13-222.131.171.244-554-4570-6	222.131.171.244	4570	192.168.0.13	554	6	26/05/2019 10:06:51 PM	1658	0	2	...	0.0	0.0	0.0	1658.0
625780	192.168.0.13-192.168.0.16-9020-52739-6	192.168.0.16	52739	192.168.0.13	9020	6	11/07/2019 01:29:09 AM	77	1	1	...	0.0	0.0	0.0	77.0
625781	192.168.0.13-192.168.0.16-9020-49784-6	192.168.0.13	9020	192.168.0.16	49784	6	20/05/2019 05:00:29 AM	240	2	1	...	0.0	0.0	0.0	120.0
625782	192.168.0.13-192.168.0.16-10000-10101-17	192.168.0.13	10000	192.168.0.16	10101	17	25/07/2019 03:25:10 AM	198	2	1	...	0.0	0.0	0.0	99.0

625783 rows × 86 columns

Fig. 6.8 Command to read and display dataset

Fig. 6.9 Count instances of attacks and normal data

```
df.Cat.value_counts()
```

```
Mirai                415677
Scan                  75265
DoS                   59391
Normal                40073
MITM ARP Spoofing     35377
Name: Cat, dtype: int64
```

features, there are 80 features available. Out of 80 features, 79 features work as an input, and the last five work as an output. Before applying the ML model, we should also check for null values and data type. Figure 6.10 shows the code and output for null values.

```
df_Mirai = df[(df['Cat']=='Mirai')]
df_Mirai = df_Mirai.sample(n=None, frac=0.05, replace=False, weights=None, random_state=None, axis=0)
df_Scan = df[(df['Cat']=='Scan')]
df_Scan = df_Scan.sample(n=None, frac=0.4, replace=False, weights=None, random_state=None, axis=0)
df_DoS = df[(df['Cat']=='DoS')]
df_DoS = df_DoS.sample(n=None, frac=0.4, replace=False, weights=None, random_state=None, axis=0)
df_Normal = df[(df['Cat']=='Normal')]
df_Normal = df_Normal.sample(n=None, frac=0.4, replace=False, weights=None, random_state=None, axis=0)
df_MITM_ARP_Spoofing  = df[(df['Cat']=='MITM ARP Spoofing')]
df_MITM_ARP_Spoofing  = df_MITM_ARP_Spoofing.sample(n=None, frac=0.5, replace=False, weights=None, random_state=None, axis=0)
```

```
df[df.isnull().any(axis=1)]
```

Src_Port	Dst_Port	Protocol	Flow_Duration	Tot_Fwd_Pkts	Tot_Bwd_Pkts	TotLen_Fwd_Pkts	TotLen_Bwd_Pkts	Fwd_Pkt_Len_Max	Fwd_Pkt_Len_Min	...	Fwd_Seg_Size_Min	Active_Mean	Active_Std

0 rows × 80 columns

Fig. 6.10 Data sampling and null value checking

There is no null value in the dataset. One of the features of Python is that we don't have to mention the data types; they are automatically assigned according to the nature of the data. Training a model on heterogenous data is not possible. First, the data should be converted into the same format. In this study, all the data is converted into the float64 type. A normalization technique is used to convert the data of the same type or within the range. Min–max normalization was used in this study. All feature instances are converted with this normalization except the last feature used as an output class. Figure 6.11 shows the code for min–max normalization.

After normalization, a label encoder is used to convert the target class into the form of a numeric value. This could be done with the label encoder method of the sklearn library. The Label encoder assigns the numeric value to each cat feature, as shown in Table 6.2. The code for the label encoder is shown in Fig. 6.12.

```
# Min-max normalization
numeric_features = df.dtypes[df.dtypes != 'object'].index
df[numeric_features] = df[numeric_features].apply(
    lambda x: (x - x.min()) / (x.max()-x.min()))
```

Fig. 6.11 Data normalization

Table 6.2 Assigning numerical value to attacks and normal class

Category label	Assigned value
DoS	0
MiTM	1
Mirai	2
Normal	3
Scan	4

```
labelencoder = preprocessing.LabelEncoder()
df.iloc[:, -1] = labelencoder.fit_transform(df.iloc[:, -1])
X = df.drop(['Cat'],axis=1).values
y = df.iloc[:, -1].values.reshape(-1,1)
y=np.ravel(y)
X_train, X_test, y_train, y_test = train_test_split(X,y, train_size = 0.7, test_size = 0.3, random_state = 0,stratify = y)
```

Fig. 6.12 Label encoding and train test split

In Fig. 6.12, 'X' stores the data frame of input features, and 'y' stores the output class. X_train and y_train are used to train the dataset, and X_test and y_test is used to evaluate the performance of the trained model. In this study, 70:30 data was used for training and testing, respectively.

DT, SVM, and neural network models were applied to the dataset, and the performance of each model was evaluated based on precision, recall, accuracy, F1 score, and confusion matrix. We also calculate the time for the execution of each model. Time plays a very important role in machine learning. As we know, more data can provide better results, but if it takes a long time to achieve accuracy, the model is not ideal. So, we also calculate the execution time of each model. To calculate the time, we use the time library. It will provide time in seconds for each model. Table 6.3 displays the model's performance matrix.

DT outperforms in all the performance metrics and takes the least time for execution. On the other hand, SVM takes a huge amount of time for execution and does not provide good results. Neural networks mostly provide good results in deep learning models, but here we have used just one hidden layer, so it could not get the advantage of multiple layers. Figure 6.13 shows the confusion matrix for all three models.

It is clear from Fig. 6.13 that decision trees provide a good classification of classes, whereas both SVM and neural networks misclassify many data instances.

IoT devices transfer huge amounts of data and can be easily attacked due to weak security. So, to make them secure, we can implement an ML model. But the time taken by the proposed models was high. We should reduce the time without affecting the performance matrix. To reduce the execution time, we used a feature selection technique. Based on the feature selection technique, the number of features selected for training was reduced from seventy-nine to four. Now we trained our models on these four features. The same performance metrics and execution time

Table 6.3 Performance evaluation of applied ML models

ML algorithm	Precision	Recall	Accuracy	F1 score	Time (in s)
DT	98.77	98.77	98.77	98.77	1.3399
SVM	79.83	78.60	78.60	78.63	601.4
NN	82.30	81.04	81.04	81.04	16.883

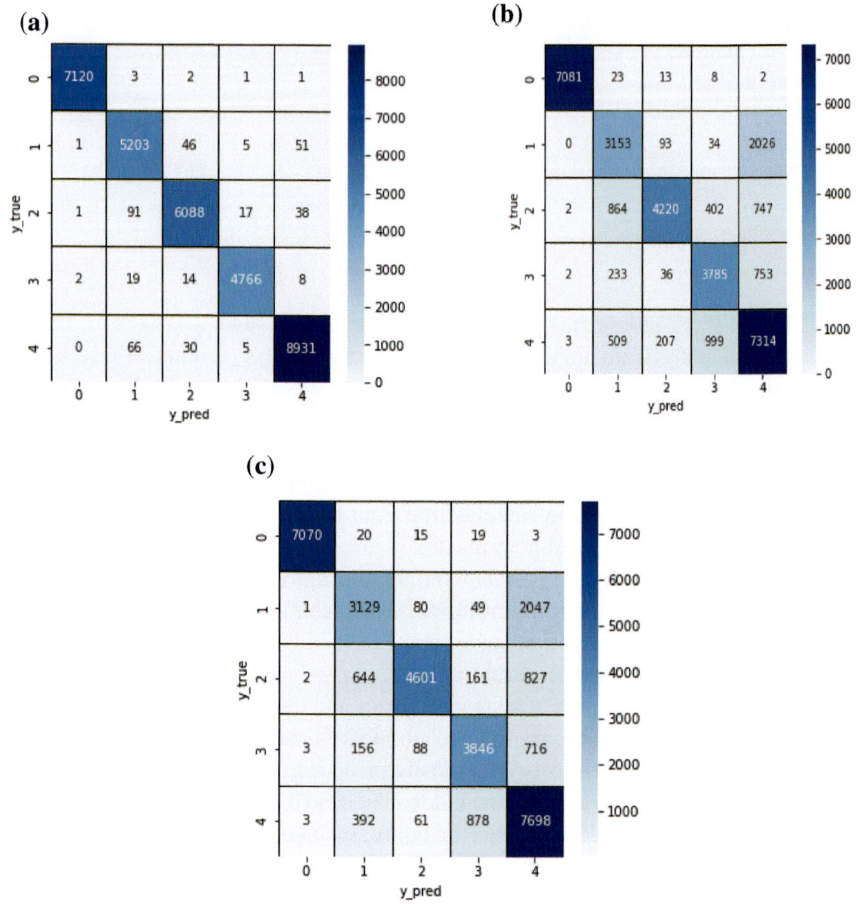

Fig. 6.13 Confusion matrix **a** decision tree **b** SVM **c** NN

were calculated. The table provides the details of performance metrics and execution time (Table 6.4).

Execution time has been reduced for all the models after feature selection. The execution time of DT has been reduced to 0.12 s from 1.13 s. SVM and NN execution times have been reduced, but there is a degradation in performance metrics.

Table 6.4 Performance evaluation of applied ML models with feature selection

ML algorithm	Precision	Recall	Accuracy	F1 score	Time (in s)
DT	98.90	98.90	98.90	98.90	0.1257
SVM	67.92	65.98	65.98	62.06	295.45
NN	82.37	79.20	79.20	78.79	12.209

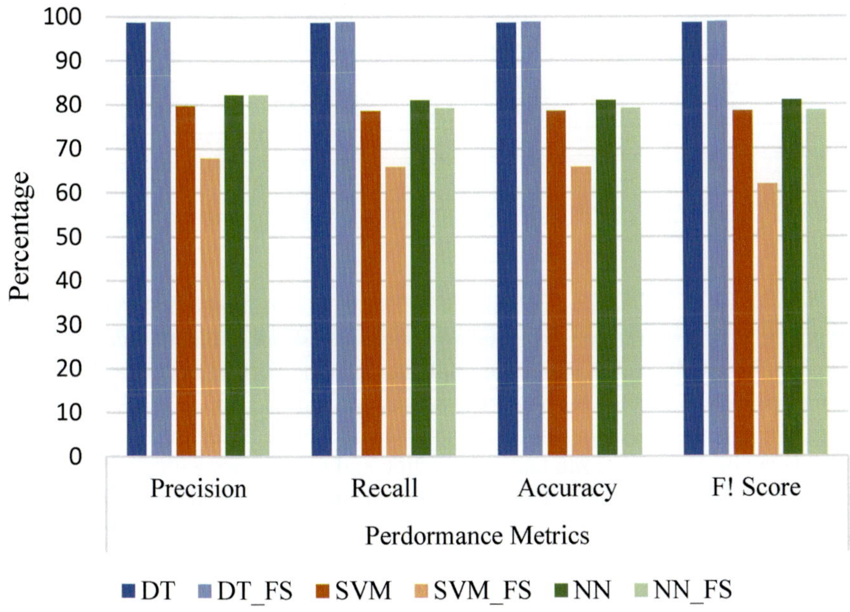

Fig. 6.14 Performance metrics

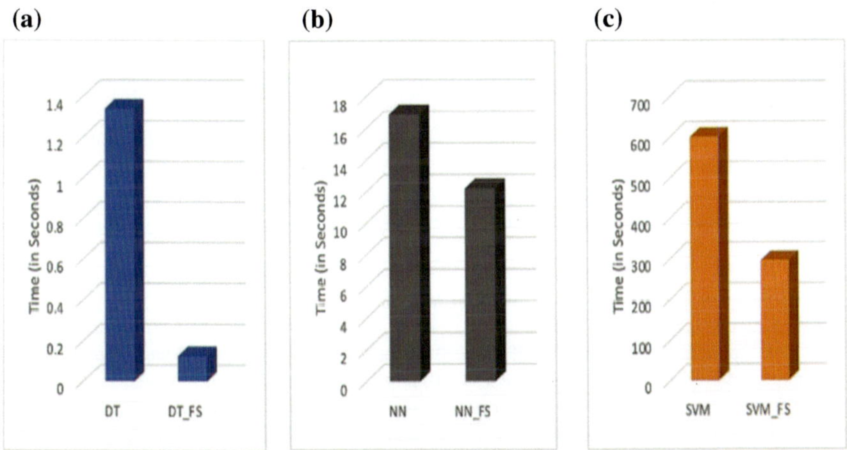

Fig. 6.15 Execution time of ML model with and without feature selection **a** decision tree **b** neural network **c** SVM

Figure 6.14 shows the bar graph representation of each model with and without feature selection. The decision tree was able to detect attacks accurately as compared to SVM and NN. Figure 6.15 shows the execution time of each model with all features and model trained after features selection.

ML approach is effective in detecting a variety of different types of intrusions. However, it is important to note that the effectiveness of the detection depends heavily on the quality of the data used for training the algorithm. It is important to pick high-quality data to obtain desirable outcomes.

6.5 Conclusions

Machine learning can be used effectively for intrusion detection in IoT systems. ML is a powerful technique that can be used to detect and respond to threats automatically. Three different ML techniques were implemented, and performance was evaluated on the IoTID20 dataset. This dataset is generated from the raw network file, so we first apply data preprocessing and then train ML models on the dataset. Decision trees provide the best result in both feature selection and model train with all features. To further enhance the scope of the study different ML models will be applied on the dataset and the DL model will also be implemented to enhance the performance.

References

Anaconda | Anaconda Distribution. https://www.anaconda.com/products/distribution. Accessed 27 July 2022.

Cui L, Yang S, Chen F, Ming Z, Lu N, Qin J (2018) A survey on application of machine learning for internet of things. Int J Mach Learn Cybern 9:1399–1417. https://doi.org/10.1007/s13042-018-0834-5

Das ML (2015) Privacy and security challenges in internet of things. In: Lecture Notes in Computer Science (including Subseries Lecture Notes in Artificial Intelligence and Lecture Notes in Bioinformatics), vol 8956. pp 33–48. https://doi.org/10.1007/978-3-319-14977-6_3/COVER

IoT Network intrusion dataset (2020) https://sites.google.com/view/iot-network-intrusion-dataset/home. Accessed 30 July 2022.

Kang H, Ahn DH, Lee GM, Jeong DY, Park KH, Kim HK (2019) IoT network intrusion dataset. IEEE Dataport. https://doi.org/10.21227/q70p-q449

Khan R, Khan SU, Zaheer R, Khan S (2012) Future internet: the internet of things architecture, possible applications and key challenges. In: Proceedings-10th International Conference on Frontiers of Information Technology, FIT 2012. pp 257–260. https://doi.org/10.1109/FIT.2012.53.

Omar M, George L (2021) Toward a lightweight machine learning based solution against cyber-intrusions for IoT. In: Proceedings-Conference on Local Computer Networks, LCN. pp 519–524. https://doi.org/10.1109/LCN52139.2021.9525002.

Outchakoucht A, Abou A, Kalam E, Es-Samaali H, Benhadou S (2020) Machine learning based access control framework for the internet of things. (IJACSA) Int J Adv Comput Sci Appl 11: 332–340. https://doi.org/10.14569/IJACSA.2020.0110243.

Punithavathi P, Geetha S, Karuppiah M, Islam SH, Hassan MM, Choo KKR (2019) A lightweight machine learning-based authentication framework for smart IoT devices. Inf Sci 484:255–268. https://doi.org/10.1016/j.ins.2019.01.073

Rathore S, Park JH (2018) Semi-supervised learning based distributed attack detection framework for IoT. Appl Soft Comput 72:79–89. https://doi.org/10.1016/j.asoc.2018.05.049

Sawadogo LM, Bassolé D, Koala G, Sié O (2021) Intrusions detection and classification using deep learning approach. In: Lecture Notes of the Institute for Computer Sciences, Social-Informatics and Telecommunications Engineering, LNICST. 400 LNICST. pp 40–51. https://doi.org/10.1007/978-3-030-90556-9_4/COVER.

Scikit-learn: Machine Learning in Python. https://jmlr.csail.mit.edu/papers/v12/pedregosa11a.html. Accessed 02 Aug 2022.

Sharma R, Sharma N (2022) Applications of artificial intelligence in cyber-physical Ssystems. Cyber Phys Syst 1–14. https://doi.org/10.1201/9781003202752-1.

Ullah I, Mahmoud QH (2020) A scheme for generating a dataset for anomalous activity detection in IoT networks. In: Lecture Notes in Computer Science (including subseries Lecture Notes in Artificial Intelligence and Lecture Notes in Bioinformatics). 12109 LNAI. pp 508–520. https://doi.org/10.1007/978-3-030-47358-7_52/COVER.

Verma A, Ranga V (2019) Machine learning based intrusion detection systems for IoT applications. Wirel Pers Commun 111:4. 111: 2287–2310. https://doi.org/10.1007/S11277-019-06986-8

Vijayanand R, Devaraj D, Kannapiran B (2017) Support vector machine based intrusion detection system with reduced input features for advanced metering infrastructure of smart grid. In: 2017 4th International Conference on Advanced Computing and Communication Systems, ICACCS 2017. https://doi.org/10.1109/ICACCS.2017.8014590.

Wireshark Download, https://www.wireshark.org/download.html. Accessed 24 July 2022.

Xiao L, Wan X, Lu X, Zhang Y, Wu D (2018) IoT security techniques based on machine learning: how do IoT devices use AI to enhance security? IEEE Signal Process Mag 35:41–49. https://doi.org/10.1109/MSP.2018.2825478

Chapter 7
Employability of Decision Support System in Data Forecasting for Internet of Things Networks

Shefali Bajaj, Sujay Bansal, Monika Mangla, Sourabh Yadav, and Rahul Sachdeva

7.1 Introduction

Time series forecasting is understanding the time series data which will then be used to protect the future values at any given interval of time (Chae et al. 2003). These data are analyzed using various statistical approaches and are trained to make estimations that can help in decision-making. It can be univariate or multivariate (Sadovnick et al. 1992).

However, the decision that the time forecasting system makes may not be the exact prediction because at different times the data may fluctuate a lot so show the outcomes are more or less likely to occur sometimes but the analysis gives the idea to explore trends in data values changing over time. The prerequisite of time series forecasting is that the data should be stationary otherwise no concept of time series model can be applicable to it (Lim and Zohren 2021).

S. Bajaj · S. Bansal
B. R. Ambedkar National Institute of Technology, Jalandhar, India
e-mail: shefalib.cs.21@nitj.ac.in

M. Mangla (✉)
Dwarkadas J Sanghvi College of Engineering, Mumbai, India
e-mail: manglamona@gmail.com

S. Yadav
Mobile Computing Lab at UNT, University of North Texas, Denton, TX 76205, United States

R. Sachdeva
Indira Gandhi Delhi Technical University for Women, Delhi, India

N. Sharma et al. (eds.), *Big Data Analytics in Intelligent IoT and Cyber-Physical Systems*, Transactions on Computer Systems and Networks,
https://doi.org/10.1007/978-981-99-4518-4_7

For data stationarity, there are three main requirements:

1. The mean of the time series should not depend upon time.
2. Variance—In this, the data should not vary with the time intervals, i.e., Variance should not vary with time. There shouldn't be any increase or decrease in data.
3. Covariance—In this, the previous data value should not vary from its own previous data value.

Data Analysis refers to breaking down the whole data into small activities for evaluation (Ahmed et al. 2010). Data analysis is a process through which raw data are gathered and get converted into useful information, which is further used to make decisions. Data are gathered and examined to give answers to many types of questions, testing the hypothesis, etc. The process of data analysis includes many activities such as requirement of the data, its collection, the processing of the data, finally its analysis using modeling techniques and algorithms.

It is a way to examine data to conclude. Data scientists and many analysts use DA techniques (Tang et al. 1991) to make several decisions. Also, businesses had a major dependency on data analysis that can help companies to understand their customers, evaluate themselves, create strategies, and improve products. Businesses use data analytics to boost business outcomes and improve their business performance. Analytics is a technique to access the data, which gives us better insight.

7.2 Applications

Time series analysis can be found in almost every stream they are dealing with (Siskos and Matsatsinis 1993). It widely ranges from economics to medical, physical sciences to finances, etc. Some of the applications of time series forecasting are as follows:

1. **Health Facilities**: It is used for blood pressure tracking, cholesterol measurement, heart rate monitoring, etc.
2. **Economics**: Used in GDP predictions.
3. **Physical Sciences**: Forecasting the weather and temperature, pollution levels.
4. **Social Sciences**: It is also used in birth rates, population, migration data, political indicators, etc.

In estimating the stock market prices, the time series forecasting is very useful and to have a recent example and can also consider the COVID cases coming etc. Data analytics means the process of surveying or examining the data to perform some tasks like predicting (Matthies et al. 2007). It comes into use in any kind of business.

1. Analysis of Fraudulence: Fraud can be in any form like the banking sector, healthcare, government policies, etc. In the above-mentioned areas, the fraud is identified by doing data analysis.
2. Traveling: Data are analyzed based on the seasons going on. According to the seasons, the hotel managers, the transportation companies got the idea of how

many bookings are done so this amount of tourist car is required on required dates.
3. Internet surfing: All search engines work on data analytics. Every time you type something on Google, Yahoo, etc. they are using many of the inbuilt data analytic algorithms to perform your task and give you the related answer within a few milliseconds.
4. Military: Here data analytics works to make the interconnection between information received and the steps of investigation to be done.
5. Education: Getting the details of the students in each class. This depends on the requirement whether you need to analyze the data of one specific class or the whole school or college or any tuition center.

7.3 Time Series Data

Time series data refer to the data that are recorded in specific and constant intervals of time. These intervals can be of constant dates, month, days, hours, or weeks. In these, the present and past values are related to each other. Every data point is dependent upon its past data point (Roselli et al. 2019).

7.3.1 *Difference from Other Forecasting Data*

What changes time series data from other data is that it is collected over an interval of time so that how the variables are changing with respect to that particular time period can be seen. Time is a crucial variable that will affect the final results of the forecasting (Yadav and Sharma 2019).

7.3.2 *Combining Time Series Data and Analysis*

Now the time series values are usually taken at a constant interval of time. If the data are not regular then pre-processing of data needs to be done manually using human interference. Usually, the gaps in the data values are filled by their previous data values. Once the preprocessing of the data is done then the first requirement to apply the time series model is to check whether the data are stationary or not. To do the same, Dicky-Fuller test is used. In the case of non-stationary data, it has to be made stationary first and then apply to the model (Nelson et al. 1999).

After the stationarity, the data are further divided into two groups: the training data and testing data. The model is trained using training data and testing data are used to forecast the future values. And at last, the accuracy of the model depends upon how accurately it predicts the values.

7.4 Decision Support System

Decision support system (DSS) is a computerized system that supports determination, judgments, and courses of action in a business organization as shown in Fig. 7.1. It has differed from ordinary operation applications whose only task is to collect data, whereas decision support systems gather data and analyze it, synthesizing it to produce comprehensive information summary reports, which may project revenue, sales, or inventory management. With the integration of multiple variables, a DSS may produce a number of different outcomes based on current and historical data. DSS can be customized for any industry or domain including the government, medical field, agricultural concern, corporate project operations, etc. (Yadav et al. 2018).

7.4.1 Dataset

This dataset is describing the Life Expectancy of 15 years from 2000 to 2015. The data are from 193 countries. This dataset shows Country, Year, Status, Life Expectancy, Adult Molarity, Hepatitis B, Diphtheria, GDP, population, and many more. We consider only nine of all from the real dataset because of not getting highly chaotic in the Excel file (Perna et al. 2010).

1. Country: In total, there are 193 countries.
2. Year: Total of 15 years' data is in CSV file from 2000 to 2015.

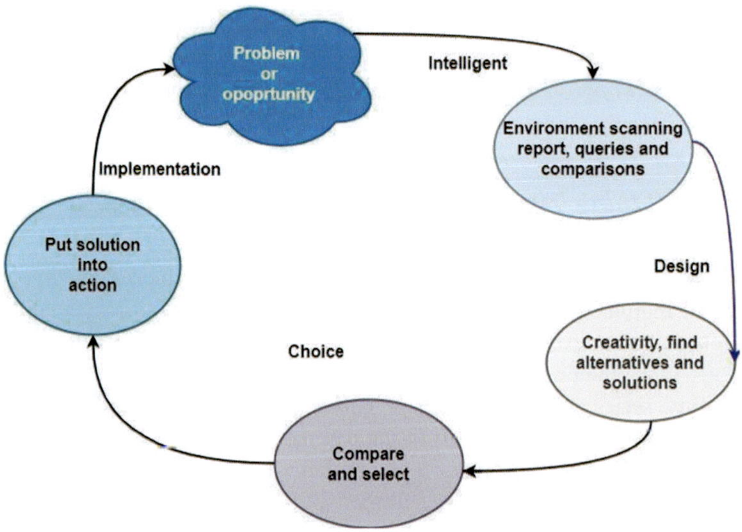

Fig. 7.1 Life cycle of DSS

3. Status: It is about whether the country is already developed or they are in progress.
4. Life Expectancy: It is calculated in terms of age.
5. Adult Molarity: These rates are calculated for both sexes by taking the probability of dying people between the ages of 15 and 60 over 1000 inhabitants.
6. Hepatitis B: Immunization rate in 1-year-old children.
7. Diphtheria: Diphtheria tetanus toxoid and pertussis immunization rate all over 1-year-old children.
8. GDP: Gross Domestic Product per unit of population.
9. Population: Total number of inhabitants in that particular country.
10. Alcohol: The consumption (in li) of intake in 15 + age groups.
11. % Expenditure: How much expense takes place on health over the overall GDP per capita?
12. Measles: How many cases encountered over 1000 inhabitants?
13. BMI: Taking the average body mass index of all the inhabitants considered in the dataset.
14. Polio: Calculating immunization rate among 1-year-old children. (in %).
15. Total expenditure: It is the total overheads done by the government on health on overall expenses.
16. HIV/AIDS: The number of inhabitants dead per 1000 births under 0–4 years.

7.5 Methodology

For any data analysis, first data need to be present for analysis so it is required to collect the data of various time ranges, i.e., it is preferred to have a minimum of at least more than 5 years of data. Now the second step is to convert that data in a single variant form, this is to make it easier for analysis, which is followed by time format required for the data, continuous data, pre-processing of the data, sorting, etc. as shown in Figs. 7.2 and 7.3 for time series analysis of BSE power.

The Life Expectancy data tell us about the health factors and the health status of many people from different countries. This dataset was made online to the public for analysis. It has been tracked and monitored by the WHO organization and was observed by Global Health Observatory. The steps to be followed for statistical analysis are shown in Fig. 7.4.

Machine will predict the future prices of the S&P Power index using the past data of about 5 years. To achieve this objective, we have to apply some steps, which are as follows.

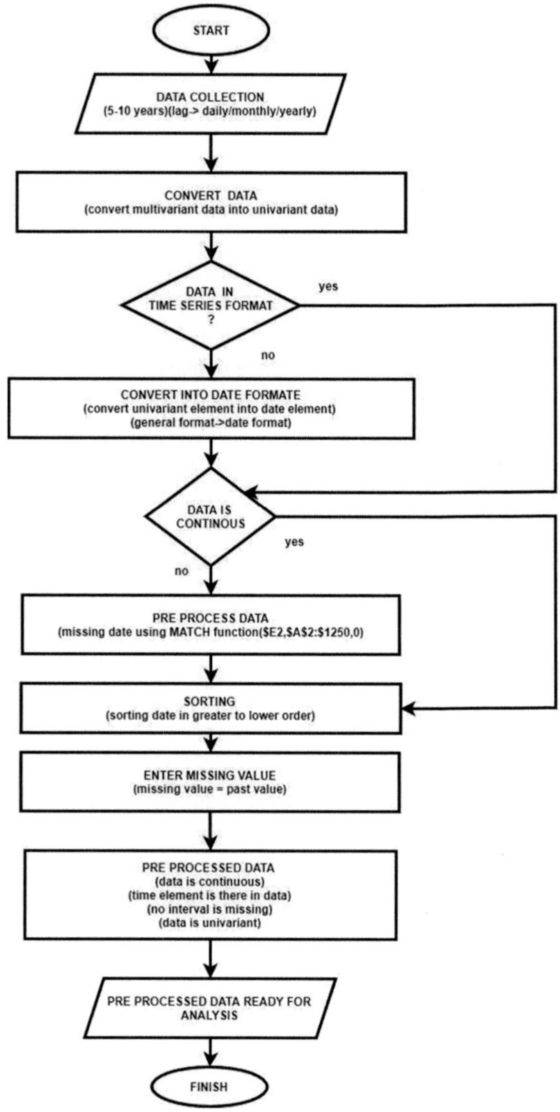

Fig. 7.2 Steps performed to make data clean and pre-processed

7.5.1 *Data Collection*

For Time Series Data, the data of the S & P Power Index are available on the BSE official website. From that we took the data as a CSV file. The data are from the last 5 years, i.e., 20 September 2016 to 01 October 2021. The data of the life expectancy are available on Kaggle at https://www.kaggle.com/datasets/kumarajarshi/life-expectancy-who?select=Life+Expectancy+Data.csv. From that, we took the data as a

Fig. 7.3 Steps are performed to make the prediction

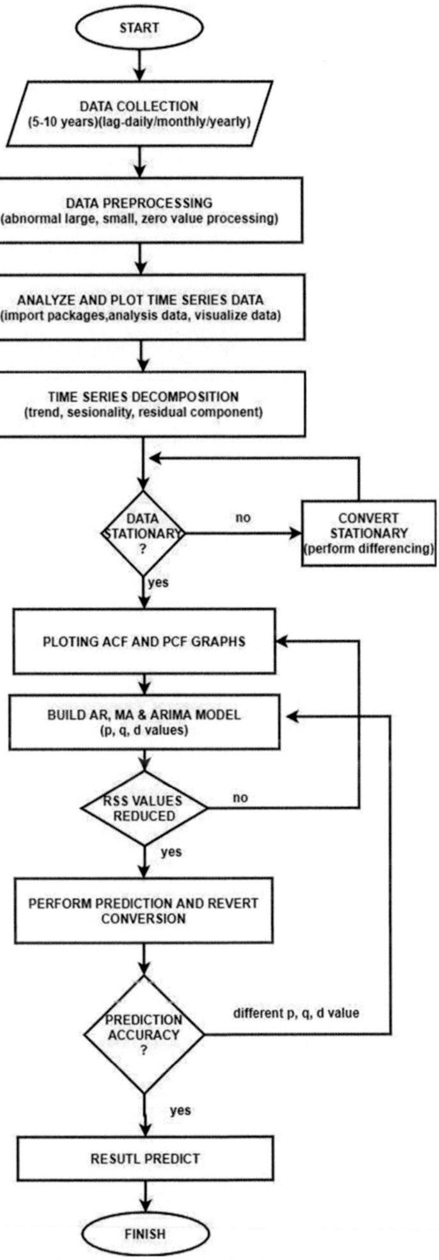

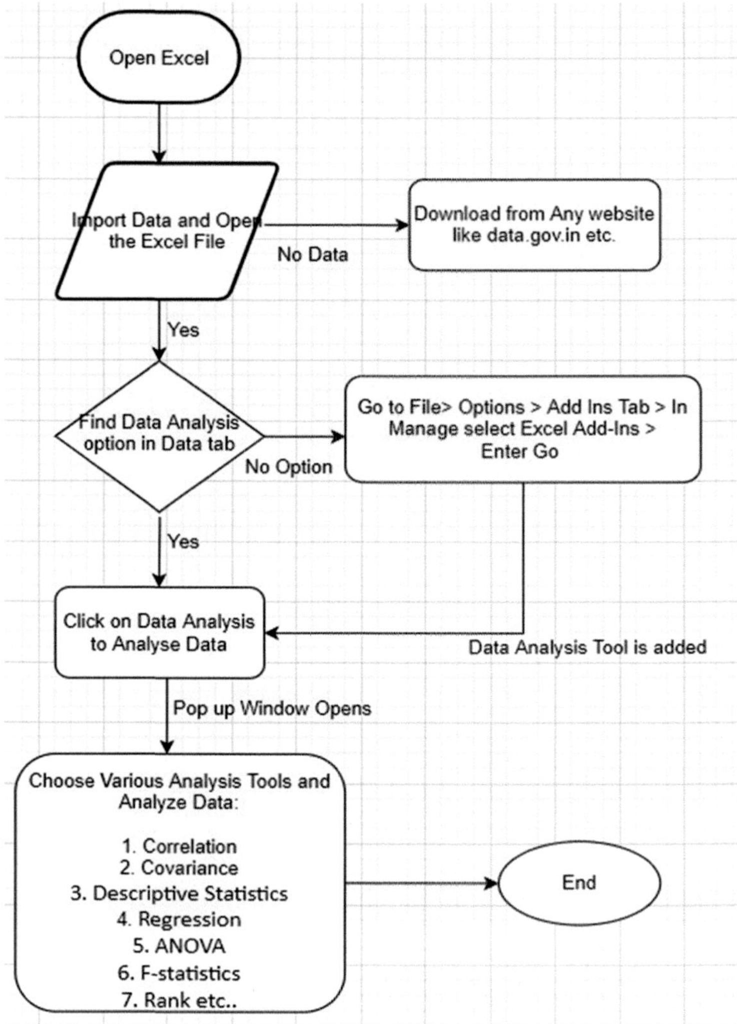

Fig. 7.4 Flowchart showing the steps to perform data analysis

CSV file. The data are for 15 years, i.e., from 2000 to 2015 in every country those are 193 in total.

7.5.2 Data Preprocessing

The data of time series come along with the open, high, and close values of the stock market. But we need only the close value because for predicting the most relatable values in future, we need to be confined about data. Stock market remains close at Saturdays and Sundays, so to fill these voids in the data we often use the previous day value. Because stock market was close at the value in the previous day. So, we start the next day with the close value of the previous day (Sharma et al. 2021a).

The Dataset for life expectancy is downloaded from Kaggle and it is unprocessed data so we need to convert it into preprocessed data (Sharma 2018). What exactly does data preprocessing mean? As the name suggests preprocessing means rectifying something. So, by all it means, for getting the understandable data, we have to change some of the data, which is in an inappropriate form. Also, we found some blank spaces in the data, so we have filled them with its previous data for ease of calculations and analysis (Najafabadi et al. 2015). Our major factors will be Country, Year, Status, Life Expectancy, Adult Molarity, Hepatitis B, Diphtheria, GDP, and Population. The empty data can be seen from some lesser-known countries like Vanuatu, Tonga, Togo, Cabo Verde, etc. Searching for all data for such countries was difficult and so decided to exclude these countries from the resulting data model.

Actually, before doing preprocessing, we have to clean our data but when we download the data, it's already a cleaned one so we proceed to the next step. For preprocessing, we have to fill in the missing values and also need to treat the NULL values. For filling the values, we have to put related values, so for that, the previous year's data will be the best. Because it's better to fill the previous accuracy than to fill it as zero. For example, if by mistake the population of a country is filled as NULL, so it is not possible to fill it as zero because we don't know the exact value. To solve this problem, we are using the previous year value because that would be more accurate and significant for the data to be shown by graphs and also for analysis (Lachtermacher and Fuller 1995).

7.5.3 Data Analysis and Visualization

Analyzing means to see or to work on. Data that we download so far need to be read or write or edit. Visualization relates to the visuals means how the data look like. To check the visuals, graph is the best way to represent any kind of statistical data. By plotting Date v/s Close value data on the graph, we get to know the trend, seasonality, residual, and observed values (Zhang and Kline 2007).

Analyzing means seeing or working on. Data that we download so far need to be read or written or edited. Visualization relates to the visuals and means how the data looks like. To check the visuals, graphs are the best way to represent any kind of statistical data. By plotting Date v/s Close value data on the graph, we get to know the trend, seasonality, residual, and observed values.

Is Data Stationary? Our data are based on time, so the model that is to be used is the Time Series Model. In this model, we can't use non-stationary data. Hence, Dickey-Fuller test is performed to check stationarity of the data. If data is found to be non-stationary, differencing part starts. It depends on the time period from one to another. It is carried out in three ways: first difference, second difference, and seasonal difference. The main point is to check the difference in p-value.

Use ACF and PACF: ACF means Auto-Correlation function and PACF means Partial Auto-Correlation function, which is a calculation of average between the present and the previous values for long “lag length”. PACF calculates the differences between short “lag length”.

ARIMA Model: Auto-Regressive Integrated Moving Average. This converts the data to stationary first if it is not stationary. ARIMA consists of three words AR+I+MA. These three are assigned as p, d, and q, respectively. Mostly used or dealing variable are p and q, which are regressive value and moving average parameter. SARIMAX is the main function of ARIMA model to check the periodicity (Mahajan et al. 2020).

Present and Future Values Prediction: To check whether the prediction will be OK for all values or not. First, we will predict the values on those data, which is already given to us. If the lines of predicted values overlap the original values, then our model is fine to go for the future prediction. And then we are ready to move to predict the future values, i.e., from 02 October 2021 to 12 October 2022.

To see the difference between the predicted values and the actual values, we used two different colors, i.e., orange and blue.

7.5.4 Statistical Techniques for Analysis

Various statistical values have been used as follows.

7.5.4.1 Covariance

This name comes from two words, i.e., jointly and difference, combining both can be considered as how the quality or characteristics vary together at the same time. Here, authors calculate the variance of more than two characteristics together. In general language, Variance is the closest value to MEAN (Maravall 1983). It can also be used to understand the intensity or strength of a relationship among variables. The higher the value, the more strongly the variable is dependent. Its value can be between $-$ infinity to $+$ infinity (Cao et al. 2019). Positive correlation tells that slope of the graph is positive while negative correlation signals that slope of graph is negative. But it doesn’t signify whether the data points are closer or farther from the slope. Similarly, zero variance that there is no relationship between the two variables and is given by the formula:

$$Cov(x, y) = \frac{\sum (x_i - \overline{x})(y_j - \overline{y})}{n}$$

	Life expectancy	Adult Mortality	Hepatitis B	Diphtheria	GDP	Population
Life expect	90.9168156					
Adult Morta	-824.1090362	15478.48745				
Hepatitis B	78.2108416	-581.5838777	832.1433794			
Diphtheria	107.3808587	-811.8183988	388.4056775	571.8292635		
GDP	56180.1446	-477104.5294	7684.322909	55851.27363	188406519.1	
Population	1637693.882	-251954926.1	-96094509.84	-9821117.766	**-10587407339**	**3.16859E+15**

Fig. 7.5 Illustration for zero covariance

The lowest value is −10,587,407,339 so covariance is lowest for GDP and population and the max value is 3.16859E + 15, with the population itself as shown in Fig. 7.5.

7.5.4.2 Correlation

Correlation means how two or more characteristics are interconnected. Are they strongly related or weakly related or not related to each other? If they are strongly related, we can choose any one attribute and neglect one for data analysis.

The C coefficient should have a value between −1 and + 1, which tells us about how strongly the two values are dependent on each other. It finds out the strength of the relationship between two variables, i.e., the level of change of one variable due to change in another. It is a pure value so it is not measured in units and is given by:

$$p(x, y) = \frac{Cov(x, y)}{SD(X) * SD(Y)}$$

where p is the correlation of x and y variables, cov is the covariance of the two, SD is the standard deviation of the individual variables x and y.

The correlation is said to be positive when both the variables move in the same direction, that is an increase in one variable leads to an increase in the other. E.g.—more time on the treadmill burns out more calories. On the other hand, correlation is negative when one variable moves in the positive direction and the other variable moves in the opposite direction or vice versa. E.g.—the increase in the speed of the vehicle may decrease the time of reach. In Excel, the CORREL function or the Analysis Tool pack add-in can be used to find the correlation coefficient between these. The correlation is illustrated in Fig. 7.6. As we can see the highest value is 0.56, so BMI and Life Expectancy have the maximum correlation between them (Singh 2000).

	Life expectancy	*Adult Mortality*	*Hepatitis B*	*BMI*	*Diphtheria*	*GDP*	*Population*
Life expectancy	1						
Adult Mortality	-0.693449117	1					
Hepatitis B	0.285032643	-0.161630623	1				
BMI	**0.568503188**	-0.398388652	0.22046511	1			
Diphtheria	0.472653844	-0.279369787	0.564399907	0.283732632	1		
GDP	0.429175079	-0.280235142	0.018902802	0.258695924	0.169782783	1	
Population	0.002030727	-0.03588164	-0.059444844	-0.056183704	-0.007363985	-0.01397969	1

Fig. 7.6 Negative correlation illustration

7.5.5 *Descriptive Statistics*

DS helps to organize the data and focuses on the main characteristics of the data so that it becomes easy to understand. It provides a summary of the data numerically or graphically.

It is given by the properties like mean, median, mode, skewness, and Standard deviation, etc. as illustrated in Fig. 7.7. Readers may refer (Singh 2000) and (https://www.kaggle.com/datasets/kumarajarshi/life-expectancy-who?select=Life+Expectancy+Data.csv) for further details regarding descriptive statistics. Various descriptive statistics for attribute Diphtheria are illustrated in Fig. 7.8.

7.5.5.1 Regression

Regression means predicting the value of one variable (dependent variable) based on other variables (independent one). It is used to predict the relation between the data like one finds a combination that appropriately fits the data and can be seen in mathematical equations. Regression analysis is primarily used for forecasting data, and further can be helpful in time series modeling and to detect the relationship in different variables. Regression can be classified into linear and non-linear regression. Further linear regression can be classified into simple linear regression

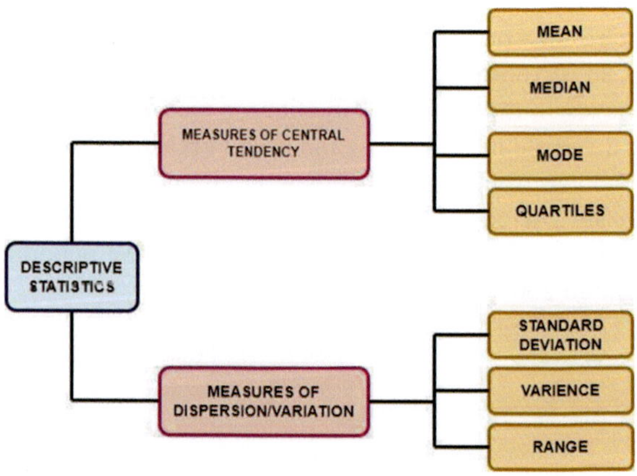

Fig. 7.7 Properties of descriptive statistics

Fig. 7.8 Descriptive statics table

Diphtheria	
Mean	82.07522124
Standard Error	0.441246441
Median	93
Mode	99
Standard Deviation	23.91702243
Sample Variance	572.0239619
Kurtosis	3.399088171
Skewness	-2.039531815
Range	97
Minimum	2
Maximum	99
Sum	241137
Count	2938

(involving single independent variable) and multiple linear regression (involving multiple independent variables).

The regression analysis for the dataset is carried out and the results are illustrated in Fig. 7.9.

R Square here is 0.48, which is not a very good fit (Torres et al. 2021) which tells how Adult Mortality and Life Expectancy are interlinked. When the value is closer to 1, the better the regression line fits the data.

Also, the f significance needs to be less than 0.05, here it is 0 so we are good to go, but if it would have been more then we need to stop using independent variables and thus need to delete variable with more p-value greater than 0.05. Here, P-value is less than 0.05 so we are good to go (Sharma et al. 2021b).

Regression Statistics	
Multiple R	0.694701615
R Square	0.482610334
Adjusted R Square	0.482434111
Standard Error	6.86086474
Observations	2938

ANOVA

	df	*SS*	*MS*	*F*	*Significance F*
Regression	1	128911.7857	128911.7857	2738.64	0
Residual	2936	138201.8185	47.07146407		
Total	2937	267113.6042			

	Coefficients	*Standard Error*	*t Stat*	*P-value*	*Lower 95%*	*Upper 95%*	*Lower 95.0%*	*Upper 95.0%*
Intercept	77.9708419	0.21009343	371.1246088	0	77.55889652	78.38278728	77.55889652	78.38278728
Adult Mortality	-0.05324222	0.001017393	-52.33201521	0	-0.055237096	-0.051247344	-0.055237096	-0.051247344

RESIDUAL OUTPUT

Observation	*Predicted Life expectancy*	*Residuals*
1	63.968138	1.031861998
2	63.54220024	-3.642200241
3	63.7019269	-3.801926901
4	63.48895802	-3.98895802
5	63.32923136	-4.12923136
6	63.11626248	-4.316262479
7	63.00977804	-4.409778039

Fig. 7.9 Summary of regression table

7.5.5.2 ANOVA

Analysis of variance is a test used to check how different the means of two or more groups are from each other as illustrated in Fig. 7.10. It did so by comparing the values. The samples to be compared are random and independent. We have one-way and two-way ANOVA.

To reject the null hypothesis, we need F > F crit. So in our analysis, 160.7 > 2.21 and we reject the null hypothesis meaning the mean of these is not equal. However, the ANOVA will not tell us where the difference lies.

7.5.5.3 F-statistics

F-statistics is the ratio of two variables that are dispersed from the mean that is their variances. It is the ratio of two variances. A larger value means more scattered the variables are. It may be given by "The variance between the samples means/variance within the samples". The F-statistics for the dataset is illustrated in Fig. 7.11.

This formula is used in one-way ANOVA. It is used to determine the equality of means. It is the ratio of two things that are expected to be nearly equal in the null hypothesis giving the value of F stats to be 1(approximately).

Groups	*Count*	*Sum*	*Average*	*Variance*
Life expectancy	2938	203296.8	69.19564329	90.94777127
Adult Mortality	2938	484231	164.8165419	15483.75762
Hepatitis B	2938	222359	75.6837985	832.4267105
Diphtheria	2938	241137	82.07522124	572.0239619
GDP	2938	21963294.04	7475.593613	188470668.4
Population	2938	38689654885	13168704.86	3.16967E+15

ANOVA

Source of Variation	*SS*	*df*	*MS*	*F*	*P-value*	*F crit*
Between Groups	4.24476E+17	5	8.48952E+16	**160.7014709**	1.4E-167	**2.214607**
Within Groups	9.30933E+18	17622	5.28279E+14			
Total	9.7338E+18	17627				

Fig. 7.10 ANOVA table

Fig. 7.11 F-statistics

	Hepatitis B	*Diphtheria*
Mean	75.71570304	82.11460828
Variance	832.9913584	572.4574286
Observations	2923	2923
df	2922	2922
F	1.455114943	
P(F<=f) one-tail	2.56226E-24	
F Critical one-tail	1.06275844	

If the f-value is less than the graph shows the group having low variability concerning the mean within the group. If the f-value is higher than it shows a case where the group means to have high variability with respect to the mean within the group. For null hypothesis rejection, we require a high f-value.

7.5.5.4 Moving Average

This is the GDP of France from year 2000 to 2015, Forecasted GDP is shown in Fig. 7.12 as it is calculated by moving average of the actual values.

Further, ranking helps in ranking all the data and percentile helps in categorizing the data into percentiles such as the top 10%, top 35%, etc.

7.5.5.5 Sampling

It is done by randomly choosing data from the dataset. Here, our data are showing the rank of life expectancy and its percentile with respect to the POINT. POINT has a direct link to our main data and with further technique and functions, we can get rank and percentile of whole data, respectively. Samples from all our main data sets have been picked randomly from our main data using a sampling tool.

7.5.6 *Descriptive Statistics*

Visual Representation of Adult morality of all the countries is illustrated in Fig. 7.13. It is evident that Adult Morality of Libya is highest in this year as shown in Fig. 7.13. Visual representation of Life Expectancy, Adult Morality, Hepatitis B,

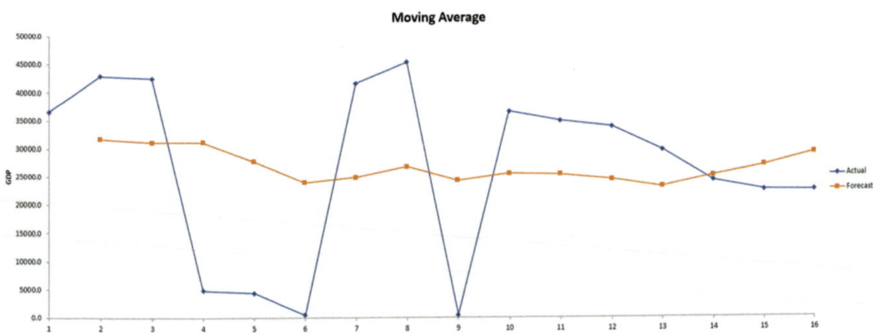

Fig. 7.12 Moving average plot

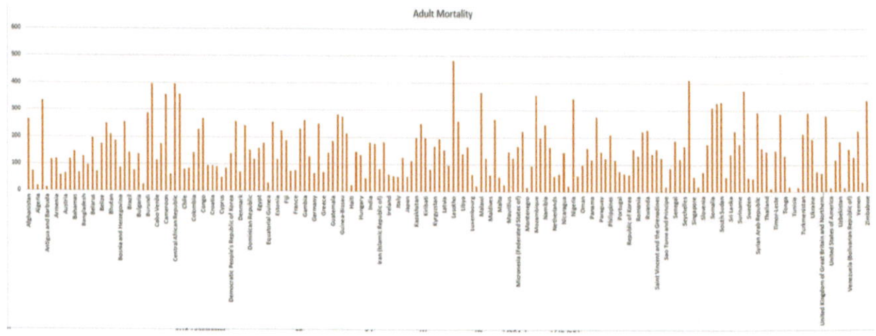

Fig. 7.13 Adult morality plot

and Diphtheria of every country in a specific year is illustrated in Fig. 7.14. Similarly, correlation and covariance are graphically represented in Figs. 7.15 and 7.16, respectively.

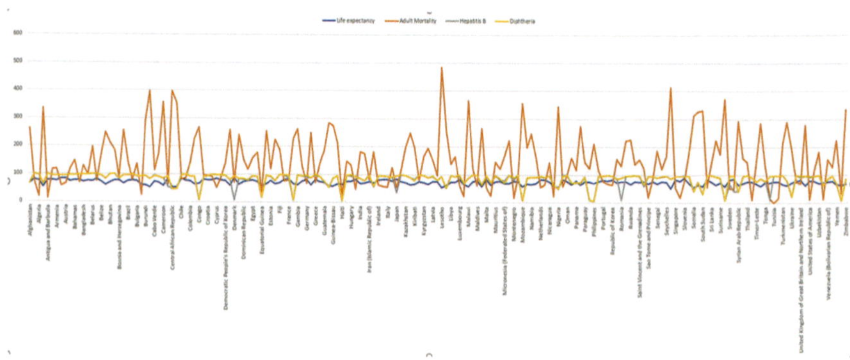

Fig. 7.14 Representation w.r.t year

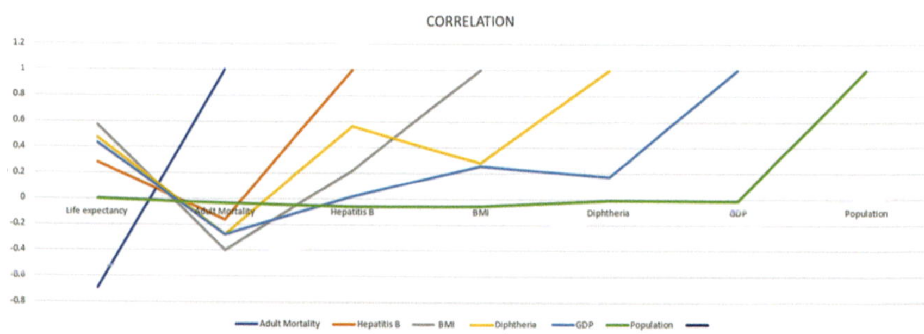

Fig. 7.15 Correlation plot

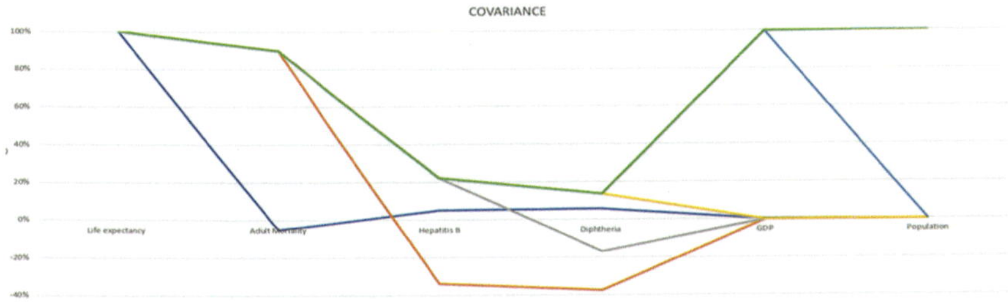

Fig. 7.16 Covariance plot

7.6 Implementation

For visualization, we are using matplotlib and to implement the ARIMA model, the stats model is used. Also, NumPy and Pandas are imported for handling data. Column names are changed here so that the person can memorize that easily. “Date” column is converted into a Date Time index. Data is visualized to check accuracy of the model as shown in Fig. 7.17. Further, mean and standard deviation are evaluated to determine the quantum of fluctuation in the data as illustrated in Fig. 7.18. Data are decomposed into various components, namely, Trend, Residual, and Seasonality. Each component of the data is shown in Fig. 7.19.

Thereafter, data are checked for the presence of stationarity by Dickey-Fuller test. If the p-value is less than 0.05, data are considered to be stationary else non-stationary. Now as the p-value is 0.872 approximately, data are non-stationary. Hence, differentials are applied and again p-value is checked for all differentials. P-value for first differential is 9.6755e-20. Further, the resultant p-value for second and third differentials are 1.008e-26 and 9.445e-08, respectively. From all these values, it is evident that data are not stationary. The data are also represented for Auto-Correlation Functions (ACF) and Partial ACF (PACF) as shown in Fig. 7.20.

ARIMA (Autoregressive Integrated Moving Average), a time series forecasting model starts and inputs will be in the format of (p, d, q). The model is implemented for prediction. The residual plot after implementation of ARIMA is illustrated in Fig. 7.21.

Fig. 7.17 Visualization of data

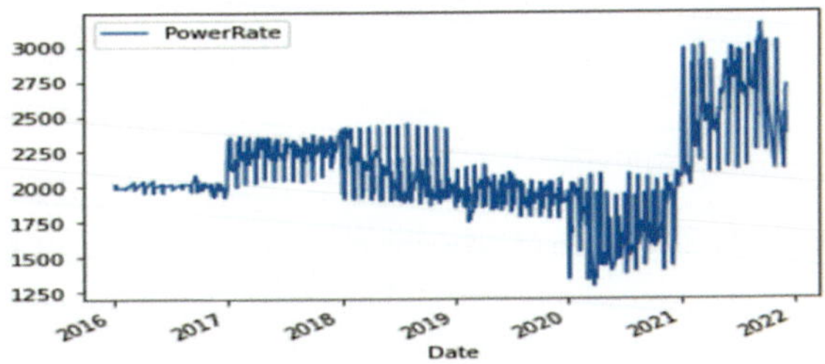

```
<matplotlib.legend.Legend at 0x7fc3d871d590>
```

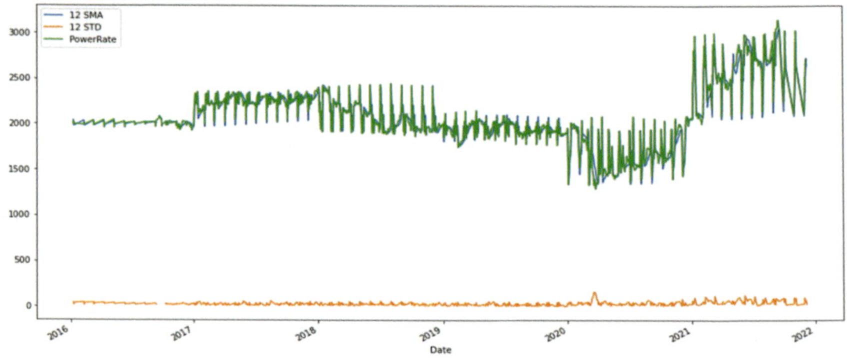

Fig. 7.18 Standard mean, standard deviation, and power rate

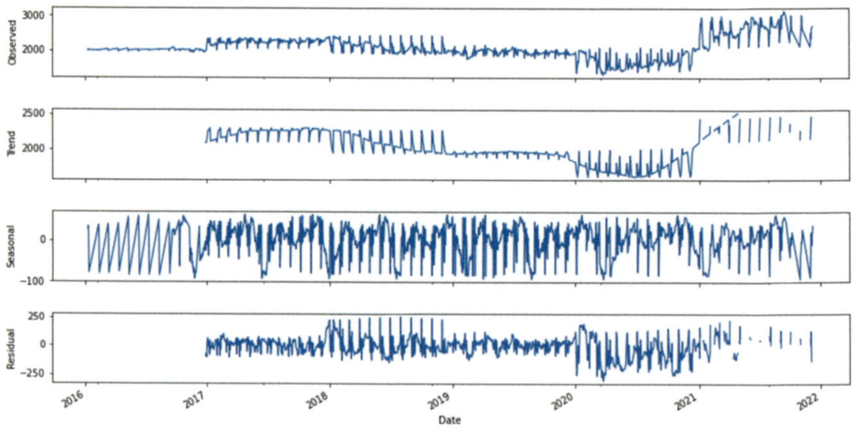

Fig. 7.19 Residual, sessional, trend, and observed data

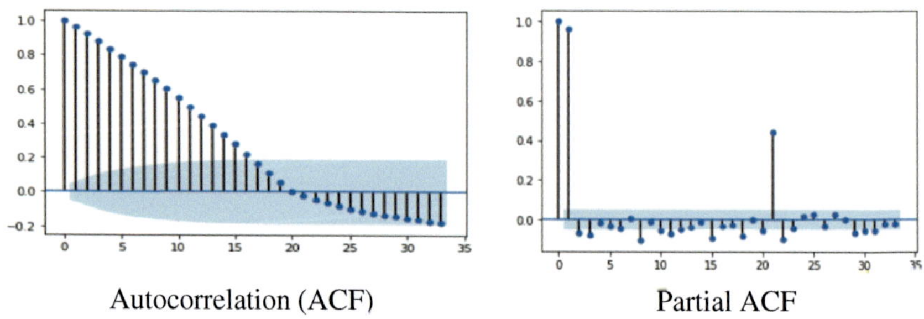

Fig. 7.20 Illustration of autocorrelation (ACF) and partial ACF

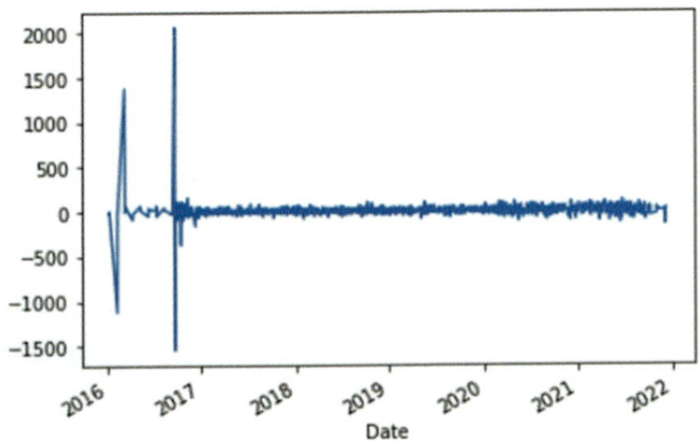

Fig. 7.21 Residual plot

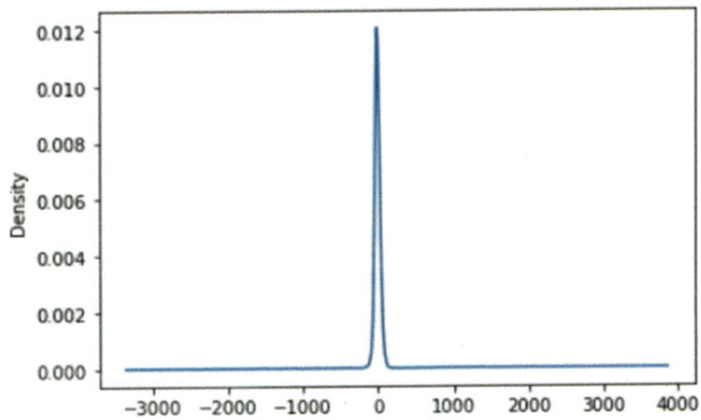

Fig. 7.22 KDE plot

To see the error by naked eyes, we can use KDE (Kernel Density Estimate) plotting. If the density is more at 0 then it will be a good sign or the green signal to the prediction. KDE plot for the data is illustrated in Fig. 7.22.

As discussed earlier, the prime objective of the current work is prediction. The plot for predicted and actual values is illustrated in Fig. 7.23. From the plot, it is evident that prediction model is performing pretty well establishing the effectiveness of the prediction model. Hence, the model can be successfully implemented in real world.

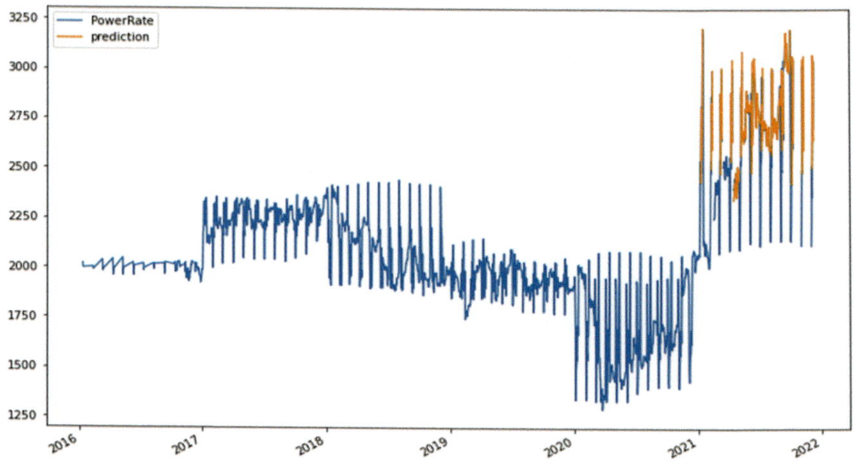

Fig. 7.23 Plot for actual and predicted values

7.7 Conclusion

In this work, authors used the S&P BSE power data after the required preprocessing. Once the data are preprocessed, it is checked for stationarity using Dickey-Fuller test and then converted the non-stationary data to stationary. Finally, ARIMA model is used to predict the values and the predicted values are compared with actual values to check for accuracy. The same is performed using Microsoft Excel as a Data Analysis Tool. Authors also explored many functions like Regression, Covariance, Correlation, ANOVA, etc. Finally, graphs and charts are also employed to visualize the data to understand the trend of GDP.

References

Ahmed NK, Atiya AF, Gayar NE, El-Shishiny H (2010) An empirical comparison of machine learning models for time series forecasting. Economet Rev 29(5–6):594–621

Cao J, Li Z, Li J (2019) Financial time series forecasting model based on CEEMDAN and LSTM. Physica A 519:127–139

Chae YM, Kim HS, Tark KC, Park HJ, Ho SH (2003) Analysis of healthcare quality indicator using data mining and decision support system. Expert Syst Appl 24(2):167–172

https://www.kaggle.com/datasets/kumarajarshi/life-expectancy-who?select=Life+Expectancy+Data.csv

Lachtermacher G, Fuller JD (1995) Back propagation in time-series forecasting. J Forecast 14(4):381–393

Lim B, Zohren S (2021) Time-series forecasting with deep learning: a survey. Phil Trans R Soc A 379(2194):20200209

Mahajan A, Rastogi A, Sharma N (2020) Annual rainfall prediction using time series forecasting. In: Soft computing: theories and applications. Springer, Singapore, pp 69–79

Maravall A (1983) An application of nonlinear time series forecasting. J Bus Econ Stat 1(1):66–74
Matthies M, Giupponi C, Ostendorf B (2007) Environmental decision support systems: current issues, methods and tools. Environ Model Softw 22(2):123–127
Najafabadi MM, Villanustre F, Khoshgoftaar TM, Seliya N, Wald R, Muharemagic E (2015) Deep learning applications and challenges in big data analytics. J Big Data 2(1):1–21
Nelson M, Hill T, Remus W, O'Connor M (1999) Time series forecasting using neural networks: should the data be deseasonalized first? J Forecast 18(5):359–367
Perna L, Thien-Seitz U, Ladwig KH, Meisinger C, Mielck A (2010) Socio-economic differences in life expectancy among persons with diabetes mellitus or myocardial infarction: results from the German Monica/Kora study. BMC Public Health 10(1):1–11
Roselli LRP, de Almeida AT, Frej EA (2019) Decision neuroscience for improving data visualization of decision support in the FITradeoff method. Oper Res Int J 19(4):933–953
Sadovnick AD, Ebers GC, Wilson RW, Paty DW (1992) Life expectancy in patients attending multiple sclerosis clinics. Neurology 42(5):991–991
Sharma N (2018) XGBoost. GRIN Verlag, The extreme gradient boosting for mining applications
Sharma N, Mangla M, Yadav S, Goyal N, Singh A, Verma S, Saber T (2021a) A sequential ensemble model for photovoltaic power forecasting. Comput Electr Eng 96:107484
Sharma N, Mangla M, Mohanty SN, Pattanaik CR (2021b) Employing stacked ensemble approach for time series forecasting. Int J Inf Technol 13(5):2075–2080
Singh S (2000) Pattern modelling in time-series forecasting. Cybern Syst 31(1):49–65
Siskos Y, Matsatsinis N (1993) A DSS for market analysis and new product design. J Decis Syst 2(1):35–60
Tang Z, De Almeida C, Fishwick PA (1991) Time series forecasting using neural networks versus Box-Jenkins methodology. SIMULATION 57(5):303–310
Torres JF, Hadjout D, Sebaa A, Martínez-Álvarez F, Troncoso A (2021) Deep learning for time series forecasting: a survey. Big Data 9(1):3–21
Yadav S, Sharma N (2018) Homogenous ensemble of time-series models for indian stock market. In: International conference on big data analytics. Springer, Cham, pp 100–114
Yadav S, Sharma N (2019) Forecasting of Indian Stock Market using time-series models. In: Computing and network sustainability. Springer, Singapore, pp 405–412
Zhang GP, Kline DM (2007) Quarterly time-series forecasting with neural networks. IEEE Trans Neural Netw 18(6):1800–1814

Chapter 8
IoT-Enabled Fuzzy Inference System for Heart Disease Monitoring

Janpreet Singh and Dalwinder Singh

8.1 Introduction

A clinical illness known as heart failure (HF) happens due to abnormalities in the basic functionality of the myocardial, which further create issues in the filling of ventricular as well as blood ejection (Inamdar and Inamdar 2016). The reduction in the functionality of the left ventricular myocardial is the most frequent reason for heart failure, but other possible causes include impairment in heart valves, endocardium, myocardium, pericardium, and great vessel, either individually or in conjunction. Genetic mutations, accelerated apoptosis, extreme or insufficient extracellular matrix proliferation, abnormal myocyte calcium cycling, elevation in neuro-humoral stimulation, ventricular remodeling, ischemia-related dysfunction, and increased hemodynamic overload are a few of the main pathogenic mechanisms causing heart failure (Dassanayaka and Jones 2015). Heart diseases are the most significant cause, due to which a plethora of individuals are dying across the world, as stated by WHO (2020). Many factors such as abnormal pulse rate, diabetics, cholesterol level, and high blood pressure are contributing symptoms which make the prediction of cardiovascular diseases process challenging and difficult (Hamo et al. 2022). Additionally, the signs of these diseases are also varied by the sex of an individual. It is analyzed that a man suffering from heart disease has more probability of having pain in the chest, and in contrast, women having this disease encounter shortness of breath, immense fatigue and nausea rather than having pain in the chest (Mayo Clinic 2022). Figure 8.1 demonstrates the normal and heart failure image.

J. Singh · D. Singh (✉)
School of Computer Science and Engineering, Lovely Professional University, Phagwara, Punjab, India
e-mail: dalwinder.singh@lpu.co.in

J. Singh
e-mail: janpreet.singh@lpu.co.in

N. Sharma et al. (eds.), *Big Data Analytics in Intelligent IoT and Cyber-Physical Systems*, Transactions on Computer Systems and Networks,
https://doi.org/10.1007/978-981-99-4518-4_8

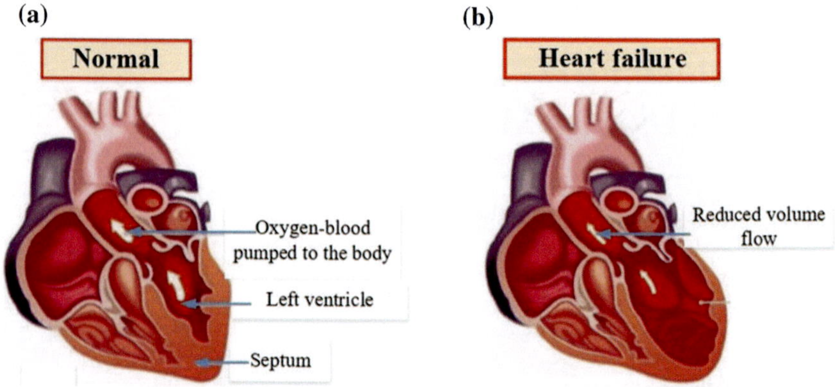

Fig. 8.1 **a** Normal heart **b** Heart failure (Junejo et al. 2019)

Heart diseases influence approximately 26 million people in a global pandemic and it makes its rank among the most common diseases (Savarese and Lund 2017). The treatment of these diseases is very expensive, and the price of diagnosis is rapidly rising due to the increase in population. Even though the treatment cost of these disorders is high, but the quality of life is low. Regardless of the significant advancement in prevention measures and medications, morbidity, as well as death, have been considerable due to heart failure. Additionally, it is found that the rate of prevalence, incidence, mortality, and morbidity varies according to the different clinical traits and etiologies detected in patients with heart diseases (Lippi and Sanchis-Gomar 2020). Contrary to Europe, no significant research has been conducted to examine the burden of heart failure in developing nations like India, leaving the epidemiology of heart failure as an unfinished agenda. Practicing internists and cardiologists are already aware of how significant this burden is expected to be, given that India is a nation in which 16% of the world's population resides and has 25% of the burden of coronary heart disease. Moreover, in this nation, 120M individuals are suffering from hypertension and a significant number of people with RHD (Chaturvedi et al. 2016).

To deliver timely, cost-effective care, a quick assessment of risk as well as prediction of heart failure are crucial (Raju et al. 2022). The accuracy of traditional risk prediction techniques is limited, and the complexity of heart failure pathogenesis and therapy generates a large amount of information that is difficult for physicians and researchers to analyze (Olsen et al. 2020). In the medical field, predicting the development of heart failure is a significant task because it calls for more expertise and information. There is a huge lack of readily accessible, experienced, and specialized doctors since all experts cannot be extremely talented in every specialism and at various locations (Shabeena 2020; Singh et al. 2020, 2021). Recent developments in data science and telemonitoring technology open the door to more effective remote patient monitoring. These solutions recommend individualized treatment plans. They contribute to preventing patient health deterioration through prompt and preventive measures. To continue treatment of individuals in the home from the hospital and back again, disease prediction systems are essential components to monitor patient hazards (Amadou Boubacar et al. 2021). The number of methodologies has been determined by different authors and researchers to anticipate this deadly disorder. However, it is the task of predicting this disease at its introductory phase is not easy because numerous factors influence it, such as accuracy, execution time, and complexity of methodology (Chicco and Jurman 2020). But, if a technique is able to predict it in an early stage, then effective diagnosis and treatment can save a significant number of lives of the patients (Choi et al. 2020; Ishaq et al. 2021).

IoT is becoming a methodology that cannot be avoided since it provides transparent and dispersed services. This technology has the capability to connect all the smart devices, such as sensors, smartphones, etc. with each other and assist in sharing of data between these gadgets (Cui et al. 2018). The main vision of IoT devices is to make the lifestyle of individuals more convenient and automated today by progressively supporting it (Adi et al. 2020). The Internet of Things generates an enormous amount of data, which is a major challenge to manage. In addition, the current network infrastructure is limited and unable to handle real-time sensitive applications, so Software Defined Networking is anticipated to be an appropriate network infrastructure for such applications (Askar 2016; Hamad and Askar 2021).

Machine learning is inspired by knowledge, which gives it intelligence. Being dependent on human input can fail. Therefore, in order to become completely resistant and compatible with human error, a system needs machine learning support (Failed 2020). In an integrated environment, human errors are easily remedied by algorithms through a process-optimizing feedback approach (Dharinya and Ephzibah 2019). ML constructs and recognizes patterns from previous behavior in order to show support for upcoming events and behavior. IoT data hidden experiences are provided by ML for quick, automated reactions and wise decision-making (Sarker 2021). IoT devices may use ML approaches to broaden existing patterns, spot abnormalities, and increase intelligence by ingesting images, audio, and video (Aldahiri et al. 2021).

8.2 Review of Literature and Research Gap Identification

A literature review has been done to highlight the benefits of machine learning and IoT methodologies for the detection of heart failure. Hence, this section contains a review of numerous medical diagnostic systems that other researchers have presented by utilizing either machine learning or IoT or a combination of both to diagnose heart failure.

8.2.1 Comprehensive Review

Chang et al. (2022) demonstrated the way in which AI technology can be utilized the determination if someone would get heart disease. The development of a system for detecting heart illness utilizing AI and ML algorithms is the main topic of the study. An application is developed using Python language by the authors since it is more dependable and aids in tracking and establishing various kinds of health monitoring applications. A random forest classifier method is created to more accurately detect heart conditions. This application outperforms training data by about 83% accuracy. The random forest classifier algorithm is examined on the basis of tests and findings. As a result, this algorithm improves the accuracy of diagnosis, and the development costs of this application are incredibly minimal.

Chicco and Jurman (2020) examined data from 299 heart failure patients, which were acquired in 2015. The authors utilized several machine learning classifiers such as Gradient Boosting, Naive Bayes, kNN, SVM, Artificial Neural Network, Decision Tree, Random Forest, and Linear Regression to rank the features that correlate to the most significant risk variables as well as anticipate the patients' survival. By using traditional biostatistical tests, this paper also conducts a different feature ranking analysis as well as compares the outcomes produced by the ML algorithms. As a result, the random forest classifier has the highest accuracy, i.e., 74%, among all other classifiers. Additionally, it is analyzed that the considered two models of feature demonstrate not only that ejection fraction and serum creatinine are adequate for identifying the survival of an individual suffering from HF from his or her history but also that the use of these selected features individually can lead to better identification of heart failure.

Kumar et al. (2021) evaluated recent developments in the discipline of quantum-enhanced machine learning and their importance in the early identification of heart failure using a data set having 14 variables. The amount of qubits in this study has been standardized according to the characteristics of the HF data using standard scalar, PCA, and min–max, and the pipelining approach has also been further utilized for optimization purposes. In this research work, it is demonstrated that quantum-enhanced machine learning algorithms, like quantum Gaussian Naive Bayes (QGNB), Quantum Decision Tree (QDT), Quantum k Nearest Neighbor (QKNN), and Quantum Random Forest (QRF), outperform conventional ML approaches in recognition of HF. The quantum random forest classifier beat all other classifiers considered in this respective work and has 89% accuracy for classification. The precision, recall, and F1-score results for the quantum random forest classifier were also excellent, coming in at 89, 93, and 88%, respectively. Moreover, there has also been a comparison done between the execution times of conventional and quantum-enhanced ML approaches. As a result, it is found that the quantum random forest has the shortest time of execution, i.e., 150ms. This paper, thus, offers a methodology to measure the differences between quantum-enhanced and conventional ML algorithms to choose the best technique for heart failure detection.

Umer et al. (2022) used Internet-of-Things (IoT) and cloud technologies to introduce a smart health care platform that enhances heart failure patients' survival estimates without taking into account manual feature engineering. The presented method explored deep learning models for identifying patients with heart failure who are still alive or who have passed away. The intelligent IoT-based framework keeps track of patients based on real-time data and offers heart failure patients prompt, efficient, and high-quality healthcare services. The model utilizes Internet of Things (IoT) sensors to collect information and transmit it to a cloud web server to process it. Deep learning algorithms further process these signals to ascertain the health of the patients. A medical practitioner who will give emergency assistance if necessary has to access to the patient's medical history and processing outcomes. The Heart Failure Clinical Records repository at UCI was explored to fetch the dataset for this research work, which has a total of 13 characteristics. The experimental findings showed that the CNN model, with an accuracy value of 92.89%, is better than other machine learning and deep learning methods.

A modified deep convolutional neural network is offered by Khan (2020) as part of an IoT framework to assess heart disease more precisely. An electrocardiogram, as well as blood pressure, is tracked using the patient's linked heart monitor and smartwatch. The MDCNN is utilized to categorize the dataset that has been gathered by sensors into abnormal and normal conditions. By contrasting the logistic regression and existing deep learning neural networks with the proposed MDCNN, the performance of the developed system is measured. The outcomes indicate that the proposed system outperforms competing methodologies. Hence, it is observed that the classification accuracy of the proposed methodology is 98.2% for the largest number of records which beats the traditional classifiers.

Ziryawulawo et al. (2022) provided a thorough analysis of methods for evaluating and determining heart disease utilizing machine learning and the IoT. IoT will enable individuals and medical professionals to assess cardiovascular ailments at an earlier stage. Various IoT methodologies in which ML approaches for the prediction and diagnosis of heart failure are implemented are compared. The comparison analysis was then evaluated according to the different performances of the developed models. Classification error, accuracy, F1-score, recall, and precision are distinct parameter measures that are taken under consideration in the comparative study. According to the final outcome of this paper, the authors concluded that among all the selected, a model created utilizing MSO—ANFIS is able to detect and continuously monitor patients' heart problems with a classification accuracy of 99.45%.

In order to protect the patients, Sandhiya and Palani (2022) presented a unique system that is utilized in the monitoring of heart disorders. This system incorporates the IoT with deep learning technology. In this research work, a new feature selection approach is used to improve classification performance using a deep learning algorithm. In the presented monitoring of heart disease, the severity of the condition is tracked in accordance with the data provided to the Internet of Things (IoT) devices. Additionally, it classifies the data of patients on the basis of many types of heart diseases as well as their severity. Furthermore, using the inputs given, it sends patients an alarm or message based on the type of heart condition. The investigations have shown and established that the developed model is more accurate in its predictions, as it has 95.12% prediction accuracy.

Kaur (2021) provided a framework for predicting heart disorders using ML approaches and IoT. In this work, the sensors of IoT have been linked to the body of a patient, which fetches the required information of the patient and makes a dataset. After that, different kinds of machine learning approaches of classifications are implemented, which assist in predicting whether the given data related to a patient imply to a patient normal or abnormal. The different ML classifiers that are executed in this study are Bayes Net, Random Forest, Simple Logistic, Decision Tree, k-star, Zero R, Naïve Bayes, and j48. Later, these classifiers are compared with one another according to the obtained classification accuracies. As a result, it is observed that the algorithm in which the J48 classifier was used has the highest accuracy, i.e., 92.56. Hence, this developed system produces positive results, aids in keeping track of heart disease patients' health and provides reliable illness data.

Sarmah (2020) developed a platform that assists in the patient monitoring having heart illnesses. This platform uses an IoT-centered Deep Learning Modified Neural Network to aid heart disorder medication administration and diagnosis. In this paper, three stages are followed by the author to successfully complete the proposed framework. These stages are authentication, encryption, and then classification. Initially, an individual having heart disease at any particular healthcare center is verified by using SHA-512 and cypher (SC). The implanted wearable Internet of Things sensor in the body of a patient will sense the information and then transfer it to the cloud. After that, the encryption process is performed on this sensed data by using the PDH-AES method and again sent to the cloud. Furthermore, the decryption of the encrypted data is done, and by using the DLMNN classifier, the classification of the acquired dataset is performed. The outputs are divided into two categories: normal and abnormal data. It indicates the condition of the heart of a patient, and a text message alert is sent to the doctor if the result is abnormal, so they can treat the individual. According to estimated research results, the DLMNN for the recognition of heart disorders performs better than existing techniques. Moreover, after contrasting to the existing AES, it is found that the presented approach provides 95.87% security, which can be considered as security's maximum level and also offers secure transmission of data. Additionally, it does so in the shortest amount of time for both encryption and decryption.

A CPS-based health care architecture is described in this research by Akter et al. (2021). Additionally, a technique for processing real-time information in order to make future strategic decisions in a unique manner employing various ML algorithms is presented. To comprehend the proposed strategy clearly in practical application, a case study involving heart disease is taken into consideration. As a result, Kaggle resources are utilized to gather a dataset on heart illness. After that, a heart disease prediction model is created by utilizing the gathered data set and implementing various machine learning classifiers. Binomial Logistic Regression, Decision Tree Classifier, Adaptive Boosting, Naive Bayes, K-Nearest Neighbors, and Random Forest are some of the classifier models that are employed. The prediction

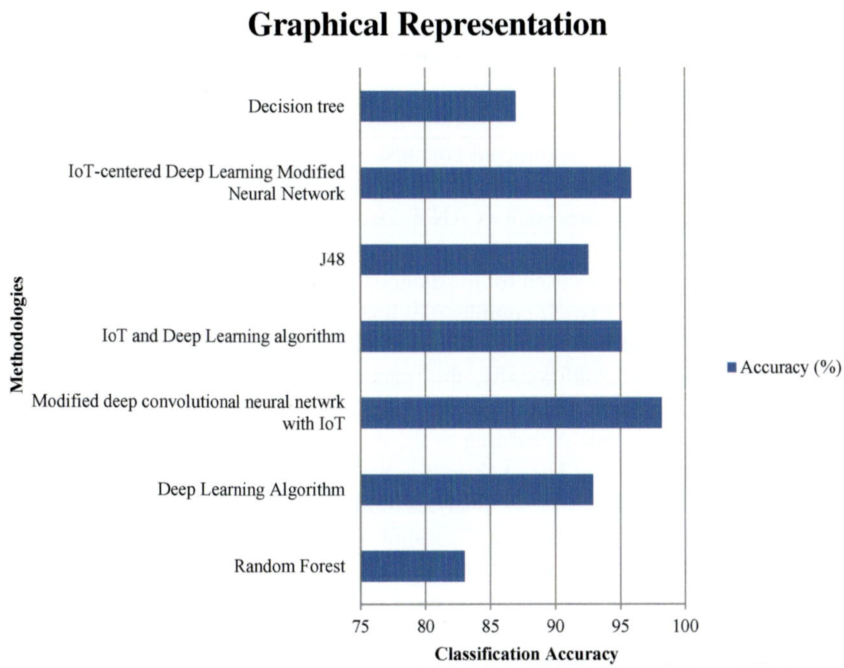

Fig. 8.2 Comparison of several heart disease detection approaches using a graphical representation

model's findings and the empirical studies correspond fairly well, and the decision tree classifiers' accuracy is about 87%. When compared to other current-generation models, the proposed model performs more accurately at estimating future decisions. Moreover, the decision tree classifier had the highest accuracy among the classifiers for predicting heart disease. It will be quicker, more accurate, and assist patients in anticipating heart issues to use this model for diagnosis. This model performs better than other existing models as well when compared to them (Fig. 8.2).

8.3 Considered Journal Comparison

Marimuthu et al. (2018) presented a summary of already published research work by using various machine learning as well as data mining techniques for the prediction of heart disease. The considered machine learning algorithm in this survey are SVM, Naïve Bayes, kNN, Fuzzy logic, Decision Tree, and ANN. This paper highlights the importance and need for a complex and combinational system, which will lead to enhancement in determining heart disease earlier and with high accuracy. Additionally, the author stated that the intelligence of a system is entirely dependent on the data stored in a system's database. The more amounts of data are stored or given to the model; it will make the system intelligent and accurate.

Seh (2019) summarized the recently published research work for the recognition of heart disease. This paper compares the finding of selected papers and draws a conclusion analytically. A significant conclusion provided by the author is that the accuracy of systems that assist in identifying heart disease can be improved by utilizing the ML approaches such as ANN, Decision Trees, and Naïve Bayes with the combination of genetic algorithms. Hence, this review majorly covers the ML and data mining techniques used for the diagnosis of heart disease and differentiates those papers according to their complexities and accuracies.

Limbitote (2020) overviewed the procedures and systems utilized for the prediction of heart disease. Additionally, this paper presents a thorough analysis of frequently used ML approaches by other researchers in their work for the identification of heart disorders. According to the conducted survey by this author, it is concluded that the Random Forest algorithm has the highest accuracy for the detection of heart disease. Moreover, the analysis of this work displays that the predictive systems for the said illness only achieve minimal accuracy. Thus, more complicated systems are required to enhance the predicting accuracy of the model.

Pandita (2021) examined numerous recent research and developments that are being done utilizing ML to recognize heart disorders. This paper stated that the Cleveland Heart Disease data set has been used by various researchers, which has 76 attributes. However, due to the missing information in some attributes, the authors have used only 14 attributes among them. Hence, after examining the association between different features and their impact on the model's accuracy, an adequate feature selection approach should be selected and used to reduce the number of features that are required in the development of an accurate system. Additionally, it is also analyzed that the neural networks and KNN are widely used approaches in numerous studies and are highly reliable at predicting heart disorders.

After noticing that the identification of the signs and symptoms of heart disease is a challenging and difficult task, Ziryawulawo et al. (2022) presented a study in which the development of the predictive system is comprehensively reviewed in which IoT is utilized for fetching the accurate data and ML approaches for the classification of acquired inputs. According to this study, the only limitations of using IoT in the detection of heart disease are processing capability and limited storage. However, this limitation can also be overcome if the researchers grab the benefits of cloud computing and integrate them into a developed model. Hence, an integration model in which ML, IoT, and cloud computing are utilized will make an advance predictive model for the detection of heart illnesses in their early stages.

8.4 Advantages of Using IoT and ML Techniques for the Detection of Heart Disease

In the medical domain, the patient's records are stored in the form of EHR, and due to the development of advanced health care systems, these records are also available on websites due to big data and further, these data are used to build a diagnostic model for any illness such as heart diseases (Nashif et al. 2018). Machine learning algorithms here assist in discovering meaningful information from this huge amount of dataset by evaluating it from a variety of angles (Hazra et al. 2017). Additionally, the intelligence of machine learning is dependent on the knowledge or data provided to the system. Hence, there are chances of making a mistake while providing the knowledge to the system as the inputs of the system are dependent on the human (Singh et al. 2023). However, the probability of making an error can be diminished by using IoT devices. These smart devices will automatically fetch the inputs from the patient without any human interference and then send the acquired data to the machine learning classifier for the prediction of the disorder (Khan 2020).

8.5 Research Methodology

A methodology is referred to a procedure that is utilized by various authors or researchers in order to accomplish an examination by briefly scrutinizing the literature and then proposing enhanced solutions for a particular problem with an appropriate result and decision. The evidence-based research approach is the most significant methodology in undertaking any research work in the medical discipline as it aids the practitioners and experts with numerous treatment plans and decision-making systems with the highest accuracy. Therefore, this research work is done by performing a systematic and statistical analysis of the various evidence from the literature review, which enlightens the use of IoT and Ml techniques for the detection of heart disease.

8.5.1 Literature Search

During the time of a decision about whether a particular research work should be included in the conducted study or not, always consider a factor known as credibility. Moreover, different databases should be explored in order to search for a research work, which offers suitable and pertinent data. In this research work, the papers that highlight the development of diagnostic models for heart disease by using IoT and ML and published from 2018 to mid of 2022 are taken user consideration. Also, these papers are searched by using some phrases and keywords such as "ML for heart disease prediction", "IoT for heart failure prediction", "Review papers for heart diseases using IoT and ML", and "IoT empowered ML systems for heart failure detection", etc. The databases that are explored in the search for perfectly suitable literature related to the chosen topic are Research Gate, IEEE, Hindawi, and Pubmed.

8.5.2 *Choosing Eligible Papers and Studies*

Before starting any research work, it is necessary to gather the literature, which is most suitable and relevant to the conducted study. Hence, eligibility criteria are made, and all the papers were first evaluated on this criterion. This criterion consists of the year of publication, must be related to heart disease, and development of any model by using either ML or IoT or a combination of both. Furthermore, while choosing any research study, it was first evaluated on the basis of its abstract and title and then went through the considered eligibility test. If a paper met the entire requirements, only then it was selected for the study.

8.6 Results and Findings

8.6.1 *Distribution of Selected Papers on the Basis of Their Year of Publication*

A total of 11 papers are considered, which have been published in the period 2018 to mid of 2022. Figure 8.3 shows the distribution of selected papers on the basis of their year of publication. As a result, it is noted that the year 2021 has a maximum number of publications in which machine learning and IoT both approaches are utilized effectively in order to propose a model which can assist experts in the diagnosis of heart failure.

To highlight the fruitful benefits of machine learning empowered by IoT, a study has been conducted. Hence, the main intent of the study is to analyze the different research work presented by various authors on the development of inference systems using ML and IoT statistically.

The analysis of various fruitful benefits and significance of using ML incorporated with IoT has been done in this complete study. The research papers that emphasized the creation of heart disease diagnostic models utilizing IoT and ML and were published between 2018 and mid-2022 are taken into account in this study. A number of scholarly articles are read and gone through eligibility criteria before selecting them in this conducted work. Moreover, the selected papers are distributed on the basis of the approaches of machine learning that is used with IoT in the development of models, database providers, and year of publications. Decision Trees and Random Forests are the two most popular ML techniques utilized in conjunction with IoT technology. Analysis shows that models created recently utilizing a combination of

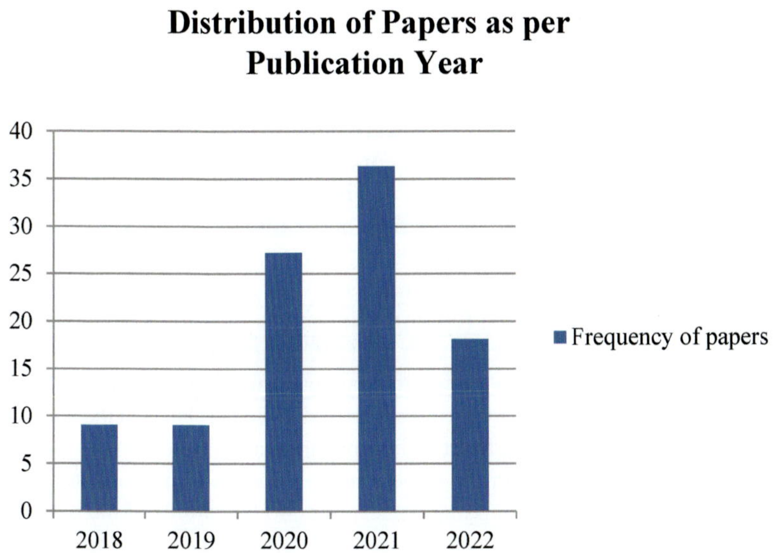

Fig. 8.3 Distribution of selected papers on the basis of their year of publication

IoT and ML approaches outperform those models that are deployed using either IoT technology or ML approach. Hence, after reading this review paper, the audience will assist in examining the significance of the ML techniques with IoT technology for the development of a diagnostic system for the detection of heart disease. Additionally, the frequency of the ML approaches along with IoT devices or alone has been demonstrated in Fig. 8.4.

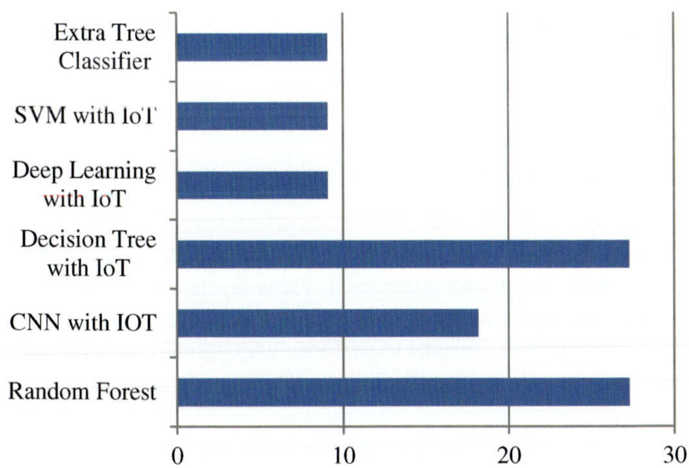

Fig. 8.4 Frequency of the ML approaches along with IoT devices or alone

8.7 Research Gap

Approximately 26 million individuals are affected by heart failure in a global pandemic, and it is becoming more common (Savarese and Lund 2017). Contrary to Europe, no significant research has been conducted to examine the burden of heart failure in developing nations like India, leaving the epidemiology of heart failure as an unfinished agenda. Practicing internists and cardiologists are already aware of how significant this burden is expected to be, given that India is a nation in which 16% of the world's population resides and has 25% of the burden of coronary heart disease. Moreover, in this nation, 120M individuals are suffering from hypertension and a significant number of people with RHD (Chaturvedi et al. 2016). Additionally, it is a very difficult task to predict if a patient suffering from heart failure will survive and assist the doctors in making accurate decisions related to the survival of an individual. Proper care of a heart failure patient can only be done if the doctor has the necessary experience in the same domain. Various limitations in the diagnostic procedure, data having missing values, inaccurate features of the patient and the inability of medical professionals to determine effects and causes related to the disease all contribute to uncertainty.

To remove this limitation, smart devices and mobile technologies are widely utilized in the domain of health and have a major influence on people's health worldwide. The numerous benefits of such approaches are extensively being used by various health professionals, which lead to a significant enhancement in the healthcare sector. In this research work, machine learning approaches and IoT techniques are effectively employed in the designing and development of a concurrent health monitoring system based on IoT and ML technology. These systems further assist in the prediction of heart failure in its earlier stages so that a patient can treat successfully before his or her condition gets worse. Hence, this integration of two approaches will fulfill the need of detection of heart failure in early stage and also the probability of having mistakes or errors while fetching the input data will also diminished by using sensory devices.

8.8 Proposed Methodology

In the development of an IoT-enabled fuzzy inference system for monitoring of heart diseases, four different layers are considered. These layers include the sensory layer, preprocessing layer, prediction layer, and performance layer, as shown in Fig. 8.5. Every layer has unique tasks or jobs to perform, which finally generate an outcome that assists in the monitoring of heart diseases. The brief description of each layer is as follows:

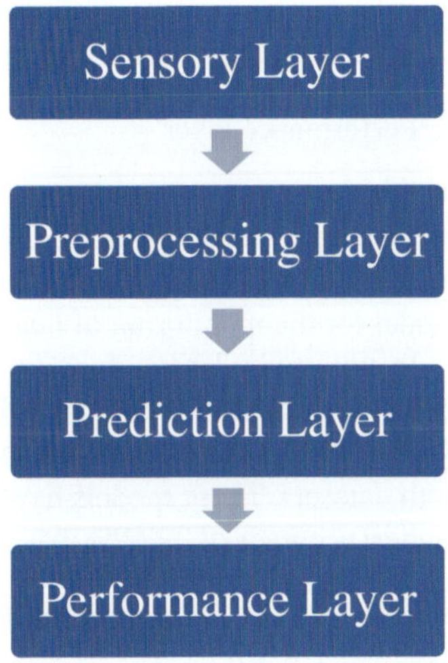

Fig. 8.5 Layers of the proposed model

- Sensory Layer

In the sensory layer, the values of selected inputs related to heart disease are grabbed by using various sensors. For example, the pulse sensor is used to monitor the irregular pulse of a patient; the persistent cough can be detected by using Doppler radar, etc. After gathering all input data, these data will transfer to the database via the Internet of Things.

- Preprocessing layer

As the collection of input data has been done by wireless communication, it might lead to raw data, which may include missing values, incomplete or inaccurate data and noise. Therefore, the handling of all null values and noisy data will be done in the pre-processing layer by using different methodologies such as moving average and normalization.

- Prediction layer

After performing the preprocessing on the collected data set, the accurate and correct dataset will be delivered to the prediction layer. The fuzzy inference system is implemented in this layer to predict heart disease adequately. Moreover, in this layer, with the usage of fuzzy membership functions, semantic points and linguistic variables, the collected input is organized and altered in a set that is in the form of fuzzy values. In the following point, the system makes fuzzier adjustments. After that, the fuzzification process of FIS will be executed if the data gathered from the sensory layer is correct. Here, the acquired data of inputs are in the crisp set and by using fuzzy membership functions and linguistic variables, they are converted into fuzzy sets. The control system of the FIS will use rules and facts and make an outcome in a set of crisp values. Furthermore, this set of crisp values is moved to a de-fuzzifier, which alters the set into fuzzy values. Finally, an output is generated, and the current health status of the patient is predicted by FIS.

- Performance layer

In this layer, the performance of the system will be calculated by using various performance measures. Basically, the classification accuracy, precision, specificity, sensitivity, and F-score will be evaluated to measure the performance of the proposed model for the monitoring of heart diseases.

In this study, I have done fine-tuning on deep learning models and proposed a small-sized CNN for the classification of cancer images in LC25000 dataset and KimiaPath24 dataset. Comparison of Five deep learning models has been done on both datasets. These models have achieved accuracies ranging from 93 to 99%. The highest accuracy of 99.69% was achieved in NasNet Large in LC25000 dataset which has the size of 343 MB and has 88.9M parameters and is 533 layers deep and on Kimia Path 24 highest accuracy of 96% was achieved in Efficient Net V2L which is the latest model of the Efficient Net family i.e. Efficient NetV2L which was developed in 2021 has been used in this study. I have saved h5 files of the architecture and the weights of the models used in this study. This proposed study has given better results in terms of accuracy than most cancer classification methods used in the literature survey. The computer vision-based techniques and deep learning models can assist pathologists to diagnose the different types of cancer at low cost and time. In future work, I would like to extend our work on different datasets of histopathological images and will try to improve the performance in the classification of cancers.

The extended form of these layers is also shown in Fig. 8.6 in which every layer is briefly illustrated, and their jobs are demonstrated in an effective way. All layers have equal importance, as the following layer is dependent on the previous layer. For example, if the sensory layer grabs the correct data and then preprocessing layer handles it properly, then according to the inputs provided by these two layers, the prediction layer will perform computation and offers a correct result. Moreover, if the prediction layer is able to deliver the accurate prediction of heart disease, only then the performance layer can evaluate the capacity to perform a task of the proposed model adequately and appropriately. Hence, in this way, all layers are dependent on each other, and every layer should perform its tasks in an effective way so that a correct result will be generated.

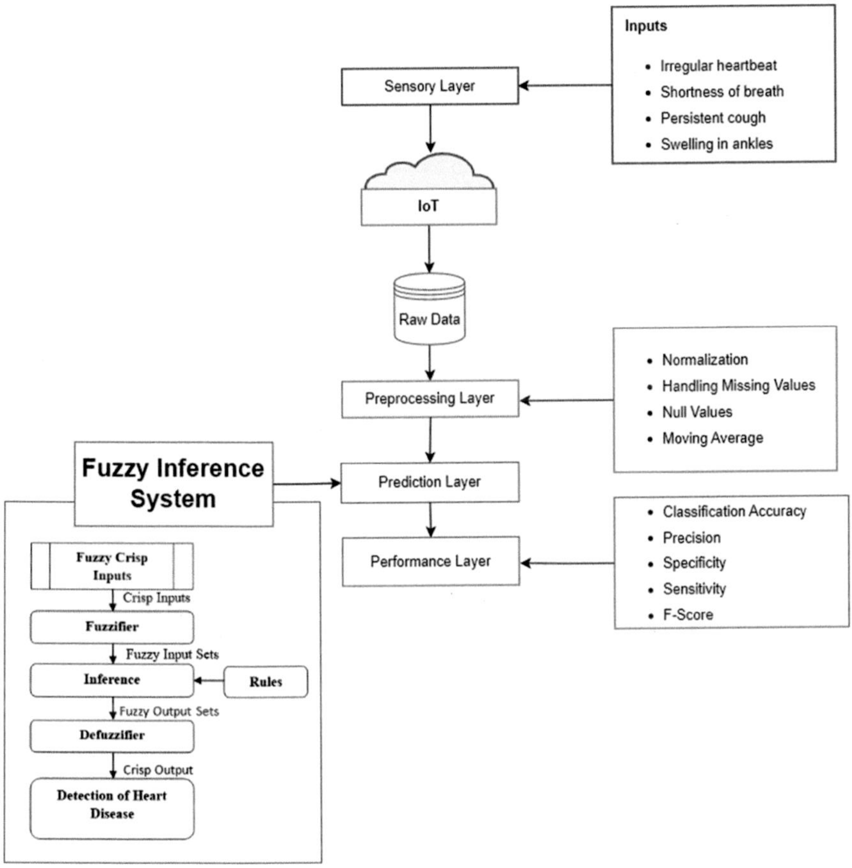

Fig. 8.6 Proposed methodology

References

Adi E, Anwar A, Baig Z, Zeadally S (2020) Machine learning and data analytics for the IOT. Neural Comput Appl 32(20):16205–16233. https://doi.org/10.1007/s00521-020-04874-y

Akter F, Kashem MA, Islam MM, Chowdhury MA, Rokunojjaman M, Uddin J (2021) Cyber-physical system (CPS) based heart disease's prediction model for community clinic using machine learning classifiers. J Hunan Univ Nat Sci 48(12)

Aldahiri A, Alrashed B, Hussain W (2021) Trends in using IOT with machine learning in health prediction system. Forecasting 3(1):181–206. https://doi.org/10.3390/forecast3010012

Amadou Boubacar H, Rahim M, Al-Hamoud G, Montesantos S, Delval C, Bothorel S, Ramirez-Gil JF (2021) Heartpredict algorithm: machine intelligence for the early detection of heart failure. Intell Based Med 5. https://doi.org/10.1016/j.ibmed.2021.100044

Askar S (2016) Adaptive load balancing scheme for data center networks using software defined network. Sci J Univ Zakho 4(2):275–286. https://doi.org/10.25271/2016.4.2.118

Chang V, Bhavani VR, Xu AQ, Hossain MA (2022) An artificial intelligence model for heart disease detection using machine learning algorithms. Healthc Analytics 2. https://doi.org/10.1016/j.health.2022.100016

Chaturvedi V, Parakh N, Seth S, Bhargava B, Ramakrishnan S, Roy A, Saxena A, Gupta N, Misra P, Rai SK, Anand K, Pandav CS, Sharma R, Prasad S (2016) Heart failure in India: the Indus (India Ukieri Study) study. J Pract Cardiovasc Sci 2(1):28–35. https://doi.org/10.4103/2395-5414.182988

Chicco D, Jurman G (2020) Machine learning can predict survival of patients with heart failure from serum creatinine and ejection fraction alone. BMC Med Inf Decis Making 20(1). https://doi.org/10.1186/s12911-020-1023-5

Choi DJ, Park JJ, Ali T, Lee S (2020) Artificial intelligence for the diagnosis of heart failure. NPJ Digital Med 3(1). https://doi.org/10.1038/s41746-020-0261-3

Cui L, Yang S, Chen F, Ming Z, Lu N, Qin J (2018) A survey on application of machine learning for internet of things. Int J Mach Learn Cybern 9(8):1399–1417. https://doi.org/10.1007/s13042-018-0834-5

Dassanayaka S, Jones SP (2015) Recent developments in heart failure. Circ Res 117(7):58–63. https://doi.org/10.1161/CIRCRESAHA.115.305765

Dharinya SS, Ephzibah EP (2019) Machine intelligence and automation: deep learning concepts aiding industrial applications. Internet Things Ind 4:237–248. https://doi.org/10.1007/978-3-030-32530-5_15

Hamad ZJ, Askar S (2021) Machine learning powered IoT for smart applications. Int J Sci Bus 5(3):92–100. https://doi.org/10.5281/zenodo.4497664

Hamo CE, Kwak L, Wang D, Florido R, Echouffo-Tcheugui JB, Blumenthal RS, Loehr L, Matsushita K, Nambi V, Ballantyne CM, Selvin E, Folsom AR, Heiss G, Coresh J, Ndumele CE (2022) Heart failure risk associated with severity of modifiable heart failure risk factors: the Aric Study. J Am Heart Assoc 11(4). https://doi.org/10.1161/JAHA.121.021583

Hazra A, Mandal SK, Gupta A, Mukherjee A, Mukherjee A (2017) Heart disease diagnosis and prediction using machine learning and data mining techniques: a review. Adv Comput Sci Technol 10(7):2137–2159

Inamdar AA, Inamdar AC (2016) Heart failure: diagnosis, management and utilization. J Clin Med 5(7):62. https://doi.org/10.3390/jcm5070062

Ishaq A, Sadiq S, Umer M, Ullah S, Mirjalili S, Rupapara V, Nappi M (2021) Improving the prediction of heart failure patients' survival using smote and effective data mining techniques. IEEE Access 9:39707–39716. https://doi.org/10.1109/ACCESS.2021.3064084

Junejo A, Shen Y, Laghari AA, Zhang X, Luo H (2019) Notice of retraction: molecular diagnostic and using deep learning techniques for predict functional recovery of patients treated of cardiovascular disease. IEEE Access 7:120315–120325. https://doi.org/10.1109/ACCESS.2019.2937290

Kaur B (2021) IOT framework for heart diseases prediction using machine learning. Int J Adv Trends Comput Sci Eng 10(3):2036–2041. https://doi.org/10.30534/ijatcse/2021/781032021

Khan MA (2020) An IOT framework for heart disease prediction based on MDCNN classifier. IEEE Access 8:34717–34727. https://doi.org/10.1109/ACCESS.2020.2974687

Kumar Y, Koul A, Sisodia PS, Shafi J, Verma K, Gheisari M, Davoodi MB (2021) Heart failure detection using quantum-enhanced machine learning and traditional machine learning techniques for internet of artificially intelligent medical things. Wirel Commun Mob Comput 2021:1–16. https://doi.org/10.1155/2021/1616725

Laha S, Chowdhury N, Karmakar R (2020) How can machine learning impact on wireless network and IoT?—a survey. In: 2020 11th International conference on computing, communication and networking technologies (ICCCNT), pp 1–7. https://doi.org/10.1109/ICCCNT49239.2020.9225652

Limbitote M (2020) A survey on prediction techniques of heart disease using machine learning. Int J Eng Res 9(06). https://doi.org/10.17577/IJERTV9IS060298

Lippi G, Sanchis-Gomar F (2020) Global epidemiology and future trends of heart failure. AME Med J 5:5–15. https://doi.org/10.21037/amj.2020.03.03
Marimuthu M, Abinaya M, Hariesh KS, Madhankumar K, Pavithra V (2018) A review on heart disease prediction using machine learning and data analytics approach. Int J Comput Appl 181(18):20–25. https://doi.org/10.5120/ijca2018917863
Mayo Clinic (2022) Cardiovascular Diseases and Cardiac Surgeryhttps://www.mayoclinic.org/medical-professionals/cardiovascular-diseases. Accessed 24 Sep 2022
Nashif S, Raihan MR, Islam MR, Imam MH (2018) Heart disease detection by using machine learning algorithms and a real-time cardiovascular health monitoring system. World J Eng Technology 06(04):854–873. https://doi.org/10.4236/wjet.2018.64057
Pandita A (2021) Review paper on prediction of heart disease using machine learning algorithms. Int J Res Appl Sci Eng Technol 9(VI):2937–2940. https://doi.org/10.22214/ijraset.2021.35626
Singh D, Rakhra M, Aledaily AN, Kariri E, Viriyasitavat W, Yadav K, Dhiman G, Kaur A (2023) Fuzzy logic based medical diagnostic system for hepatitis B using machine learning. Soft Comput 1–17
Raju KB, Dara S, Vidyarthi A, Gupta VMNSSVKR, Khan B (2022) Smart heart disease prediction system with IOT and fog computing sectors enabled by cascaded deep learning model. Comput Intell Neurosci 2022:1–22. https://doi.org/10.1155/2022/1070697
Sandhiya S, Palani U (2022) An IOT enabled heart disease monitoring system using grey wolf optimization and deep belief network. Research Square. https://doi.org/10.21203/rs.3.rs-1058279/v1
Shabeena T (2020) IoT based heart disease prediction using higher order Boltzmann deep belief neural network. Int J Sci Res (IJSR) 9(10):44–48. https://doi.org/10.21275/SR20831111725
Sarker IH (2021) Machine learning: algorithms, real-world applications and research directions. SN Comput Sci 2(3). https://doi.org/10.1007/s42979-021-00592-x
Sarmah SS (2020) An efficient IOT-based patient monitoring and heart disease prediction system using deep learning modified neural network. IEEE Access 8:135784–135797. https://doi.org/10.1109/ACCESS.2020.3007561
Savarese G, Lund LH (2017) Global public health burden of heart failure. Card Fail Rev 03(01):7–11. https://doi.org/10.15420/cfr.2016:25:2
Seh AH (2019) A review on heart disease prediction using machine learning techniques. Int J Manage IT Eng 9(4):2018–2224
Singh D, Verma S, Singla J (2020) A comprehensive review of intelligent medical diagnostic systems. In: 2020 4th International conference on trends in electronics and informatics (ICOEI) (48184), pp 977–981. https://doi.org/10.1109/ICOEI48184.2020.9143043
Singh D, Verma S, Singla J (2021) A neuro-fuzzy based medical intelligent system for the diagnosis of Hepatitis B. In: 2021 2nd International conference on computation, automation and knowledge management (ICCAKM), pp 107–111. https://doi.org/10.1109/ICCAKM50778.2021.9357765
Umer M, Sadiq S, Karamti H, Karamti W, Majeed R, Nappi M (2022) IOT based smart monitoring of patients' with acute heart failure. Sensors 22(7):24–31. https://doi.org/10.3390/s22072431
World Health Organization (2020) The top 10 causes of death. https://www.who.int/news-room/fact-sheets/detail/the-top-10-causes-of-death
Ziryawulawo A, Ogare AC, Ayebare F, Sinde R (2022) Application of IOT and machine learning techniques for heart disease prediction and diagnosis: a comprehensive review. Int J Adv Sci Res Eng 08(07):76–85. https://doi.org/10.31695/IJASRE.2022.8.7.7

Chapter 9
Implantable and Wearable Devices for IoT Applications—A Prototype of Integrated Multi-Feature Smart Shoes and Glass for the Safe Navigation of Blind People

Jothimanivannan Yuvanesh, S. Sherine, and I. Kala

9.1 Introduction

It is mentioned in Tamil Literature that "More than gifted to born as a human being, it is a rare gift to born without any disease, blind and deaf". Blindness means lacking of vision. Mostly, blindness resorts to complete vision loss that is no perception of light by people who suffers blindness. Blindness may occur instantaneously by sudden adverse actions that damage eyes or over an ambit of time due to some illness or diseases in our body. There are multifarious attributes that roots in the occurrence of blindness in people. Total vision loss owes to severe injury in eye, final stage glaucoma, total detachment of retina, stroke in the eye called vascular occlusion, intense infection in the eye, clinically called Endophthalmitis, and so on. The WHO says that blindness affects the people over the age of 50 in majority. As reported by World Health Organization, blind people have visual acuity worse than 3/60. In other words, it is the inability to perceive objects at a distance of 3 m. Blindness critically affects standard of living among adult populations. Adults with blindness have high levels of anxiety and depression and low levels of participation in labor force and

J. Yuvanesh (✉) · I. Kala
Department of Computer Science and Engineering, PSG Institute of Technology and Applied Research, Coimbatore, India
e-mail: d20z113@psgitech.ac.in

I. Kala
e-mail: kala@psgitech.ac.in

S. Sherine
Computer Science and Engineering, Karunya Institute of Technology and Sciences, Coimbatore, India

N. Sharma et al. (eds.), *Big Data Analytics in Intelligent IoT and Cyber-Physical Systems*, Transactions on Computer Systems and Networks,
https://doi.org/10.1007/978-981-99-4518-4_9

productivity. In the case of old aged people, blindness contributes to higher risks of falls and fractures, difficulty walking, social isolation and a greater probability of early entry into care homes. The mainstream problem for a blind person is to navigate around places independently. They encounter various obstacles on their way while walking on roads and they need someone to assist them in day to day life outside their place of residence. Otherwise they can use the help of guide dogs which in return needs regular training and maintenance. The use of walking canes might be suggested but it is also disadvantageous at a certain point when it breaks while hitting on hard objects. In order to avoid obstacles in outdoor places, it is very much mandatory for many of the blind people to keep a clear mental image or mapping of their surrounding paths if they need to walk independently on the roads. The most predominant necessity for a blind person is to obtain independence while moving across places. A blind person can lead such ante independent life with certain specifically designed assistive devices for them. They must feel safe while walking on the roads. The system proposed in this paper helps them achieve the same.

9.2 Literature Survey

The numerous research and journal papers on IoT-based solutions for assisting blind people are presented in chronological order.

In 2012, Ziad et al. reported on the design and development of rehabilitative shoes and spectacles for the blind (Abu-Faraj et al. 2012). It was developed with the notion of contributing to ameliorate the life of persons with vision impairment. A part of the system uses three pairs of ultrasonic sensors mounted on different sides of each shoe so as to detect ground-level obstacles of various heights, ground pits, and holes. The other part of system has a pair of ultrasonic transducers mounted above bridge of spectacles to identify head level obstacles.

In 2015, Alam et al. developed a model that aimed at directing the visually impaired people and avert them from unnecessary crashes with the impediments via prerecorded voice commands. This system has two arrangements, a shoe and a cane, integrated together to work as a single unit by Bluetooth technology. The shoe unit has three IR sensors and the cane also has an IR sensor to ascertain the presence of hindrances above the ground level. The cane comprises a pressure switch also to alert the user if they lose hold of the cane and LED lighting system on the cane helps alert the people surrounding the blind person about their presence such that to mitigate the risk of injury (Alam et al. 2015).

In the year 2019, Rutuja et al. proposed smart shoes to analyze different functionality of gait while walking. It also combines discrete working modules such as pedometer, health tracking system (Rutuja Anil Shinde et al. 2019).

In 2020, Roy et al. reported the design and implementation of smart shoes for secure movements to aid blind and visually impaired. The prototype was designed to detect obstacles and wet floors. Even though the implemented prototype has electrical safety systems to limit the false alarms, it has a false detection rate of 5.3% when the battery level reaches its minimum (Zeid Daou et al. 2020).

In the same year, Uma Maheshwari et al. came up with an idea to develop assistive device to enhance the quality and comfort in the lives of visually challenged people. The proposed system includes three ultrasonic sensors to locate an obstacle and find its distance from the user and an IR sensor to find out the depth of a place in case of pits using IR sensing technology (Failed 2020).

The next year, in 2021, Chava et al. proposed IoT-based smart shoe for the blind. The proposed system incorporates smart shoes and a pair of smart glasses. Sensors have been used to sense the obstacles and vibrators on peripheral sides of each shoe buzz if any obstacles are identified in that specific direction. Smart glasses have been designed to sense if any obstructions occur at head level of the user (Chava et al. 2021).

During the same period, Anisha et al. propounded a design of low-cost smart shoe for visually impaired (Anisha et al. 2021). The system uses ultrasonic sensors to measure the distance of the obstacles, and there is a buzzer to alert and assist the visually challenged person to traverse.

In the same year, Pradeep Kumar et al. recommended a prototype of assistive shoes for people who suffer from vision loss. It consists of a phone module and a shoe module. The phone app makes use of Google Maps to guide the user by giving information about the path. The shoe module helps to detect obstacles and moist place (Kumar et al. 2021).

Also in 2021, Porkodi et al. presented a review of smart shoes for blind using IoT. It discussed about various features of smart shoes, different technologies to design, develop, and implement a smart shoe (Conference for Convergence in Technology (I2CT) 2021)

9.3 Proposed System

The system presents a prototype of smart shoes and glass in order to guide the user to safely navigate from one place to another without the aid of any guide dog, any person or a cane. Hence, the design proposed includes integrated smart assistive shoes and glasses for visually impaired people which will detect obstacles at ground and head levels in their path and notify them through audio feedback, to detect ground pits, to identify wet surface in order to prevent falling of the user, to detect and notify the location of the user to the person whose mobile is paired with the Bluetooth, when the user encounters with a fall if any and to utilize the energy harnessed while walking to power up the entire system.

Advantages

- There are no existing systems that included multiple features to ensure safe navigation of visually challenged people that allows them to lead their life independently outside their home.

- Hence, the model proposed will ensure the safety of the user while walking out doors by incorporating multiple features like detection obstacles, wet surfaces, and pits, informing the concerned person about the location of the user when he/ she meets with a fall.

9.4 System Implementation

9.4.1 Shoe Module

This section presents the assembly of the sensors and other components on the shoe and working of the prototype developed. The power source for the entire system is a 12 V lead acid battery, which is placed on the wooden plank and an ON–OFF switch is used to ON and OFF the circuit when needed to prevent wastage of power.

- As for ultrasonic sensors

The ultrasonic sensors used here are to detect objects at ground level of the user. One ultrasonic sensor is placed on the toe cap and its echo and trig pins are connected to the digital pins 2 and 3 of the micro-controller, respectively. The two other ultrasonic sensors are fixed to the right and left of the shoe, in which the echo and trig pins of left ultrasonic sensor are coupled with the Arduino mega's digital pins 4 and 5 accordingly and the echo and trig pins of the ultrasonic sensor to the right of the shoe are connected to the digital pins 6 and 7 of the Arduino mega micro-controller. The HC-SR04 fixed at the toe cap of the shoe identifies the obstacle in-front of the user and ultrasonic sensor on each side of the shoe is used to detect obstacles on the right and left side of the user. The threshold distance set for the HC-SR04 sensor arranged at the toe cap, left, and right sides of the shoe are 25 cm, 10 cm, and 40 cm, respectively. This means that if the ultrasonic sensor at the toe cap detects an object at a distance less than 25 cm an audio alert about the same saying "Obstacle detected" will be given to the user. Similarly, if objects are found at a distance less than 10 and 40 cm by ultrasonic sensors on the left and right of the shoe, then audio feedback will be given to the user. The data about the distance identified by all the three sensors are transferred via Bluetooth to the mobile and also and text alert saying "obstacle detected" can be viewed using Blue Serial App which can be downloaded from the Android Play Store (Figs. 9.1 and 9.2).

A REL_35 Water Level Sensor is fixed to the right side of the shoe. The signal pin of the sensor is connected to analog input pin A1 of the micro-controller. When the sensor detects water, pin A1 on Arduino becomes HIGH and then the buzzer is turned on and gives a beep to alert the user to avoid the wet surface. The data about the water detection are transferred via Bluetooth to the mobile and text alert saying "water detected" can be viewed using Blue Serial App, which can be downloaded from the Android Play Store.

Fig. 9.1 Integrating ultrasonic sensors, IR sensor to the toe cap and Bluetooth and Voice module on the vamp of the shoe

Fig. 9.2 Mounting of water sensor and fall detection sensor on the shoe

- As for Infrared sensor

An IR sensor is mounted upside down on the toe cap of the shoe to detect ground pits and holes. The OUT pin of the IR sensor is connected to the Arduino mega 25600P development board's digital pin 10. When the IR sensor detects pits and holes, the LED indicator lights up, giving a low-level output signal in the OUT pin of the sensor and the buzzer gives a beep to alert the user about the pits identified and the same data are transmitted to the phone through HC05 Bluetooth module where an alert saying "pits detected" could be seen in the Blue Serial mobile application installed from Google Play Store.

- As for piezoelectric sensors

15 piezoelectric plates are lined up on the bottom sole of the shoe. They have two output pins. One is positive pin and the other one is negative pin which means

Fig. 9.3 Piezoelectric plates lined up on the sole of the shoe

ground. Connect positive potentials of piezoelectric sensors to the analog input pin A3 of the ATmega 2560P Arduino micro-controller. Depending on the pressure applied, a 35 mm piezoelectric disc could generate 4 V–6.2 V. The energy harnessed while walking, that is by applying pressure on the piezoelectric plates in which the mechanical energy gets converted to electrical energy is enhanced by using the buck boost converter (XL6009). This energy then is stored in capacitors and then used to power the whole system (Fig. 9.3).

- As for fall detection sensor and GPS module

ADXL335 triple axis accelerometer is fixed on the right side of the shoe. This sensor is used to detect a fall when the user encounters if any. Any of the analog OUT pin (Xout, Yout, Zout) of the MEMS sensor can be connected to the analog input pin A0 of the Arduino mega micro-controller. The threshold angle is fixed as 268 and if there is any change of angle measured from the threshold angle the buzzer will be turned on. The latitude and longitude of the user will be located using GPS NEO 6 m module whose RX/TX pins are connected to the 16 and 17 UART digital pins of the micro-controller and the data will be sent to mobile via Bluetooth HC05, which can be viewed in Blue Serial application (Fig. 9.4).

- As for Bluetooth and voice recorder and buzzer modules

The Bluetooth module HC05 is mounted on the vamp of the shoe and its RX/TX pins are connected respectively to the UART digital pins 18 and 19. The range of HC05 Bluetooth module is up to 10 m. The Bluetooth technology is a wireless technology that is used to transfer data from the system to the mobile phone. Whenever ultrasonic sensors sense obstacles, water sensor detects water, IR sensor finds pits, fall detection sensor identifies fall of the user, the distance of the obstacle identified, the latitude and longitude of the user are sent to the mobile connected through the Bluetooth. Alerts for the above-mentioned issues are also shown on the mobile and also voice alert for obstacle detection will be given through voice recorder and playback module

Fig. 9.4 Image of alerts in Blue Serial mobile application

which is coupled with a 3W speaker and a beep sound for water and pit detection will be provided by the buzzer fitted on the wooden plank.

9.4.2 Glass Module

The glass module is designed in such a way to identify obstacles at head level of the user. An ultrasonic sensor is mounted on the bridge of the glass whose echo and trig pins are soldered to the digital pins 8 and 9 of the same micro-controller unit fixed on the wooden plank, which is used to control the sensors and other components placed on the shoe module (Fig. 9.5).

9.4.3 Mobile Application Module

Once when the user falls and a fall is detected, the latitude and longitude of the user location located using GPS are sent over to the mobile phone via Bluetooth HC05.

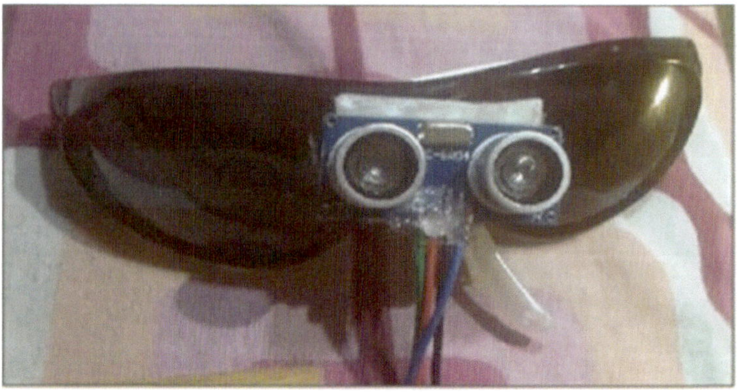

Fig. 9.5 Image of HC-SR04 sensor fixed on the bridge of the glass

- Blue Serial mobile application

Blue Serial is an open source application available in the Google Play Store, which could be easily downloaded and installed. The application is used for connecting an Android device to a serial Bluetooth-enabled device such as Arduino and other devices. It has been tested extensively to work.

- With the JY-MCU module and should work for a wide range of devices

  Steps to connect with the Bluetooth module.

Step 1: Install the Blue Serial Application from the Google Play Store.

Step 2: Turn on the Bluetooth on your mobile.

Step 3: Then pair your mobile to HC05 Bluetooth module in your Bluetooth settings.

Step 4: While pairing the device the application will ask for password which is usually 0000 or 1234. Type in the password.

Step 5: Then open the Blue Serial application and there will be a button with label "Search for paired devices" on the bottom left corner of the window. Click it to see the paired devices on the screen.

Step 6: After that click the "Connect" button on the lower right-hand corner of the mobile screen.

Step 7: Once after the connection message alerts can be seen on the screen (Figs. 9.6, 9.7 and 9.8).

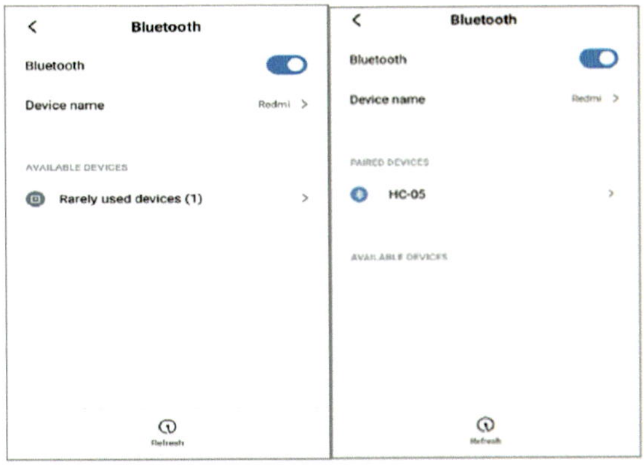

Fig. 9.6 Images of Bluetooth setting

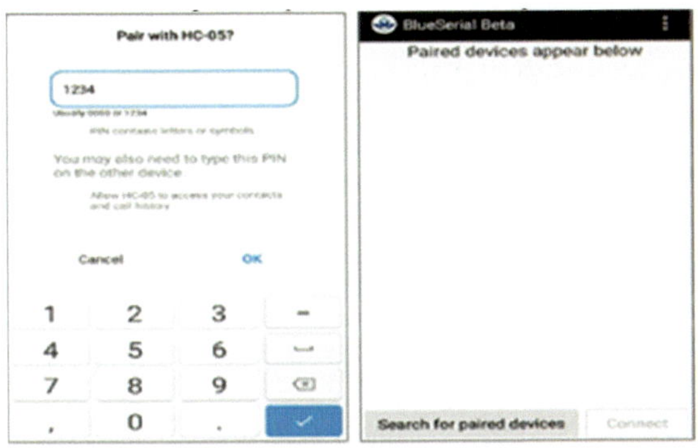

Fig. 9.7 Pairing of mobile device and Bluetooth HC05

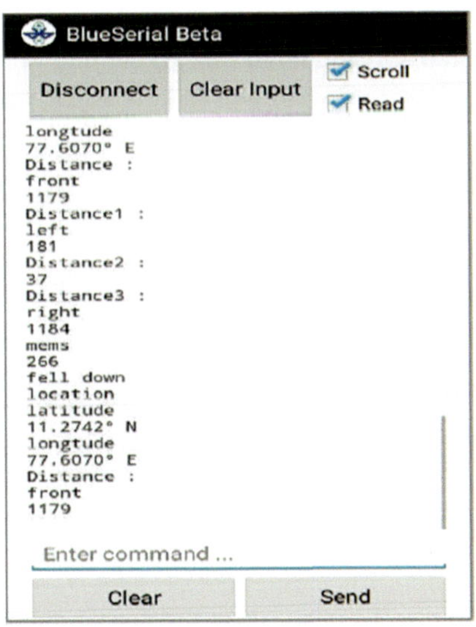

Fig. 9.8 Image of message alerts and location

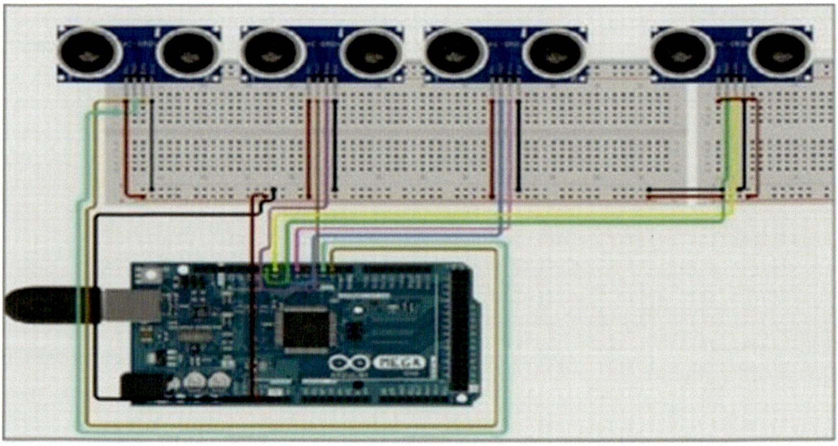

Fig. 9.9 Interfacing of Ultrasonic sensors (HC-SR04) with Arduino mega (ATmega2560P)

9.5 System Design and Components Interfacing

9.5.1 Interfacing HC-SR04

Step 1: Connect Vcc of all sensors into +5V of Arduino.

Step 2: GND of all sensors will go into GND of Arduino.

Step 3: Trig Pin of first sensor into Pin # 3 of Arduino.

Step 4: Echo pin of first sensor into Pin # 2 of Arduino.

Step 5: Trig Pin of second sensor into Pin # 5 of Arduino.

Step 6: Echo pin of second sensor into Pin # 4 of Arduino.

Step 7: Trig Pin of third sensor into Pin # 7 of Arduino.

Step 8: Echo pin of third sensor into Pin # 6 of Arduino.

Step 9: Trig Pin of fourth sensor into Pin # 9 of Arduino.

Step 10: Echo pin of fourth sensor into Pin # 8 of Arduino (Fig. 9.9).

9.5.2 Interfacing IR Sensor

Step 1: The first pin is the OUT pin connected to the digital pin 10 of the Arduino Mega 2560.

Step 2: The second pin is the GND pin connected to the GND of the micro-controller.

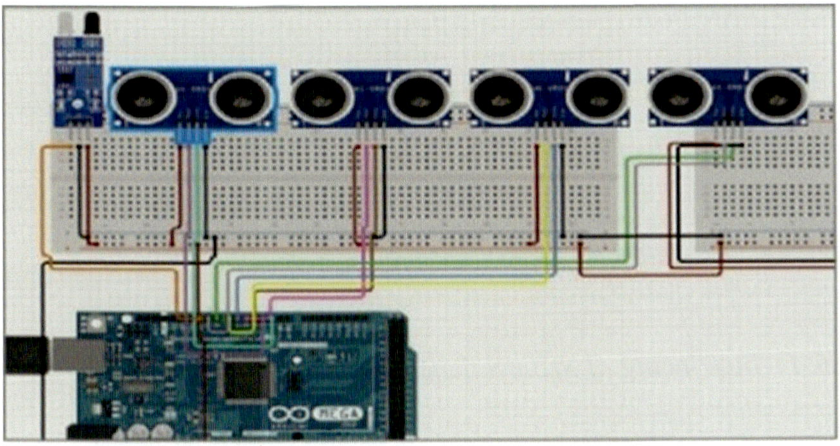

Fig. 9.10 Interfacing of Infrared sensor with Arduino mega (ATmega2560P)

Step 3: The third pin is the Vcc pin connected to the +5V pin of the Arduino Mega (Fig. 9.10).

9.5.3 *Interfacing Water Level Sensor*

Step 1: Connect Vcc of water level sensor to 5 V of Arduino mega board.

Step 2: Connect GND of water sensor to GND of the micro-controller.

Step 3: Connect Signal pin of the water level sensor to Arduino mega A1 analog input pin (Fig. 9.11).

Fig. 9.11 Interfacing of water sensor with Arduino

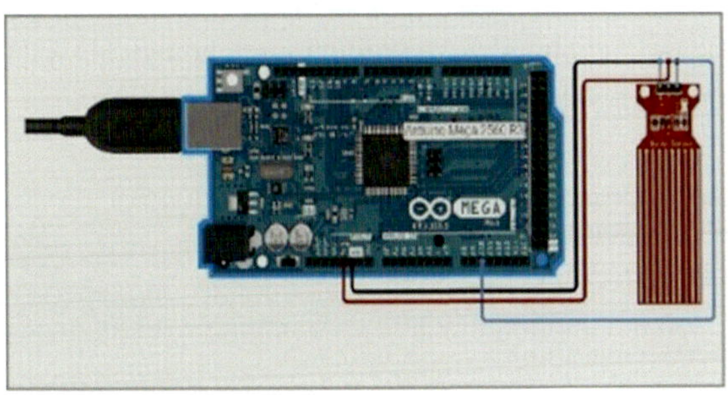

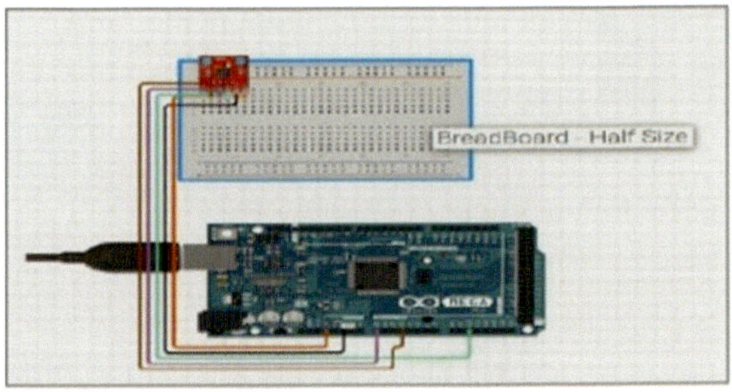

Fig. 9.12 Interfacing of tri-axis accelerometer ADXL335 with Arduino

9.5.4 *Interfacing of MEMS Sensor*

Connect any one of the analog output pin of the accelerometer to the analog input pin A0 of the micro-controller. Figure 4.5 depicts how MEMS sensor should be interfaced with the Arduino mega micro-controller (Fig. 9.12).

9.5.5 *Interfacing of Bluetooth and GPS Modules*

Step 1: Connect Vcc pin of GPS Neo6m with Arduino mega 3.3V and connect Vcc pin of Bluetooth HC05 module with 5V of Arduino.

Step 2: Connect GND of both Bluetooth and GPS modules to the GND of the micro-controller.

Step 3: Connect RX pin of GPS to UART/TX of the Arduino which is digital pin 16.

Step 4: Connect TX pin of GPS module with UART/RX of the Arduino mega which is digital pin 17.

Step 5: Connect RX pin of HC05 to UART/TX of the Arduino which is digital pin 18 and TX pin of the Bluetooth module with UART/RX, which is digital pin 19 of the micro-controller (Fig. 9.13).

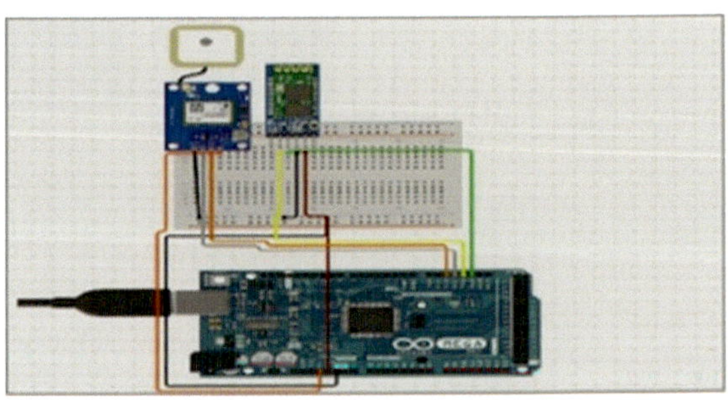

Fig. 9.13 Interfacing of Bluetooth HC05 and GPS NEO 6 m with Arduino

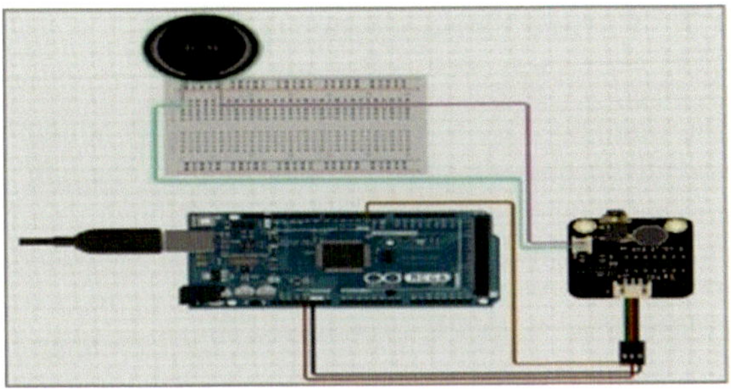

Fig. 9.14 Interfacing of APR ISD1820 along with 3W speaker to Arduino

9.5.6 Interfacing Voice Module and Speaker

Step 1: Interface voice recorder Vin pin to 5V pin of Arduino mega.

Step 2: Connect GND of voice recorder to GND of the micro-controller.

Step 3: Connect POS terminal of the speaker to SP+ of the voice module and NEG terminal of the speaker with SP- of the APR module.

Step 4: Connect PE pin of voice recorder to A8 pin of the Arduino mega, which is an analog input pin (Fig. 9.14).

9.6 Result and Analysis

The system is tested for various scenarios by placing obstacles in-front of the shoe, to the right and left of the shoe. The HC-SR04 sensors were able to detect obstacle precisely according to the threshold value given in the sketch and voice alert for the same can be heard through the speaker.

Water sensor is tested by pouring a droplet of water on it and it correctly indicated the presence of water by turning on the buzzer. The working of the accelerometer is checked by tilting it so that the change in angle goes below the threshold and at the same time the latitude and longitude acquired through GPS are sent over Bluetooth technology to the mobile, and it can be viewed after pairing our mobile device to the HC05 module through Blue Serial application that could be easily downloaded and installed from the Google Play Store. IR sensor is tested to locate pits by adjusting the sensor to identify any deep surfaces. Piezoelectric sensors are tested by pressing them and put a pressure on them. The output voltage of these sensors is measured using a multi-meter at the battery. Numerous tests have been conducted on the prototype to improve accuracy and to prevent false detection. With each and every test, the system showed improved accuracy in detecting obstacles, water, pits, and fall of the user. The overall weight of the system is 450g, initial charging of the 12V 1.3Ah battery required 8 hours. The estimated power consumption of the entire system is 760.35mA.

9.7 Future Enhancements

The prototype can be further improvised by adding much more features like gait analysis, pedometer step counting, wireless charging of mobile, Google Maps integration to help guide the blind people to their destination. The complexity of the circuit and the weight of the whole system can be reduced furthermore.

References

Abdel-Jaber H, Albazar H, Abdel-Wahab A, Amir M, Alqahtani A, Alobaid M (2021) Mobile based IoT solution for helping visual impairment users. Adv Internet of Things 11:141–152. https://doi.org/10.4236/ait.2021.114010

Abi Zeid Daou R, Chehade J, Abou Haydar G, Hayek A, Boercsoek J, Olmedo JJS (2020) Design and implementation of smart shoes for blind and visually impaired people for more secure movements. In: 2020 32nd international conference on microelectronics (ICM), pp 1–6. https://doi.org/10.1109/ICM50269.2020.9331779

Abu-Faraj ZO, Jabbour E, Ibrahim P, Ghaoui A (2012) Design and development of a prototype rehabilitative shoes and spectacles for the blind. In: 2012 5th international conference on biomedical engineering and informatics, pp 795–799. https://doi.org/10.1109/BMEI.2012.6513135

Anisha M et al (2021) Low-cost smart shoe for visually impaired. In: 2021 Third international conference on intelligent communication technologies and virtual mobile networks (ICICV), pp 1108–1111.https://doi.org/10.1109/ICICV50876.2021.9388432

Alam ST, Shrivastava S, Alam ST, Sasikala R (2015) Smart assistive device for visually impaired people. Int J Eng Res Technol (IJERT) 04(03)

Chava T, Srinivas AT, Sai AL, Rachapudi V (2021) IoT based smart shoe for the blind. In: 2021 6th international conference on inventive computation technologies (ICICT), pp 220–223. https://doi.org/10.1109/ICICT50816.2021.9358759

Garimella RC, Sastry VR, Mohiuddin MS (2015) Piezo-Gen—an approach to generate electricity from vibrations. Procedia Earth Planet Sci 11: 445–456. ISSN 1878-5220. https://doi.org/10.1016/j.proeps.2015.06.044. https://www.sciencedirect.com/science/article/pii/S1878522015000958

Haller S, Karnouskos S, Schroth C (2009) The Internet of Things in an enterprise context. In: Future internet symposium. Springer, Berlin, Heidelberg, pp 14–28

Kala I, Karthik S, Srihari K (2022) QoS-dependent and varied clustered routing (QoS–VCR) in wireless sensor network. Wireless Pers Commun. https://doi.org/10.1007/s11277-021-09441-9

Kala I, Karthik S, Srihari K (2021) Advanced hybrid secure multipath optimized routing in Internet of Things (IoT)-based WSN. Int J Commun Syst 34: e4782. https://doi.org/10.1002/dac.4782

Kumar P, Inchara KM, Lekhashree S, Likhith CN, Pavan U (2021) Real time assistive shoe for visually impaired people. In: 2021 6th international conference for convergence in technology (I2CT), pp 1–5. https://doi.org/10.1109/I2CT51068.2021.9417928

Mala NS, Thushara SS, Subbiah S (2017) Navigation gadget for visually impaired based on IoT. In: 2017 2nd international conference on computing and communications technologies (ICCCT), pp 334–338. https://doi.org/10.1109/ICCCT2.2017.7972298

Maheshwari BU, Subashini PR (2020) Sneak-sight shoes for the visually challenged. In: 2020 international conference on power, energy, control and transmission systems (ICPECTS), pp 1–3. https://doi.org/10.1109/ICPECTS49113.2020.9337030

Niveditha K, Kavya PD, Nivedha P, Pooja B, Lakshmikantha GC (2020) Virtual eye for blind using IoT, Int J Eng Res Technol (IJERT) 8(11) IETE—2020

Pradeepa R, Dr. R. Porkodi (2021) Smart shoes for blind using internet of things: a review. Int J Creative Res Thoughts (IJCRT) 9(2): 387–404. ISSN:2320-2882

Shinde RA, Nalbalwar SL, Singh S (2019) Smart shoes: walking towards a better future. Int J Eng Res Technol (IJERT) 08(07)

Tarkoma S, Katasonov A (2011) Internet of Things strategic research agenda (IoT–SRA). Finnish strategic centre for science, technology, and innovation: for information and communications (ICT) services, businesses, and technologies, Finland

Weyrich M, Ebert C (2016) Reference architectures for the Internet of Things. IEEE Softw 33(1): 2019. https://doi.org/10.1109/MS.2016.20

Whitmore A, Agarwal A, Da Xu L (2015) The Internet of Things—a survey of topics and trends. Inf Syst Front 17(2): 261–274

Part III
Application of Artificial Intelligence in Big Data

Chapter 10
A Feature Selection Technique Using Self-Organizing Maps for Software Defect Prediction

Krishna Pal Sharma, Shivam, Nonita Sharma, Ravi Sharma, and Mukesh Mishra

10.1 Introduction

Software defect prediction has become an important task in software development process to ensure that major defects are handled before the software reaches the customer. There is no upper bound to the number of defects that can be present in a large software project and hence allocating resources efficiently to handle defective modules can be especially useful. Defect prediction datasets are often composed of units called modules. Modules are further a collection of features which are used for predicting the class label for that module using various techniques. The impact each feature has on prediction accuracy varies and hence some features have high importance for prediction while some can be discarded during data preprocessing. Bharavi Mishra et al. showed that combining machine learning algorithms with feature selection can yield good performance on prediction tasks (Mishra and Shukla

K. P. Sharma (✉) · Shivam · R. Sharma
Computer Science and Engineering, Dr. B R Ambedkar National Institute of Technology, Jalandhar, India
e-mail: Kpsharma17vce@gmail.com

Shivam
e-mail: shivam.cs.19@nitj.ac.in

R. Sharma
e-mail: ravis.cs.19@nitj.ac.in

N. Sharma
Indira Gandhi Delhi Technical University for Women, Delhi, India
e-mail: nonitasharma@igdtuw.ac.in

M. Mishra
Massey University, Palmerston North, New Zealand
e-mail: m.mishra@massey.ac.nz

N. Sharma et al. (eds.), *Big Data Analytics in Intelligent IoT and Cyber-Physical Systems*, Transactions on Computer Systems and Networks,
https://doi.org/10.1007/978-981-99-4518-4_10

2011). Also, redundant software metrics can affect the performance of the classifier (Jiarpakdee et al. 2016). Hence, careful selection of meaningful features becomes an imperative task before defect prediction.

In the existing literature, many feature selection methods can be classified as filter-based, wrapper-based or hybrid techniques (Priyavrat and Sikka 2021). In filter-based methods, a predefined evaluation criterion is used to assign importance to attributes which can be used to select best feature subset for defect prediction. In wrapper-based methods, learning algorithms are used to get the desired feature subset. Wrapper-based methods usually provide better results but are also costlier than filter-based methods (Yadav et al. 2022). Therefore, methods which examine all subsets of feature set will provide us more insight into the importance of each feature and help in reducing high dimensionality for better prediction. Reducing dimensionality has been shown to reduce high computational needs (Mangla et al. 2022). Hybrid methods take advantage of benefits of both filter-based and wrapper-based methods. Therefore, it is crucial to apply the correct method for feature selection based on the given requirements.

This paper also introduces a wrapper-based technique for feature selection. If there a "N" features in a dataset, we find all dataset subsets of size "N-1" and analyze all subsets further to select one of them for the next cycle. After each cycle, one feature gets eliminated and we can continue this process any number of times until there are more features to eliminate. We also use the results of each cycle for defect prediction and after termination of algorithm, and we select the feature subset with the highest prediction accuracy. This paper has the following contributions:

- Proposed a wrapper-based feature selection technique.
- Used cost matrix to give the set some initial priorities for features.
- The technique can be run desired number of times with eliminating one feature in each iteration.
- Comparison with existing techniques on multiple datasets.

The organization of the sections ahead is as follows. Section 10.2 provides information on existing literature on feature selection in defect prediction. Section 10.3 discusses the methodology used in this paper for feature selection. Section 10.4 provides the results of the experimentation. Section 10.5 is for conclusion on this paper.

10.2 Literature Review

Zhoe Xu et al. proposed a new framework called MICHAC which uses maximal information coefficient for ranking features to remove irrelevant ones and then eliminate redundant features using hierarchical agglomerative clustering (Xu et al. 2016a). They used 11 NASA datasets and four AEEEM projects for feature selection and compared their results with five existing techniques.

Zhou Xu et al. analyzed the effectiveness of 32 existing feature selection methods (Xu et al. 2016b). They eliminated several limitations in analyzing the effectiveness by considering both noisy and clean datasets. They showed that the type of dataset and noise have insignificant impact over results. Also, Filter-based and wrapper-based methods give better performance but also take more time.

Khadijah et al. compared algorithms from the filter method and embedded method approaches, respectively. They found that embedded method SVE-RFE performed better in terms of accuracy than the filter method (Khadijah et al. 2020).

Qiao Yu et al. used feature subset selection and feature ranking approaches to analyze the effectiveness of these feature selection methods for cross-project defect prediction. Their results on NASA and PROMISE datasets show significant impact of both the selection techniques on prediction performance (Yu et al. 2019).

Shulong Liu et al. proposed a new feature selection technique based on Feature clustering using FF-Correlation measure and Feature ranking using the FC-Relevance method. They also propose a new metric to calculate the redundancy rate based on correlation (Liu et al. 2014).

Sukmawati Anggraeni Putri et al. proposed a framework that incorporates cost matrix into traditional feature selection algorithms. They showed that the cost sensitive feature selection approach helps reduce the overall cost and also addressed the class-imbalance problem (Putri and Fricyadie 2017).

Hadeel Alsolai et al. reviewed 15 Feature selection papers from journals and conferences from 2007 to 2017. Filter-based feature selection approaches are the most commonly used approaches with RELIEF being the most popular algorithm (Alsolai and Roper 2019).

Yu Qiu et al. proposed a new model called neural forest by combining Neural networks and Decision Forest algorithm. They were able to automatically explore feature representations using which, they got better results than state-of–the-art methods (Qiu et al. 2019).

Huan and Lei studied various existing feature selection techniques for both classification and clustering (Liu and Lei 2005). They further grouped these algorithms based on several factors. The authors also proposed a technique which uses integration of existing feature selection methods.

Petr and Jana worked on the problem of stability in feature selection techniques (Somol and Novovičová 2010). They studied the stability for techniques which produce varying-sized subsets. The authors also proposed stability measures for feature selection.

10.3 Methodology

In this section, we will discuss the technique used for feature selection step-by-step. This methodology has been devised with defect datasets in mind. The datasets used in this paper are JM1, KC1 and PC1 from PROMISE repository (Tong et al. 2017). All these datasets have 21 features and one class attribute which can evaluate to either true or false. Using these 21 features and class attribute as input, we will get a feature set using our technique which will be used for prediction.

10.4 Dataset Comparison

Before applying feature selection on the given datasets, it is imperative to use data preprocessing to make the datasets fit for further computation. The first step is to handle null values if present. Then, we normalize the data to put all attributes on the same scale. The next step is to shuffle the data in order to avoid having biased distribution. All these steps can be performed using Pandas library of python (Table 10.1).

The features present in these datasets are given in Table 10.2.

In this technique, we are going to consider all possible subsets of the given feature set. This is done in order to improve accuracy, but this also has a drawback because it will increase the computational needs. The time and space required for this technique will generally be more than the filter-based methods. But this trade-off is needed to ensure that we don't miss any potential feature-set. If there are "N" features in the given dataset, the first step to get all dataset subsets of size "N-1". There will be "N" such subsets because the number of ways in which we can select "N-1" classes from "N" classes is given by nCn-1 (Fig. 10.1).

The second step involves usage of self-organizing maps to cluster each subset into two clusters. The dataset used in this paper is unbalanced in nature. 80.65% of the modules are classified are defective. Therefore, it is imperative to consider the clustering algorithm which can perform well for unbalanced datasets also. Modified Kohonen self-organizing maps can perform clustering on unbalanced datasets effectively (Ahmad et al. 2017). However, we will use unmodified self-organizing maps in this paper and observe the impact on performance of prediction. The reason for choosing two clusters is that we have two classes in defect dataset i.e., "defect" and "not a defect". Therefore, unsupervised classification of such datasets into two clusters should give us an insight into how close these classification results are to the original labels. This can be done by applying labels on those clusters as "defect" or

Table 10.1 Datasets used and their properties.ont sizes of headings

Dataset	Instances	Attributes	Defective modules
JM1	10,885	22	8779
KC1	2109	22	326
PC1	1109	22	1032

Table 10.2 Features present in JM1, KC1 and PC1 datasets

Loc	v(g)	ev(g)	iv(g)	n
V	l	d	i	
B	T	lOCode	lOComment	
lOCodeAndComment	Uniq_Op	uniq_Opnd	total_Op	
branchCount	total_Opnd	lOBlank	e	

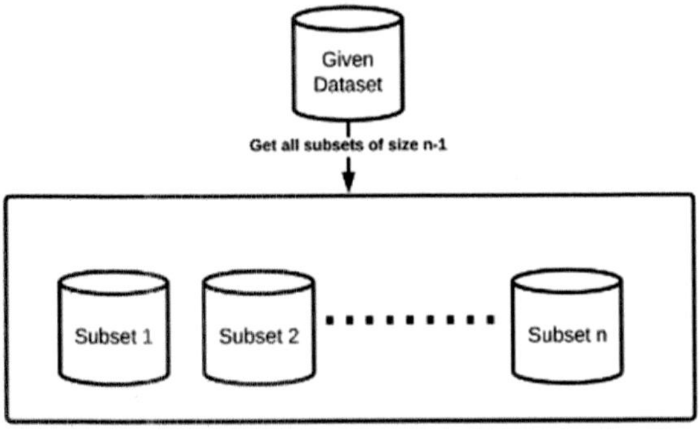

Fig. 10.1 Obtaining "N" subsets of size "N-1" from dataset

"not defect" by understanding the distribution of the data in the clusters. If majority of data points in the clusters originally lied in "defect class", we will label that cluster as "defect" and "not a defect" otherwise (Fig. 10.2).

The third step is to use Euclidean distance to measure the similarity between original labels and the labels we added using clustering. However, there may be situations where the developer or analyst may want to assign some priorities to features. This can be based on experience or some historical data. Thus, to assign

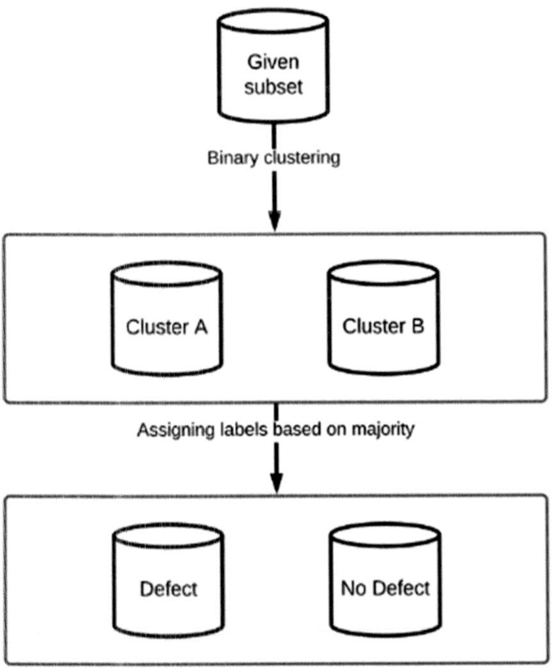

Fig. 10.2 Binary clustering of given subset

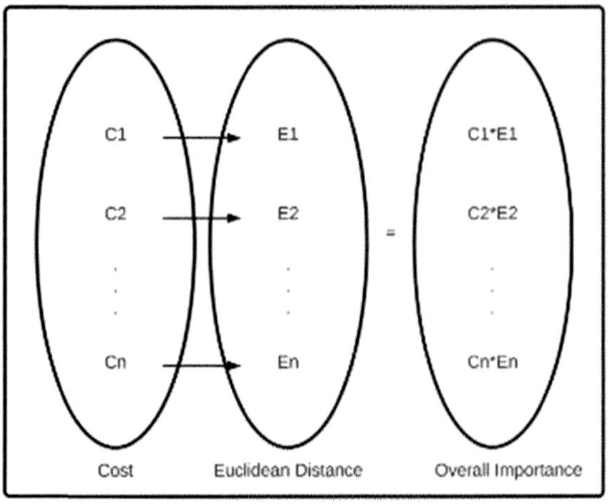

Fig. 10.3 Calculating total importance of features

importance to features, we will use cost matrix. This cost matrix can be filled by the developer by providing higher cost to more importance features. Since there are "N" features, it will be the matrix containing "N" columns and 1 row. Then, we multiply costs of all features with the corresponding Euclidean distances obtained for those features and obtain the overall feature importance value (Fig. 10.3).

Now, all subsets can be compared based on the obtained overall value. The subset with the lowest calculated value would give us the desired feature subset because it is closest to the original labels. We can either select this feature set or we can try to remove more features by repeating this process again with the newly obtained subset as the base dataset.

For comparison purpose, we will use 7 feature selection algorithms provided by Weka 3.8.5 (Frank et al. 2016; Kohavi and John 1997).

10.5 Results

The following experiments were performed on a Windows 10 system. The implementation of the technique described in this paper was done in python 9.1 (64-bit). Weka 3.8.5 was used for getting results for some existing feature selection techniques (Sharma et al. 2022).

For obtaining the cost matrix, we can fill the values based on experience, analysis or historical data. For this paper, we have chosen to use ANOVA since our features are continuous and the target variable is categorical in nature. The more the feature has impact on the target variable, the more will be the ANOVA test value for that feature (Shrivastava et al. 2022). Therefore, we need to take the reciprocal of each value to get cost of that feature since we will be multiplying the cost with the Euclidean distance values to get the overall importance value. Since the less the Euclidean distance, the better, we need inverse of the cost provided by the ANOVA test to get the actual cost to be used in this technique. We are plotting ANOVA test results for each dataset as bar plots for comparison as shown in Fig. 10.4. Since, there is no general rule for how many features are optimal, we will stick to the idea behind the wrapper-method and find the accuracy after each cycle where one feature gets eliminated.

For JM1. For experiments on the JM1 dataset, we find its cost matrix using ANOVA technique. Then we apply 8 feature selection techniques on the given dataset to compute the 3 best performing features. Further, we apply a machine learning algorithm called Random Forest for defect prediction. Then, we analyze the impact of eliminating features on computational time. We repeat this experiment for KC1 and PC1 datasets also.

Each algorithm gives a list of features after computation. Some algorithms remove features and provide a pruned list while others just provide the whole list sorted according to feature priority. Therefore, we will show the top performing features according to each algorithm for comparison (Table 10.3).

Accuracy versus #features. For each dataset, we have used 66% of data points for training purpose and rest for testing. Also, for each algorithm, 10-fold cross validation is used. After applying the algorithm once, one feature gets eliminated from the dataset. Hence, we apply CFSSOM multiple times on the dataset obtained

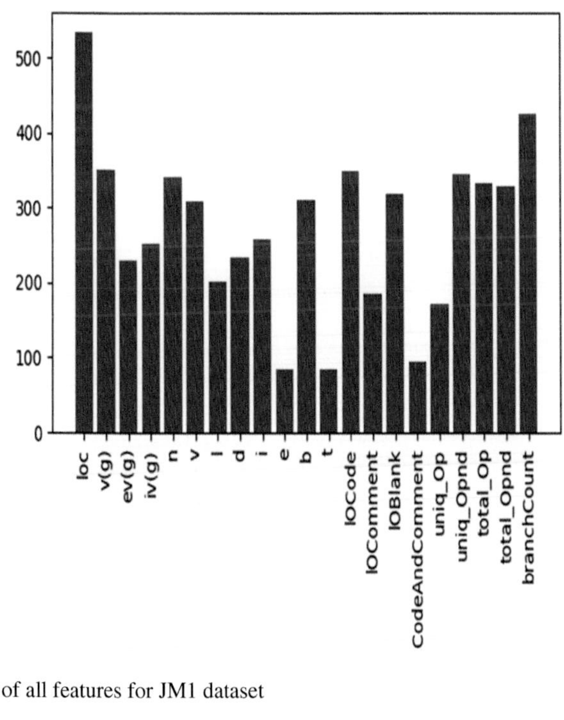

Fig. 10.4 Cost of all features for JM1 dataset

Table 10.3 Feature selection comparison using JM1 dataset

S.No	Algorithm	Rank 1	Rank 2	Rank 3
1	CFSSOM	loc	lOCode	l
2	CfsSubsetEval	loc	v(g)	ev(g)
3	ClassifierAttributeEval	branchCount	l	d
4	CorrelationAttributeEval	loc	branchCount	Uniq_Opnd
5	InfoGainAttributeEval	loc	lOCode	Uniq_Opnd
6	ReliefFAttributeEval	i	l	ev(g)
7	SymmetricalUncertAttributeEval	loc	lOCode	v(g)
8	OneRAttributeEval	v(g)	loc	branchCount

after each cycle until we have only 5 features remaining and then analyze the trend of prediction accuracy and RMSE from the obtained data. We will select features where we have the maximum prediction accuracy. As shown in Table 10.4, we obtain a maximum accuracy when number of features is 19 i.e., after eliminating "t" and "e".

Weighted average. Table 10.5 shows the trend of three important indicators. It is to be noted that these results are weighted average of class-level results. Since there are two possible classes in our data, we calculate individual precision, recall and F-measure for each class and then find weighted average for both classes.

Table 10.4 Results using JM1 per feature removal

Feature count	Feature eliminated	Correctly classified (%)	RMSE
21	None	81.5158	0.3671
20	t	81.7731	0.3656
19	e	81.9017	0.3657
18	uniq_Opnd	81.8374	0.3666
17	total_Op	81.8558	0.3655
16	total_Opnd	81.7915	0.3675
15	branchCount	81.6261	0.3675
14	lOComment	81.718	0.3685
13	lOBlank	81.4424	0.3709
12	locCodeAndComment	81.6169	0.3705
11	uniq_Op	81.1484	0.3723
10	b	81.3321	0.3729
9	d	81.0749	0.3729
8	ev(g)	80.8911	0.3754
7	i	81.0749	0.3741
6	iv(g)	80.2113	0.3798
5	n	80.1929	0.3829

Table 10.5 Precision, recall and F-measure for JM1 dataset

Feature count	Precision	Recall	F-measure
21	0.784	0.815	0.784
20	0.788	0.818	0.787
19	0.790	0.819	0.791
18	0.789	0.818	0.788
17	0.789	0.819	0.790
16	0.788	0.818	0.787
15	0.785	0.816	0.785
14	0.787	0.817	0.787
13	0.782	0.814	0.784
12	0.786	0.816	0.816
11	0.778	0.811	0.780
10	0.781	0.813	0.783
9	0.777	0.811	0.780
8	0.776	0.809	0.780
7	0.778	0.811	0.781
6	0.766	0.802	0.773
5	0.766	0.802	0.774

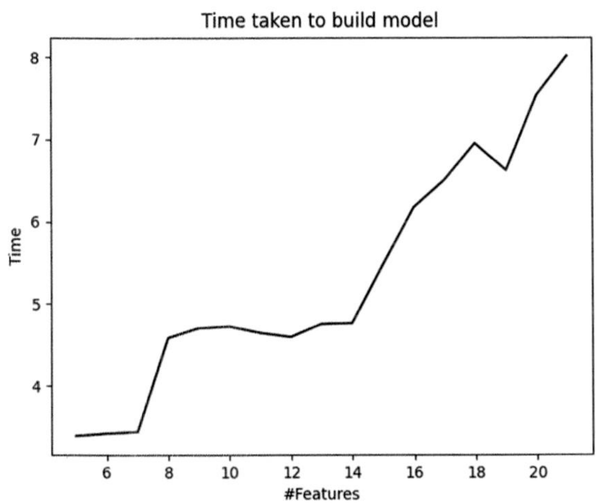

Fig. 10.5 Line plot for time taken

10.5.1 Time Taken to Build Model

See Figs. 10.5, 10.6, 10.7 and Tables 10.6, 10.7, 10.8.

For KC1.

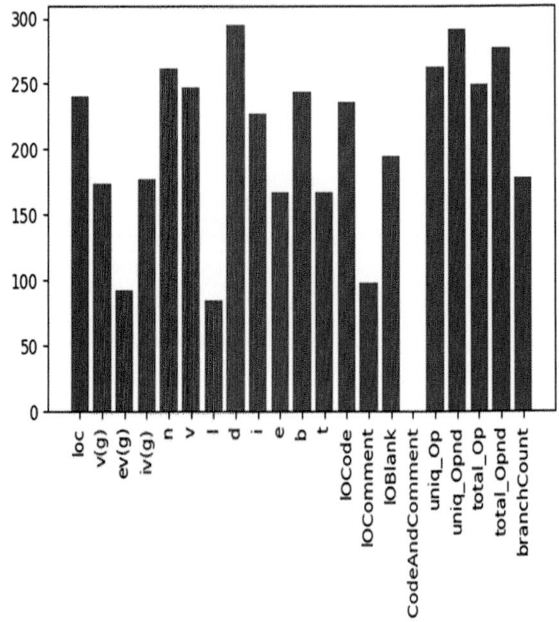

Fig. 10.6 Cost of all features for KC1

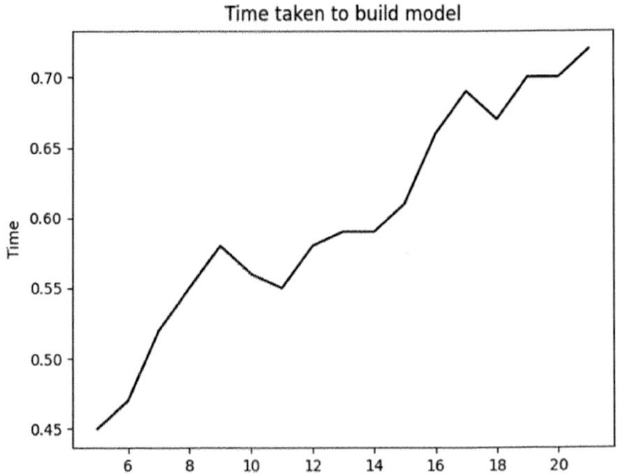

Fig. 10.7 Line plot for time taken

Table 10.6 Feature selection comparison using KC1 dataset

S.No	Algorithm	Rank 1	Rank 2	Rank 3
1	CFSSOM	b	v(g)	v
2	CfsSubsetEval	v	d	i
3	ClassifierAttributeEval	branchCount	l	d
4	CorrelationAttributeEval	d	Uniq_Opnd	Uniq_Op

(continued)

Table 10.6 (continued)

S.No	Algorithm	Rank 1	Rank 2	Rank 3
5	InfoGainAttributeEval	b	Total_Opnd	n
6	ReliefFAttributeEval	l	Uniq_Op	i
7	SymmetricalUncertAttributeEval	uniq_Opnd	d	v
8	OneRAttributeEval	uniq_Op	l	v(g)

Table 10.7 Results using KC1 per feature removal

Feature count	Feature eliminated	Correctly classified (%)	RMSE
21	None	86.4326	0.3193
20	locCodeAndComment	86.5275	0.3185
19	uniq_Op	86.4801	0.3213
18	uniq_Opnd	85.6262	0.3242
17	total_Op	86.0531	0.3222
16	total_Opnd	85.7211	0.3232
15	branchCount	86.1006	0.3208
14	l	86.5275	0.3219
13	d	85.4839	0.324
12	i	85.7685	0.325
11	e	85.6262	0.3235
10	ev(g)	84.8197	0.3273
9	t	85.389	0.3295
8	lOCode	85.0569	0.3317
7	lOComment	84.8197	0.3323
6	iv(g)	85.6736	0.3321
5	n	84.2505	0.3348

Table 10.8 Precision, Recall and F-measure for KCI dataset

Feature count	Precision	Recall	F-measure
21	0.845	0.864	0.847
20	0.846	0.865	0.847
19	0.845	0.865	0.847
18	0.834	0.856	0.838
17	0.840	0.861	0.843
16	0.836	0.857	0.840
15	0.840	0.861	0.843
14	0.846	0.865	0.848
13	0.833	0.855	0.838

(continued)

Table 10.8 (continued)

Feature count	Precision	Recall	F-measure
12	0.836	0.858	0.840
11	0.834	0.856	0.838
10	0.823	0.848	0.830
9	0.830	0.854	0.834
8	0.827	0.851	0.833
7	0.826	0.848	0.833
6	0.835	0.857	0.839
5	0.819	0.843	0.827

For PC1.

See Figs. 10.8, 10.9 and Tables 10.9, 10.10 and 10.11.

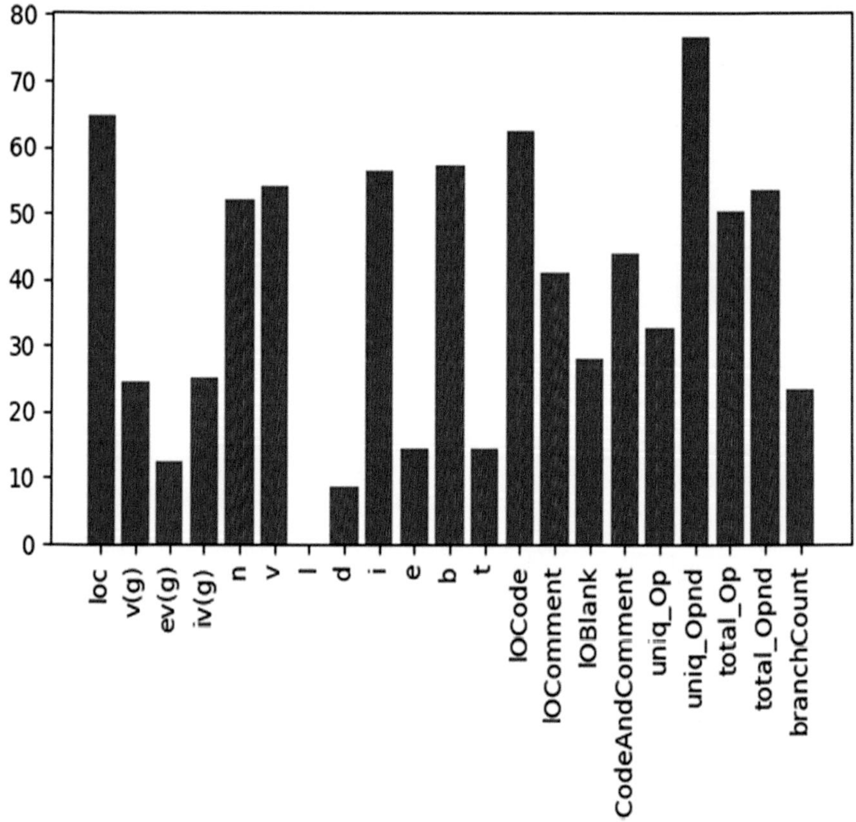

Fig. 10.8 Cost of all features for PC1 dataset

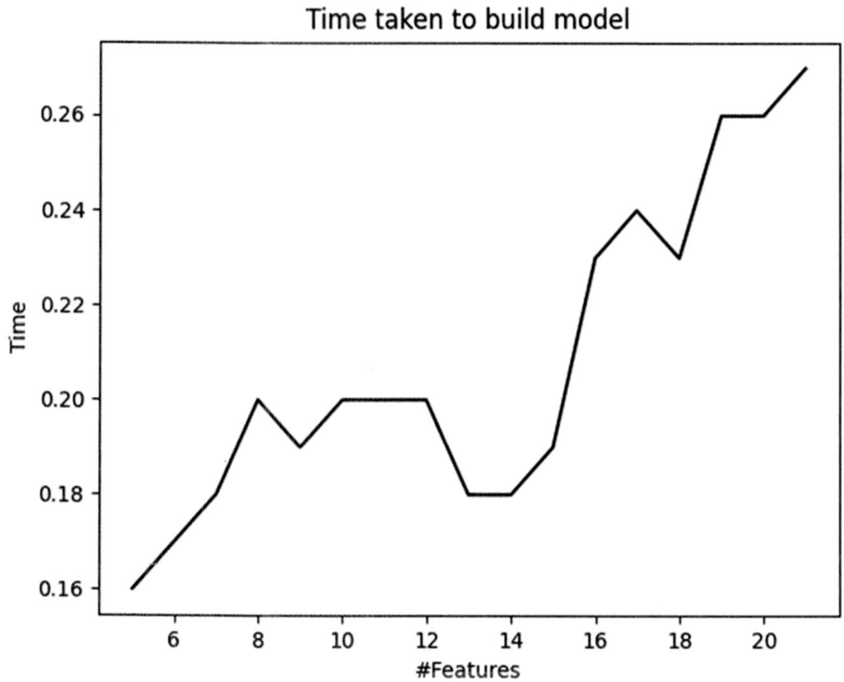

Fig. 10.9 Line plot for time taken

Table 10.9 Feature selection comparison using PC1 dataset

S.No	Algorithm	Rank 1	Rank 2	Rank 3
1	CFSSOM	loc	v(g)	n
2	CfsSubsetEval	v(g)	I	lOComment
3	ClassifierAttributeEval	branchCount	L	D
4	CorrelationAttributeEval	uniq_Opnd	lOBlank	loc
5	InfoGainAttributeEval	lOBlank	loc	uniq_Opnd
6	ReliefFAttributeEval	L	uniq_Opnd	lOComment
7	SymmetricalUncertAttributeEval	lOBlank	lOComment	uniq_Opnd
8	OneRAttributeEval	lOComment	I	T

Table 10.10 Precision, Recall and F-measure for PCI dataset

Feature count	Feature eliminated	Correctly classified (%)	RMSE
21	None	93.5018	0.2233
20	l	94.0433	0.2232
19	d	93.6823	0.2218
18	i	93.5018	0.225
17	e	93.7726	0.2255
16	b	93.6823	0.2239
15	t	94.1336	0.2215
14	lOCode	94.0433	0.2203
13	lOComment	94.0433	0.2203
12	lOBlank	93.231	0.2336
11	locCodeAndComment	93.5018	0.2346
10	uniq_Op	93.5012	0.2323
9	uniq_Opnd	93.0505	0.2343
8	total_Op	92.87	0.2427
7	total_Opnd	92.509	0.2408
6	branchCount	93.231	0.2384
5	ev(g)	93.1408	0.2383

Table 10.11 Precision, recall and F-measure for PCI dataset

Feature count	Precision	Recall	F-measure
21	0.921	0.935	0.923
20	0.929	0.940	0.930
19	0.924	0.937	0.926
18	0.921	0.935	0.924
17	0.926	0.938	0.928
16	0.924	0.937	0.926
15	0.931	0.941	0.931
14	0.930	0.940	0.931
13	0.930	0.940	0.931
12	0.917	0.932	0.921
11	0.919	0.935	0.920
10	0.920	0.935	0.923
9	0.913	0.931	0.917
8	0.907	0.929	0.912
7	0.903	0.925	0.910
6	0.915	0.932	0.918
5	0.914	0.931	0.918

10.6 Conclusion

In this paper, we have successfully proved the impact of feature selection on defect prediction by introducing a new wrapper-based technique. We observed the impact of eliminating one feature per cycle and observed an initial increase in prediction accuracy up to a few cycles and then decrease after reaching a maxima value. We have also proved that having high number of features does not specifically mean better results, but it may result in unnecessary computation. Therefore, feature selection can not only provide better prediction results but also save computational time. In future, we intend to study the optimal number of features taking both prediction accuracy and computational time under consideration.

References

Ahmad A, Yusoff R, Ismail MN, Rosli NR (2017) Clustering the imbalanced datasets using modified Kohonen self-organizing map (KSOM). Comput Conf 2017:751–755. https://doi.org/10.1109/SAI.2017.8252180

Alsolai H, Roper M (2019) A systematic review of feature selection techniques in software quality prediction. In: 2019 international conference on electrical and computing technologies and applications (ICECTA). Ras Al Khaimah, United Arab Emirates, pp 1–5. https://doi.org/10.1109/ICECTA48151.2019.8959566

Eibe F, Hall MA, Witten IH (2016) The WEKA workbench. Online appendix for "Data Mining: Practical Machine Learning Tools and Techniques". Morgan Kaufmann, 4th edn

Hall MA (1998) Correlation-based feature subset selection for machine learning. Hamilton, New Zealand

Mishra B, Shukla KK (2011) Impact of attribute selection on defect proneness prediction in OO software. In: 2011 2nd international conference on computer and communication technology (ICCCT-2011). Allahabad, India, pp 367–372. https://doi.org/10.1109/ICCCT.2011.6075151

Jiarpakdee J, Tantithamthavorn C, Ihara A, Matsumoto K (2016) A study of redundant metrics in defect prediction datasets. In: 2016 IEEE international symposium on software reliability engineering workshops (ISSREW). Ottawa, ON, Canada, pp 51–52.https://doi.org/10.1109/ISSREW.2016.30

Khadijah AA, Wirawan PW, Kurniawan K (2020) The comparison of feature selection methods in software defect prediction. In: 2020 4th international conference on informatics and computational sciences (ICICoS), pp 1–6. https://doi.org/10.1109/ICICoS51170.2020.9299022

Kira K, Rendell LA (1992) A practical approach to feature selection. In: Ninth international workshop on machine learning, pp 249–256

Kohavi R, John GH (1997) Wrappers for feature subset selection. Artif Intell 97(1–2):273–324

Kononenko I (1994) Estimating attributes: analysis and extensions of RELIEF. In: European conference on machine learning, pp 171–182

Liu H, Yu L (2005) Toward integrating feature selection algorithms for classification and clustering. IEEE Trans Knowl Data Eng 17(4): 491–502. https://doi.org/10.1109/TKDE.2005.66

Liu S, Chen X, Liu W, Chen J, Gu Q, Chen D (2014) FECAR: a feature selection framework for software defect prediction. In: 2014 IEEE 38th annual computer software and applications conference, pp 426–435. https://doi.org/10.1109/COMPSAC.2014.66.a

Mangla M, Sharma N, Mohanty SN (2022) A sequential ensemble model for software fault prediction. Innov Syst Softw Eng 18(2):301–308

Priyavrat SN, Sikka G (2021) Multimodal sentiment analysis of social media data: a review. In: Singh PK, Singh Y, Kolekar MH, Kar AK, Chhabra JK, Sen A (eds) Recent innovations in computing. ICRIC 2020. Lecture notes in electrical engineering, vol 701. Springer, Singapore. https://doi.org/10.1007/978-981-15-8297-4_44

Putri SA (2017) Combining integreted sampling technique with feature selection for software defect prediction. In: 2017 5th international conference on cyber and IT service management (CITSM), pp 1–6. https://doi.org/10.1109/CITSM.2017.8089264

Qiu Y, Liu Y, Liu A, Zhu J, Xu J (2019) Automatic feature exploration and an application in defect prediction. IEEE Access 7:112097–112112. https://doi.org/10.1109/ACCESS.2019.2934530

Robnik-Sikonja M, Kononenko I (1997) An adaptation of relief for attribute estimation in regression. In: Fourteenth international conference on machine learning, pp 296–304

Sharma N, Awasthi LK, Mangla M, Sharma KP, Kumar R (eds) (2022) Cyber-physical systems: a comprehensive guide, 1st edn. Chapman and Hall/CRC. https://doi.org/10.1201/9781003202752

Shrivastava VK, Shrivastava A, Sharma N et al (2022) Deep learning model for temperature prediction: an empirical study. Model Earth Syst Environ. https://doi.org/10.1007/s40808-022-01609-x

Somol P, Novovičová J (2010) Evaluating stability and comparing output of feature selectors that optimize feature subset cardinality. IEEE Trans Pattern Anal Mach Intell 32(11):1921–1939. https://doi.org/10.1109/TPAMI.2010.34

Tong H; Liu B; Wang S (2017) "Benchmark data sets", Mendeley Data, V1, https://doi.org/10.17632/923xvkk5mm

Xu Z, Xuan J, Liu J, Cui X (2016a) MICHAC: defect prediction via feature selection based on maximal information coefficient with hierarchical agglomerative clustering. In: 2016a IEEE 23rd international conference on software analysis, evolution, and reengineering (SANER), pp 370–381. https://doi.org/10.1109/SANER.2016.34

Xu Z, Liu J, Yang Z, An G, Jia X (2016b) The impact of feature selection on defect prediction performance: an empirical comparison. In: 2016b IEEE 27th international symposium on software reliability engineering (ISSRE), pp 309–320. https://doi.org/10.1109/ISSRE.2016.13

Yadav S, Tomar P, Nehra V, Sharma N (2022) Hybrid model for software fault prediction. In: Cyber-physical systems. Chapman and Hall/CRC, pp 85–103

Yu Q, Qian J, Jiang S, Wu Z, Zhang G (2019) An empirical study on the effectiveness of feature selection for cross-project defect prediction. IEEE Access 7:35710–35718. https://doi.org/10.1109/ACCESS.2019.2895614

Chapter 11
Lumbar Spine Disease Prediction with KNN, Random Forest and Decision Tree: A Study

Ruchi and Dalwinder Singh

11.1 Introduction

Lumbar Spine is a disease that affects the human body and causes a formation issue with the Lumbar Spine gland that is an important part of the human body. Metabolism corresponding to the body is adversely affected with Lumbar Spine related disease. These diseases cause imbalance in the regulation rate of the body's metabolism. Machine learning based mechanism plays a critical role in the process of Lumbar Spine prediction. This paper discussed the models that take the information from UCI machine learning repository. The dataset then fed into the system for training. After the training process, finalized testing is performed (Sidiq 2019).

Lumbar Spine disease can affect the human body and severe issues can be a result related to metabolism as discussed in (Jha et al. 2022). The effective classification of disease using the machine learning approaches including decision tree is discussed by (Ionita 2016). The decision tree-based mechanism generally operates on a dataset. The dataset can be derived from benchmarked websites including UCI machine learning repository, Kaggle and many more (Yasir 2020). The attributes of the dataset must be pre-processed to remove noise that ultimately increases the classification accuracy. The mode based approach as followed in (Chaubey et al. 2020). In this approach, the highest value from the attribute is replaced with the missing value. After this phase, segmentation is performed. The segmentation process divides the overall dataset into critical and non-critical parts (Gorade 2017). The last phase

Ruchi
School of Computer Science and Engineering, Lovely Professional University, Phagwara, Punjab, India

D. Singh (✉)
School of Computer Science and Engineering, Lovely Professional University, Phagwara, Punjab, India
e-mail: dalwinder.singh@lpu.co.in

N. Sharma et al. (eds.), *Big Data Analytics in Intelligent IoT and Cyber-Physical Systems*, Transactions on Computer Systems and Networks,
https://doi.org/10.1007/978-981-99-4518-4_11

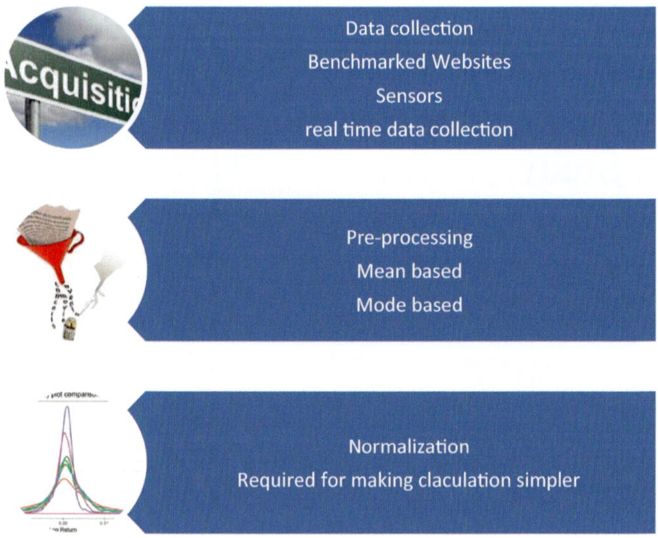

Fig. 11.1 Data collection and pre-processing-based methodology

is classification that predicts the disease. The classification process includes the neural network-based approach, decision tree, KNN, random forest and many more as discussed in (Chaubey et al. 2020). The approaches followed and the corresponding results are discussed in Sect. 11.4. Machine learning based methodology used for the data collection of Lumbar Spine related disease is given in Fig. 11.1. This is a critical phase as rest of the phases depends upon this phase (Priyadharsini and Sasikala 2022). Normalization process is used to bring the result in certain range for improving the process of calculations.

Rest of the paper is organized as follows. Section 11.2 provides the in-depth study of approaches used for Lumbar Spine related diseases. Section 11.3 gives the problem formulation that ultimately result in better techniques for prediction of Lumbar Spine related diseases. Section 11.4 gives the empirical study of the existing approaches and the last section gives the conclusion.

11.2 Related Work

The work has been done towards detecting the disease caused by Lumbar Spine. This section is discussed in multiple parts. At the first place, the role of Lumbar Spine chemicals has been discussed and in the next section the data mining mechanism used for prediction are discussed and at the end, comparative analysis of the techniques have been presented.

A. Role of Lumbar Spine chemicals

It is widely believed that both Lumbar Spondylosis and Herniated Disk can cause changes in the way that lipoproteins are organized as discussed in (Fonseca 2013). Since a few key lipoprotein transport catalysts are controlled by Lumbar Spine hormones, it is not surprising that the Lumbar Spine disease frequently causes a worsening of lipoprotein transport (Gereben 2015). By suppressing the rate-limiting protein of cholesterol amalgamation (hydroxymethylglutaryl coenzyme A [HMG CoA] reductase), intracellular free cholesterol stifles endogenous cholesterol production, enabling the bone to govern its own cholesterol level through a nearby input control framework. By activating the enzyme HMG CoA reductase, which catalyses the conversion of HMG CoA to mevalonate, Lumbar Spine chemically animates the hepatic cholesterol union once again. This results in a better cholesterol union in the Herniated Disk and a decreased one in Lumbar Spondylosis (Chakraborty 2010).

Nevertheless, because Lumbar Spine chemicals may also regulate the union and oxidation of LDL cholesterol, where blood cholesterol levels are alternately increased in Lumbar Spondylosis and decreased in the Herniated Disk. Additionally, about 3% of thyroxine (T4) is linked to lipoproteins, mostly HDL (92%), and less so LDL (6.7%). (Hood 2000). The LDL receptor detects the T4-LDL complex, which functions as a credible tool for T4 entry into cells. Finally, a Lumbar Spine hormone activates the LDL receptor, increasing the fragmented catabolic speed of apoB without changing the rate at which it is produced. Plasma HDL concentrations have been described as normal or, conversely, decreased in the Herniated Disk and normal or even raised in severe Lumbar Spondylosis (Arriagada 2015).

B. Data mining techniques used for Lumbar Spine prediction

There are a number of machine learning mechanisms that are used for the prediction of Lumbar Spine. The methodology corresponding to Lumbar Spine related disease prediction with machine learning is presented in Fig. 11.2.

Feature extraction is critical in the prediction of Lumbar Spine related disease. Correlation based feature extraction is discussed in (Snekhalatha and Gomathy 2018a). Using the correlation-based mechanism, highest correlated features or attributes from the ultrasound images are extracted from the image. The correlation based mechanism is discussed in (Vanderpump 2011) to reduce the dimension

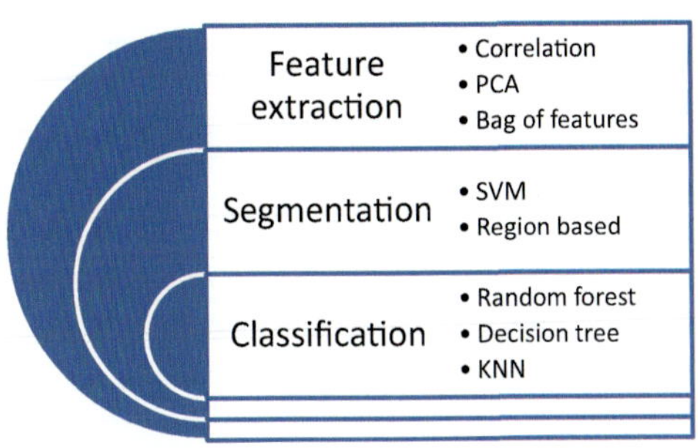

Fig. 11.2 Data mining phases for the Lumbar Spine prediction

to increase the execution speed of the overall operation. The performance gain of almost 5% was observed in terms of execution speed through the correlation-based approach (Yu et al. 2022). Furthermore, Principal component analysis reduces the dimensionality of the overall operation (Mugasa et al. 2020). PCA concentrates on extracting the features from the dataset that can contribute towards the actual output. Using PCA may reduce the overall classification accuracy.

Bag of features is another important feature extraction mechanism that is graph based (Han et al. 2022). This mechanism allows effective extraction of features from the image even if the feature is not that correlated with the result. Large and complex images may take large amount of time for feature extraction. To avoid the issue, back propagation mechanism with the feed forward approach can be used (Snekhalatha and Gomathy 2018b).

Segmentation is the next phase in the classification of the disease. The segmentation process can be achieved with the support vector machine. This mechanism is based upon the hyperplanes (Kumar 2020). The support vector machine divides the dataset into critical and non-critical parts. The critical parts are evaluated for having Lumbar Spine disease. Each hyperplane within SVM is attached with either class 0 or 1. 0 which means the disease is absent and 1 indicates presence of the disease. SVM is commonly collaborated with the convolution neural network for the Lumbar Spine image segmentation as done in (Ma et al. 2017). The issue with this approach is that limited classes are only available for prediction. The performance gain in terms of segmentation accuracy is achieved with limited classes using SVM. Another important segmentation mechanism is region based (Poudel et al. 2018). The region based approach identifies the critical regions to be retained that may be adversely affected with the Lumbar Spine disease. The issue with this approach is the high degree of misclassification.

The last phase is the classification process. Classification can be accomplished with the help of KNN, Random Forest, and decision tree. KNN is the K nearest approach that has static K values (Bai et al. 2020). KNN approach is based upon Euclidean distance calculated to determine the nearest class for prediction. The issue is of static K values that must be dynamic for better prediction accuracy. Random forest-based approach for classification is another approach for Lumbar Spine related disease (Prochazka et al. 2019). This approach is much faster as compared to the KNN approach for classification with the difference that classification depends upon the hot and trial method. The classification accuracy fluctuates in this case and hence may not be suitable for complex and large datasets. The decision tree-based classification model is the most common and simple in the prediction of Lumbar Spine related diseases (Yadav and Pal 2020a). Decision tree forms the branches (parts of disease) corresponding to the main tree (disease) and performs the classification by skipping the branches that do not correlate. The classification accuracy of the decision tree corresponding to Lumbar Spine related diseases is better as compared to the KNN and random forest-based approach.

C. Comparative analysis

This section presents the comparative analysis of different techniques used for Lumbar Spine related diseases. The comparative analysis is presented in Table 11.1.

Table 11.1 Comparative analysis of techniques used for lumbar spine disease prediction

Reference	Technique	Parameters	Description
Akbaş et al. (2013)	Multiple approaches including random forest, SVM and KNN	Classification accuracy	Ensemble based approach produced better classification accuracy in the prediction of Lumbar Spine
Ioniţă (2016)	Lumbar spine related disease prediction using machine learning	Classification accuracy	High classification accuracy in the range of 90% was achieved
Yadav and Pal (2019)	Ensemble based approach in the detection of Lumbar Spine within women	Prediction accuracy	High classification accuracy along with F-score was achieved
Liu et al. (2019)	CNN based mechanism for lumbar spine related disease	Multiple classes of prediction	Multiple classes or diseases are predicted using CNN
Yadav and Pal (2020b)	Decision tree based lumbar spine prediction	Classification accuracy	High classification accuracy with limited class of prediction
Jia et al. (2020)	Correlation based lumbar spine disease prediction	Correlation analysis, classification accuracy	High classification accuracy is achieved
Zhu et al. (2021)	Generic deep learning based mechanism in the prediction of lumbar spine related disease	Classification accuracy	Accuracy is high but execution speed is high
Agilandeeswari et al. (2022)	Lumbar spine prediction system using expert system	Classification accuracy	Classification accuracy is substantial but it is not suitable for multiple diseases

11.3 Problem Formulation

The mechanisms including the support vector machine, KNN and random forest predict the Lumbar Spine however it is not possible to predict the after effect of the Lumbar Spine. This means it is not possible to predict the future impact associated with the Lumbar Spine related diseases. Furthermore, correlation analysis can be introduced within the convolution neural network to improve the process of the prediction. The pre-processing process also has the issue as the noise handling mechanism must accommodate background subtraction and contrast enhancement that is not considered in earlier work (Prasad et al. 2016; Azar et al. 2013). The issue can be resolved using the mode-based mechanism within the existing median filtering for handling salt and pepper noise along with contrast enhancement for improving the overall classification process. Furthermore, the classification phase can be modified using the ensemble-based approach for achieving performance gain.

11.4 Empirical Study

The empirical study is based upon the result obtained corresponding to Lumbar Spine disease prediction using random forest, KNN and decision tree-based approaches. The result corresponding to these approaches is given in Table 11.2.

The result corresponding to the decision tree is better as indicated in Table 11.2. The classification accuracy however depends greatly upon the feature selection which is random within the random forest approach. This means that the random forest based approach can serve better in a certain situation due to the fluctuating nature of this approach. The plot corresponding to Table 11.2 is given in Fig. 11.3

Table 11.2 Results corresponding to KNN, Decision tree and random forest

References	Approach	Dataset	Classification accuracy (%)
Huang et al. (2020), Yadav and Pal (2020b)	Diagnostic algorithm with decision tree	Lumbar spine dataset with csv format	95
Li et al. (2020)	Deep learning with random forest	PLCO	94
Dov et al. (2021)	Supervised learning based prediction model using KNN	ODDS	91

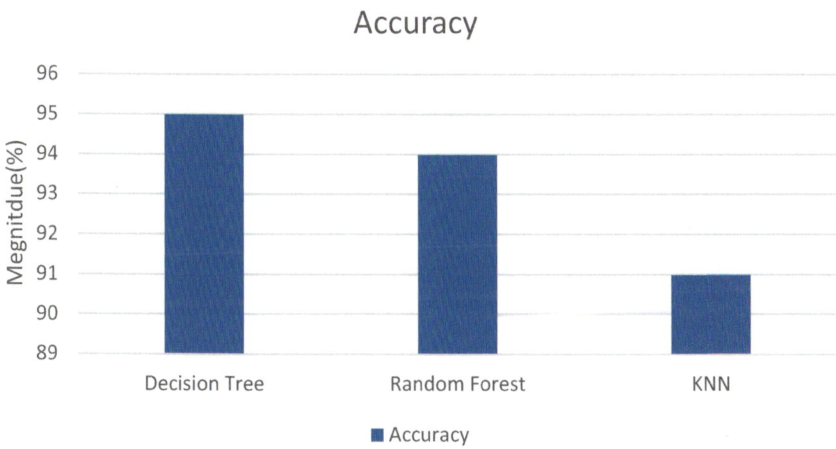

Fig. 11.3 Plots of classification accuracy

11.5 Conclusion

Lumbar Spine can impact the immune system of the body adversely. Early detection and treatment is critical in this regard. Modern methodologies corresponding to the Lumbar Spine detection are effective but further improvement is desired. The proposed work highlights the certain aspects including pre-processing and classification that can be modified to achieve the highest possible classification accuracy. The analysed result corresponding to the decision tree produced better classification accuracy in the range of 95% but with the complex dataset this might reduce. To increase the stability, the convolution neural network-based mechanism having the mode based pre-processing mechanism can reduce the degree of misclassification further. Furthermore, the process of normalization can be introduced within the CNN to reduce complexity and improve the convergence process.

References

Agilandeeswari L, Muralibabu K, Khatri I, Advani J, Nihal SM (2022a) An efficient lumbar spine disease prediction system—a study. Lecture notes in networks and systems, 417 LNNS, 544–552. https://doi.org/10.1007/978-3-030-96302-6_51/COVER

Akbaş A, Turhal U, Babur S, Avci C (2013) Performance improvement with combining multiple approaches to diagnosis of Lumbar Spine cancer. Engineering 5(10):264–267. https://doi.org/10.4236/eng.2013.510b055

Arriagada AA, Albornoz E, Opazo M (2015) Excess iodide induces an acute inhibition of the sodium/iodide symporter in lumbar spine male rat cells by increasing reactive oxygen species. Endocrinology 156: 1540–1551

Azar AT, El-Said SA, Hassanien AE (2013) Fuzzy and hard clustering analysis for lumbar spine disease. Comput Methods Prog Biomed 111(1):1–6. https://doi.org/10.1016/j.cmpb.2013.01.002

Bai Y, Kakudo K, Jung CK (2020) Updates in the pathologic classification of lumbar spine neoplasms: a review of the world health organization classification. Endocrinol Metab 35(4):696. https://doi.org/10.3803/ENM.2020.807

Chaubey G, Bisen D, Arjaria S, Yadav V (2020a) Lumbar spine disease prediction using machine learning approaches. Natl Acad Sci Lett 44(3): 233–238. https://doi.org/10.1007/S40009-020-00979-Z

Dov D, Kovalsky SZ, Assaad S, Cohen J, Range DE, Pendse AA, Henao R, Carin L (2021) Weakly supervised instance learning for lumbar spine malignancy prediction from whole slide cytopathology images. Med Image Anal 67. https://doi.org/10.1016/J.MEDIA.2020.101814

Fonseca TL, M. C.-M. M. C. (2013) Coordination of hypothalamic and pituitary T3 production regulates TSH expression. J Clin Investig 123:1492–1500

Gereben B, E. M. M. R. (2015) Scope and limitations of iodothyronine deiodinases in lumbar spondylosis. Nat Rev Endocrinol 11:642–652

Gorade SM, A. D. P. P. (2017) A study of some data mining classification technique. Int Res J Eng Technol 4(4):3112–3115

Han P, Guo J, Lai H, Song Q (2022) Construction method of knowledge graph under machine learning. Int J Grid Util Comput 13(1):11–20. https://doi.org/10.1504/IJGUC.2022.121423

Hood A, C. K. (2000) Differential effects of microsomal enzyme inducers on in vitro thyroxine (T(4)) and triiodothyronine (T(3)) glucuronidation. Toxicol Sci 55:78–84

Huang BL, Chabot JA, Lee JA, Kuo JH (2020) A stepwise analysis of the diagnostic algorithm for the prediction of malignancy in lumbar spine nodules. Surgery (United States) 167(1):28–33. https://doi.org/10.1016/J.SURG.2019.05.079

Ionita I, L. I. (2016) Prediction of lumbar spine disease using data mining techniques. Broad Res Artif Intell Neurosci 7(3):115–124

Jha R, Bhattacharjee V, Mustafi A (2022) Increasing the prediction accuracy for lumbar spine disease: a step towards better health for society. Wirel Pers Commun 122(2):1921–1938. https://doi.org/10.1007/S11277-021-08974-3/FIGURES/7

Jia M, Li Z, Pan M, Tao M, Lu X, Liu Y (2020) Evaluation of immune infiltrating of lumbar spine cancer based on the intrinsic correlation between pair-wise immune genes. Life Sci 259. https://doi.org/10.1016/J.LFS.2020.118248

Kumar HHS (2020) A novel approach of SVM based classification on lumbar spine disease stage detection. In: Proceedings of the 3rd international conference on smart systems and inventive technology, ICSSIT 2020, pp 836–841. https://doi.org/10.1109/ICSSIT48917.2020.9214180

Liu T, Guo Q, Lian C, Ren X, Liang S, Yu J, Niu L, Sun W, Shen D (2019) Automated detection and classification of lumbar spine nodules in ultrasound images using clinical-knowledge-guided convolutional neural networks. Med Image Anal 58. https://doi.org/10.1016/J.MEDIA.2019.101555

Ma J, Wu F, Jiang T, Zhao Q, Kong D (2017) Ultrasound image-based Lumbar Spine nodule automatic segmentation using convolutional neural networks. Int J Comput Assisted Radiol Surg 12(11): 1895–1910. https://doi.org/10.1007/S11548-017-1649-7

Makhdoomi SM, Rakhra M, Singh D, Singh A (2022) Artificial-intelligence based prediction of post-traumatic stress disorder (PTSD) using EEG reports. In: 2022 5th International Conference on Contemporary Computing and Informatics (IC3I), Uttar Pradesh, India, pp 1073–1077. https://doi.org/10.1109/IC3I56241.2022.10072671

Prasad V, Rao TS, Babu MSP (2016) Lumbar spine disease diagnosis via hybrid architecture composing rough data sets theory and machine learning algorithms. Soft Comput 20(3):1179–1189. https://doi.org/10.1007/s00500-014-1581-5

Priyadharsini D, Sasikala S (2022) Efficient lumbar spine disease prediction using features selection and meta-classifiers. In: Proceedings—6th international conference on computing methodologies and communication, ICCMC 2022, pp 1236–1243. https://doi.org/10.1109/ICCMC53470.2022.9753986

Singh D, Rakhra M, Aledaily AN, Kariri E, Viriyasitavat W, Yadav K, Kaur A et al (2023) Fuzzy logic based medical diagnostic system for hepatitis B using machine learning. Soft Comput 1–17

Singh D, Verma S, Singla J (2021) A neuro-fuzzy based medical intelligent system for the diagnosis of Hepatitis B. In: 2021 2nd international conference on computation, automation and knowledge management (ICCAKM), pp 107–111. https://doi.org/10.1109/ICCAKM50778.2021.9357765

Snekhalatha U, Gomathy V (2018a) Ultrasound lumbar spine image segmentation, feature extraction, and classification of disease using feed forward back propagation network. Adv Intell Syst Comput 563:89–98. https://doi.org/10.1007/978-981-10-6872-0_9/COVER

Soewu T, Singh D, Rakhra M, Chakraborty GS, Singh A (2022) Convolutional neural networks for MRI-based brain tumor classification. In: 2022 3rd International Conference on Computation, Automation and Knowledge Management (ICCAKM), Dubai, United Arab Emirates, pp 1–7. https://doi.org/10.1109/ICCAKM54721.2022.9990173

Umar Sidiq D, Aaqib SM, Khan RA (2019) Diagnosis of various lumbar spine ailments using data mining classification techniques. Int J Sci Res Comput Sci Eng Inf Technol 5(1): 2456–3307

Vanderpump MPJ (2011) The epidemiology of lumbar spine disease. Br Med Bull 99(1):39–51. https://doi.org/10.1093/BMB/LDR030

Yadav DC, Pal S (2019) To generate an ensemble model for women lumbar spine prediction using data mining techniques. Asian Pac J Cancer Prev 20(4): 1275–1281. https://doi.org/10.31557/apjcp.2019.20.4.1275

Yadav DC, Pal S (2020a) Prediction of lumbar spine disease using decision tree ensemble method. Human-Intell Syst Integr 2(1): 89–95. https://doi.org/10.1007/S42454-020-00006-Y

Yasir HS (2020) Lumbar spine disease data set|Kaggle. Kaggle. https://www.kaggle.com/datasets/yasserhessein/LumbarSpine-disease-data-set

Yu R, Tian Y, Gao J, Liu Z, Wei X, Jiang H, Huang Y, Li X (2022) Feature discretization-based deep clustering for lumbar spine ultrasound image feature extraction. Comput Biol Med 146.https://doi.org/10.1016/J.COMPBIOMED.2022.105600

Zhu YC, AlZoubi A, Jassim S, Jiang Q, Zhang Y, Wang YB, de Ye X, Du H (2021) A generic deep learning framework to classify lumbar spine and breast lesions in ultrasound images. Ultrasonics 110. https://doi.org/10.1016/J.ULTRAS.2020.106300

Chapter 12
Classification of Skin Cancer Using Dermoscopy Datasets by an Automated Machine Learning System

Puneet Thapar and Manik Rakhra

12.1 Introduction

In accordance with the WHO, among the most common causes of death globally is malignant tumor. Research and treatment are being done for several types of cancer. The problem is that SC is one of the modern world's most rapidly expanding cancers. Current statistics show that SC cases are rising more quickly than any other types of cancer. The study's objective is to create a reliable technique that would be applied in clinical practise to differentiate between benign and malignant lesions in skin lesions using dermoscopy pictures. More than 1,890 people in Australia had died from skin cancer per annum (Australia 2010). The growth rate of skin cancer published by World of Cancer is shown in Fig. 12.1 (World of Cancer n.d.).

According to Fig. 12.1, skin cancer is the third most common malignancy in both males and females, with the highest incidence occurring in Australia and New Zealand. Furthermore, during the holidays, the European states are exposed to sunlight in low-latitude areas, and the percentage of skin melanoma has increased. Lesion has the potential to be extremely invasive. The most dangerous aspect of melanoma is how it spreads throughout the body via the lymphatic and cardiovascular systems. Consequently, the key to melanoma diagnosis is early detection of the disease (Masood and Adel 2013).

Melanoma cancer develops as a result of unmanaged melanocyte reproduction. It attacks adjacent tissues as well as other body organs, putting life in danger (Pennisi et al. 2016). Melanoma initially resembles a benign mole, but unlike it, it quickly

P. Thapar · M. Rakhra (✉)
Department of Computer Science and Engineering, Lovely Professional University, Phagwara 144411, Punjab, India
e-mail: rakhramanik786@gmail.com

P. Thapar
e-mail: puneet.thapar90@gmail.com

N. Sharma et al. (eds.), *Big Data Analytics in Intelligent IoT and Cyber-Physical Systems*, Transactions on Computer Systems and Networks,
https://doi.org/10.1007/978-981-99-4518-4_12

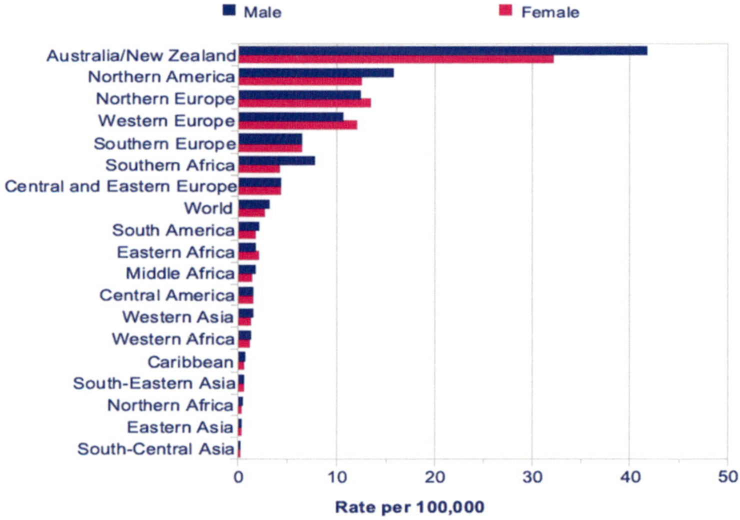

Fig. 12.1 The global skin cancer rate

becomes uneven in colour, unevenly spaced in shape, and greatly expanded in diameter. Dermatologists have traditionally performed skin cancer detection in clinics. Although dermoscopy, a non-invasive observation for inspecting lesions at higher resolution during the detection process, is used, the end result is always achieved through histological examination involving invasive intervention (Bakheet 2017).

Image processing technology uses a series of steps to diagnose skin cancer. Preprocessing, segmentation techniques, extraction of features, and diagnostic classifiers are some of these phases (Sadeghi et al. 2013).

"Biopsy" is the first step that doctors take to detect cancer. By this process, a small portion of the cell is taken from the patient's body and tested in the laboratory (Jaleel et al. 2013). This procedure takes time because the sample is sent to several laboratories. Throughout this process, the risk of cancer spreading to other parts of the body increases, making it more dangerous. To address this issue, the digital image processing techniques were developed. This technology promotes the early diagnosis of skin cancer, and clear mole images can be acquired without the application of any oil to the skin, making it the fastest and most accurate method. However, most significantly, due to its higher intensification, the use of different segmentation techniques have been applied by researchers for skin cancer detection that can avoid surplus removal of completely undamaging of skin lesions (Achakanalli and Sadashivappa 2014).

12.2 Techniques to Optimized Skin Cancer

12.2.1 Image Classification

Image classification refers to the method of categorising an image based on pixel identifiers or matrices constructed between those pixels using predefined rules. Image classification is the primary application of Deep Neural Networks (DNNs) in the analysis of medical image processing. A particular input image is accepted by the system using an image classification model, which processes it, and then produces a judgement to determine whether a disease is present (Balaji and Lavanya 2019). Table 12.1 lists various image classification models used in image processing.

12.2.2 Principal Component Analysis (PCA)

PCA can act as a pure square sum and cross-product matrix (SSCP), a covariance matrix that scaled square amounts and a cross-product matrix. The matrix for correlation is square sums and cross-products from standardized information. The analysed result for SSCP type object and Covariance is different because these objects differ only due to a global scaling factor. A correlation matrix is useful if the contrast between the individual is more, or there is a difference between units of measurement of the individual (Singh 2021).

Principal Component Analysis (PCA) reflects both common and unique variance of the variables, and may it look like a variance-focused method that aims to regenerate the correlations. It is used much more than the main factor analysis (PFA), and component location variables are frequently used. The primary components in one particular orthogonal dimension are explained as a linear combined by the originate terms.

12.2.3 Speed-Up Robust Feature (SURF)

Speed-Up Robust feature is a newly-developed methodology, mainly used for feature detection purpose of an image. This is vbecause this number of features is already stored in the score of recognition. The SURF technique is used to create a set of pairs between the entire image and each specific database object. Because of the powerful attribute of SURF algorithm, this is useful for the detection of the object in the image, including invariance of scale, invariance of translation, invariance of lighting, invariance of comparison and invariance of rotation. This algorithm composes of mainly the below four kinds:

Table 12.1 Techniques for image classification

Techniques for classifying	Advantages	Limitations and/or assumptions
Neural Network (NN)	• It has the potential to be used for classification or regression • Be able to represent Boolean functions • It is manageable for noisy input signals • Neural network instances can be classified using more than one output	• The algorithmic structure is difficult to grasp. • Having too many attributes leads to overfitting • Only through experimentation can the optimal network structure be computed
Support Vector Machine (SVM)	• It represents non-linear class boundaries • Typically, overfitting occurs • This algorithm's computational complexity results in quadratic optimization challenge • Effectively control the complexity of decision rules and the frequency of errors	• NN training is slower than Bayes and Decision Tree training • Determining the optimal parameter is difficult if the training data is not linearly separated • The algorithmic structure is difficult to grasp.
Fuzzy Logic (FL)	• A number of nonlinear relationships have been identified to describe the properties	• Previous experience is necessary for improved performance • If the decision's direction is unclear, precise solutions are not obtained
Genetic Algorithm (GA)	• This algorithm is used for both feature selection and classification problems • Primarily used in optimization • Produces mostly "good" results but not always "best" solutions • With this approach, large, complex, non-differentiable, and multimodal space can be easily managed • Searching approach that is efficient in a complex problem space • It removes irrelevant and noisy features from the features used for classification	• It is not possible to retrieve the computational or scoring function development • This method is inefficient for computing some optimal rather than global values • It is difficult to represent the training or output data

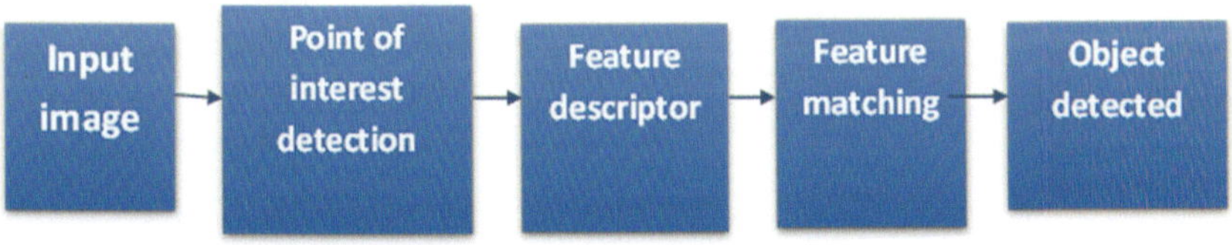

Fig. 12.2 Flow chart of SURF algorithm

- Generation of Integral Image
- Detection of Fast Hessian (i.e. detection of point of interest)
- Assignment of descriptor orientation
- Generation of descriptor.

Both subsequent aspects of the method utilized an integral image to properly change the speed. To use the integral image, it is compulsory to calculate the surface integral of the variable size from the original image each time to pick up just four values of the pixel (Fig. 12.2).

To locate the corresponding point of images, SURF employs the Hessian matrix factors that contribute. When using the SURF algorithm, all of them representative points can operate through similar weight, where it could be achieved by assigning representative points to random weights:

$$W_P = \frac{Number\,of\,detected\,images\,with\,respect\,to\,point\,P}{Number\,of\,training\,images\,in\,object} \tag{12.1}$$

12.2.4 Optimization Techniques

It is the process of either optimization or function reduction constraints. Algorithms for collaborating to evaluate enhanced clarification or uncontrolled optimum/continuous and minimal amount differentiable functions are used. Based on the presence of constrictions, the principle of the equation, the third acceptable amount of the decision variables, and the last number of optimum solution based on the three components (objective function, variable, and constraints), this could be classified as an optimization problem. Figure 12.3 depicts the various Optimization techniques.

12.3 Data Set Used

To simulate and evaluate the model, we will have to use the dermoscopic skin lesion dataset. Here are some existing datasets for detecting melanoma from skin images that we used in our research.

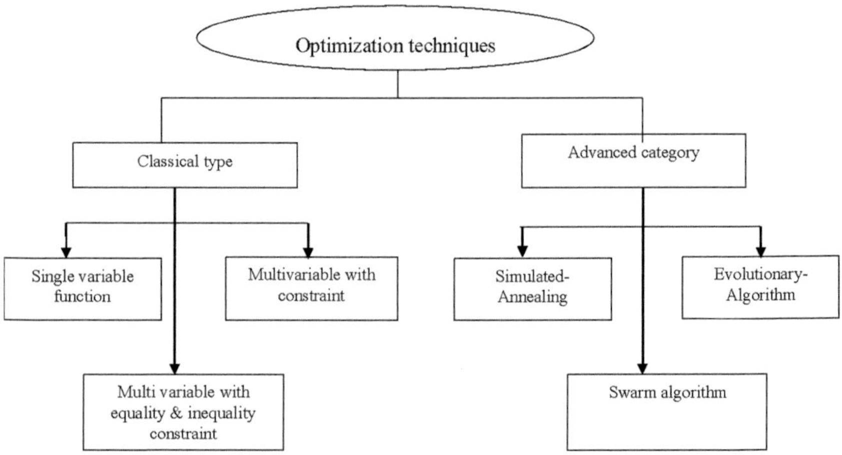

Fig. 12.3 Optimization methods

12.3.1 ISIC 2018

The collection is in the form of dermoscopic images, and it covers the examination of skin lesions for melanoma identification. Dermoscopy is an image processing method that removes the skin's surface reflection. It offers increased diagnostic precision.

12.3.2 PH-2 Dataset

The PH-2 dataset contains a large number of skin lesions that have been manually segmented for clinical diagnosis and research purposes. Dermatologists, who specialise in skin diseases, perform the identification of various skin lesion dermoscopic structures. Dermoscopic images from the PH-2 dataset will be made freely available for scientific research.

12.3.3 ISBI 2017

It is a collection of over 10,000 images of dermoscopic skin lesions used for research study and medical diagnostics. A sub-set of the skin lesion dermoscopic images has been annotated and marked up by skin cancer experts.

12.4 Literature Survey

See Table 12.2.

Table 12.2 Summary of the reviewed literature on malignant melanoma skin cancer

Author	Techniques	Results	Future scope
Rahman et al. (2022)	Anisotropic diffusion filtering and FBB	More effective than any prior ML algorithms in terms of AC (99.85%), SP (95.70%), and SE (91.55%)	An improved system for remote checkups from the comfort of the patient's home would be developed in the future, using a more promising ML algorithm
Kaur et al. (2022)	DCNN	The suggested DCNN classifier achieved 81.41%, 88.23%, and 90.42% accuracies on the ISIC 2016, 2017, and 2020 datasets, correspondingly, indicating higher implementation associated with the other state-of-the-art networks	Further multi-class categorization of the built DCNN model was possible to forecast various forms of skin malignancies
Aljohani and Turki (2022)	CNN	Google Net could get the best results on both testing and training sets (76.08% and 74.91%, respectively)	To boost prediction accuracy, it was advocating an ML/deep learning hybrid strategy
Thanh et al. (2020)	Automatic image processing	The suggested technique offers high accuracy and overall excellent presentation: 96.6%, 93.9%, and 86.7%, respectively, are the segmentation accuracy, Dice, and Jaccard scores	Using deep-learning models and the characteristics used here for melanoma skin cancer diagnosis would further enhance the accuracy of every job
Zghal and Derbel (2020)	Filtering and contrast-enhancing, ABCD	There is a 92% accuracy rate and an 87% accuracy rate for the suggested technique, which indicates its dependability	Eventually, the accuracy of the suggested system would be improved so it could be used on a different database as well
Abdulhamid et al. (2020)	Global optimization	The segmentation technique beats the current best practices but was on par with emerging methods based on deep neural networks	Segmenting melanoma in skin cancer diagnosis would be a good use for the suggested approach

(continued)

Table 12.2 (continued)

Author	Techniques	Results	Future scope
Thaajwer et al. (2020)	Image processing techniques and the SVM	Combining and applying the shape, color, and GLCM characteristics to the classifier results in high accuracy of 83%	Having dark skins was essential for further experimentation in the study
Khan et al. (2019)	SVM, KNN	In addition to the color features, the GLCM and LBP features would be used in combination to produce an impressive classification accuracy of 96%	DERMIS images should be used to test the planned method's efficacy and performance
Majumder and Ullah (2018)	ABCD	An overall 98% success rate was attained with the feature extraction block that included all parameters for 200 photos	Melanoma skin cancer could be diagnosed quickly and accurately with this new approach
Alquran et al. (2017)	PCA and SVM	A 92.1% accuracy rate was attained utilizing the full set of data (11 features) in the SVM classifier and the same AC is achieved using the PCA (five features)	The next work on the SC recognition system could be more precise and effective if the technology was deployed as a stand-alone mobile application and thus more dependable and practical
Jain and Pise (2015)	CAD	Patients and doctors alike could benefit from the suggested technology, which could help them detect skin cancer more quickly and precisely	Automatic diagnostics could be performed by using this device
Sheha et al. (2012)	MLP	According to the results, texture analysis is an effective tool for distinguishing melanocytic skin cancers from other premalignant lesions with a high degree of precision	To distinguish between Melanocytic Nevi and Malignant Melanoma, the features recovered are based on the Gray Level Co-occurrence Matrix (GLCM) and the Multilayer perceptron classifier (MLP)
Rajabi-Estarabadi et al. (2019)	OCT	The correlation between histology and HD-OCT and SV-OCT diagnoses of MIS or MM was documented time and time again in the literature	Usage of optical coherence tomography (OCT) for the diagnosis of skin cancers has to be increased and standardized

(continued)

Table 12.2 (continued)

Author	Techniques	Results	Future scope
Ciazynska et al. (2021)	BCC	It is necessary to assess the prevalence of primary BCCs and SCCs about the variables: age, gender, tumour location, and subtype of tumour	The need of enhancing NMSC registration regulations to enhance methods for the prevention and treatment of these tumors and the importance of doing
Collins et al. (2019)	SCC	In transplant patients, skin malignancies tend to be more invasive, which results in a greater morbidity and fatality rate	Strategies for the prevention and therapy of field leukoplakia as well as non-melanoma squamous cell carcinoma
Subtaweesin et al. (2018)	FTIR	Antimicrobial properties were shown using curcumin-loaded hydrogel films against Aspergillus. The films were made with curcumin solutions with values of 0.5 and 1.0 mg/mL	It seemed as if the mechanical capabilities of curcumin-loaded chitosan-coated films, especially their bending capabilities, were inferior compared to those of chitosan-coated film
Mane and Shinde (2018)	Biopsy	The results of the suggested system reveal that a supported vector machine with a linear kernel provides the highest level of accuracy	In the future to address the problems that have been outlined above, It is necessary to have computer-assisted diagnostics for skin cancer
Vijayalakshmi (2019)	SVM	The result is a wholly automatic method for detecting dermatological illness based on lesion photographs, rather than relying on medical people	The early detection of potentially lethal disorders such as melanoma is critical in establishing the likelihood of successfully undergoing treatment and emerging healthy

12.5 Proposed Methodology

The methodology is divided into two parts—training portion and classification process (Thapar 2022a) (Table 12.3).

The proposed work is segmented into three sections as follows.

The following flow diagram, depicted in Fig. 12.4, can be used to demonstrate the training portion of the work flow diagram.

The training process takes the raw data i.e. the images with the associated label as input. For each image in the list, the improved segmentation is applied. Once the segmentation process is done, the segmented part is processed for the feature extraction. A novel fitness function of the swarm intelligence behaviour is developed,

Table 12.3 Methodology section

Section	Purpose	Steps/algorithms
1	Pre-processing	Any classification mechanism's objective is to raise the true positive rate in order to improve the overall accuracy of the suggested system Modifications in this section are compulsory as the segmentation produces a significant role in the classification of the skin image
2	Feature extraction and optimization	This section aims to extract the feature of the segmented part using SURF algorithm. The extracted features are not always the best possible features which the training section will be looking for. Optimization will give a important part in the selection for appropriate features
3	Training and classification	The optimized features will be passed to the training section and the trained architecture will be used to classify the test data. Evaluation criteria like Precision, Recall, and F measure would be retrieved based on the classification of test data that performed the best

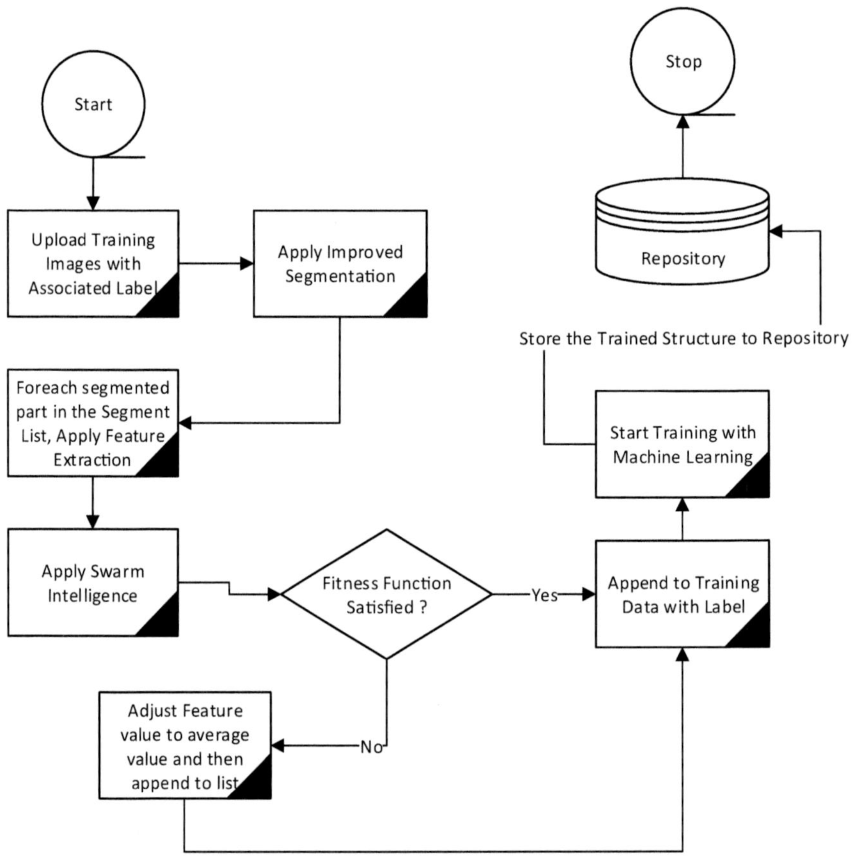

Fig. 12.4 Training process

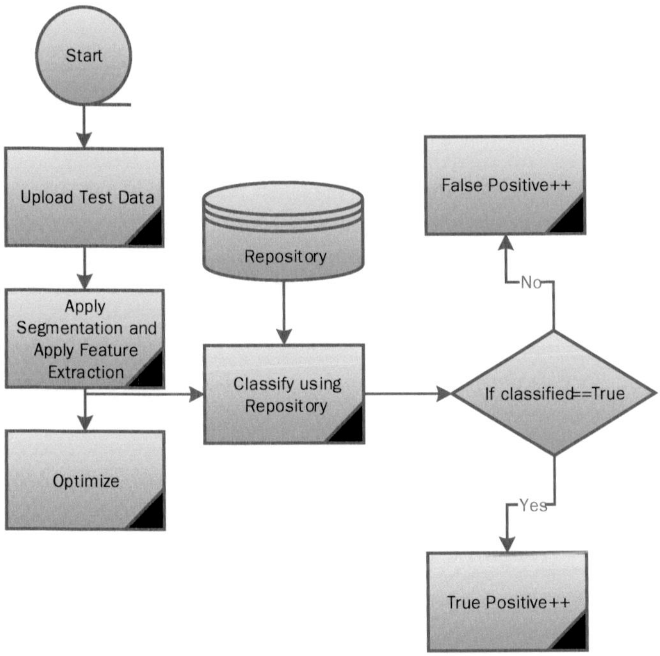

Fig. 12.5 Classification process

and optimization architecture is then performed over the retrieved characteristics (Thapar 2022b). If a feature fulfils the Swarm Intelligence's fitness function, it is given immediately to the training phase; otherwise, the average feature of the segmented region is used in its place. The trained architecture will be passed to the classification section (Fig. 12.5).

The classification process takes the trained structure as well as the test image as input and processes the same setup of steps to obtain the classified result. If the classified result matches with the target label then the true positive is incremented else false positive is incremented. Increase in false positive will decrease the overall accuracy of the proposed system. Therefore, the goal of any classification mechanism is to increase the true positive rate in order to increase the proposed system's overall accuracy (Thapar 2022c).

Expected Outcome:

(a) An optimized segmented region for better feature extraction
(b) An optimal set of features optimized by Swarm Intelligence
(c) A better classified lesion for high accuracy.

12.6 Results and Discussion

We use data from three different databases i.e. ISIC 2018, ISBI 2017, and PH-2 data sets to estimate our approach and the predicted best segmentation method. We attempt to test our system against each of these databases. All experiments are conducted on noise removal and code execution within the latest version of MATLAB. We provide the parameters Accuracy, Specificity, and Sensitivity to quantify the classification.

Table 12.4 shows our result using different paramets on these three databases.

As per the result from Table 12.4, it is clear that the accuracy percentage using PH-2 dataset is 98.7% that is higher than other two dat sets. İt is also clear from results that Specificity and Sensitivity percentage is also higher or better in case of PH-2 data set. So İt is clear from the study that PH-2 dataset is a better option to choose for designing an efficient Skin classifier system.

Figure 12.6 represents the graphical representation of different paramets on the basis of three different data sets and found that PH-2 data sets bars are high during every case or parameter evaluation.

Table 12.4 Parameters comparison using different data sets

Data set	Accuracy (% age)	Specificity (% age)	Sensitivity (% age)
ISIC 2017	97.8	96.7	95.4
PH-2	98.7	97.5	98.6
ISBI 2018	98	97.1	96.2

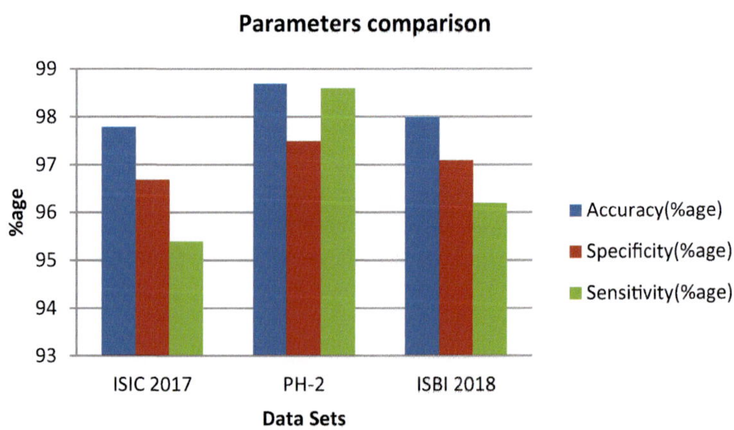

Fig. 12.6 Graphical representation of comparing different parameters using datasets

12.7 Conclusion

The general procedure for automatic skin cancer diagnosis has been covered in this study. This paper's major objective is to describe in depth the various methods for identifying Dermoscopy images. There is discussion of each technique and each of its sub techniques. Additionally, several research methodologies' work has been discussed. At the conclusion, a comparative research based on various parameters like accuracy, specificity and sesitivity using datasets, including PH-2, ISIC 2017, and ISBI 201, has been given. The PH-2 Data set with image scaling, such as Color image to Gray image, and thresholding algorithm performed much better with the greatest accuracy up to 98.7%, as shown above. Researchers who aim to develop a system for automatically detecting skin cancer will find this paper useful.

References

Abdulhamid M, Idris A, Şahiner A, Rahebi J (2020) New auxiliary function with properties in nonsmooth global optimization for melanoma skin cancer segmentation. BioMed Res Int

Achakanalli S, Sadashivappa G (2014) Statistical analysis of skin cancer image–a case study. Int J Electron Commun Eng (IJECE) 3(3)

Aljohani K, Turki T (2022) Automatic classification of melanoma skin cancer with deep convolutional neural networks. AI 3 (2):512–525

Alquran H, Qasmieh IA, Alqudah AM, Alhammouri S, Alawneh E, Abughazaleh A, Hasayen F (2017) The melanoma skin cancer detection and classification using support vector machine. In: 2017 IEEE Jordan conference on applied electrical engineering and computing technologies (AEECT), pp 1–5. IEEE

Bakheet S (2017) An SVM framework for malignant melanoma detection based on optimized HOG features. Computation 5(1):4

Balaji K, Lavanya K (2019) Medical image analysis with deep neural networks. In: Deep learning and parallel computing environment for bioengineering systems, pp 75–97. Academic Press

Ciążyńska M, Kamińska-Winciorek G, Lange D, Lewandowski B, Reich A, Sławińska M, Pabianek M et al. (2021) The incidence and clinical analysis of non-melanoma skin cancer. Scientific Reports 11(1):1–10

C. W. O. Australia, Ed., Causes of Death 2010, Australian Bureau of Statistics, Canberra, Australia

Collins L, Asfour L, Stephany M, Lear JT, Stasko T (2019) Management of non-melanoma skin cancer in transplant recipients. Clin Oncol 31(11):779–788

Jain S, Pise N (2015) Computer-aided melanoma skin cancer detection using image processing. Procedia Comput Sci 48:735–740

Jaleel, Abdul J, Salim S, Aswin RB (2013) Computer aided detection of skin cancer. In: 2013 international conference on circuits, power and computing technologies (ICCPCT), pp.1137–1142. IEEE

Kaur R, GholamHosseini H, Sinha R, Lindén M (2022) Melanoma classification using a novel deep convolutional neural network with dermoscopic images. Sensors 22(3):1134

Khan, Qasim M, Hussain A, Rehman SU, Khan U, Maqsood M, Mehmood K, Khan MA (2019) Classification of melanoma and nevus in digital images for diagnosis of skin cancer. IEEE Access 7:90132–90144

Majumder S, Ullah MA (2018) Feature extraction from dermoscopy images for an effective diagnosis of melanoma skin cancer. In: 2018 10th international conference on electrical and computer engineering (ICECE), pp 185–188. IEEE

Mane S, Shinde S (2018) A method for melanoma skin cancer detection using dermoscopy images. In: 2018 Fourth international conference on computing communication control and automation (ICCUBEA), pp 1–6. IEEE

Masood A, Adel Ali A-J (2013) Computer aided diagnostic support system for skin cancer: a review of techniques and algorithms. Int J Biomed Imaging

Nonita S, Xalikovich PA, Kumar CR, Rakhra M, Samori IA, Maquera YM, Gonzáles JLA (2022) Intelligent water drops algorithm-based aggregation in heterogeneous wireless sensor network", J Sensors vol. 2022, Article ID 6099330, 12 pages. https://doi.org/10.1155/2022/6099330

Pennisi A, Bloisi DD, Nardi D, Giampetruzzi AR, Mondino C, Facchiano A (2016) Skin lesion image segmentation using Delaunay Triangulation for melanoma detection. Comput Med Imaging Graph 52:89–103

Rahman, Mahbubur Md, Nasir MK, Nur A, Khan SI, Band S, Dehzangi I, Beheshti A, Rokny HA (2022) Hybrid feature fusion and machine learning approach for melanoma skin cancer detection

Rajabi-Estarabadi A, Bittar JM, Zheng C, Nascimento V, Camacho I, Feun LG, Nasiriavanaki M, Kunz M, Nouri K (2019) Optical coherence tomography imaging of melanoma skin cancer. Lasers Med Sci 34(2):411–420

Sadeghi M, Lee TK, McLean D, Lui H, Atkins MS (2013) Detection and analysis of irregular streaks in dermoscopic images of skin lesions. IEEE Trans Med Imaging 32(5):849–861

Singh D, Verma S, Singla J (2021) A neuro-fuzzy based medical intelligent system for the diagnosis of Hepatitis B. 2021 2nd international conference on computation, automation and knowledge management (ICCAKM), pp 107–111

Singh D et al. (2023) Fuzzy logic based medical diagnostic system for hepatitis B using machine learning. Soft Computing:1–17

Sheha MA, Mabrouk MS, Sharawy A (2012) Automatic detection of melanoma skin cancer using texture analysis. Int J Comput Appl 42(20):22–26

Subtaweesin C, Woraharn W, Taokaew S, Chiaoprakobkij N, Sereemaspun A, Phisalaphong M (2018) Characteristics of curcumin-loaded bacterial cellulose films and anticancer properties against malignant melanoma skin cancer cells. Appl Sci 8(7):1188

Thaajwer, Ahmed MA, Piumi Ishanka UA (2020) Melanoma skin cancer detection using image processing and machine learning techniques. In: 2020 2nd international conference on advancements in computing (ICAC), vol 1, pp 363–368. IEEE

Thanh, Dang NH, Prasath VB, Hieu LM, Hien NN (2020) Melanoma skin cancer detection method based on adaptive principal curvature, color normalization and feature extraction with the ABCD rule. J Digital Imaging 33(3):574–585

Thapar P, Rakhra M, Cazzato G, Shamim Hossain Md (2022a). A novel hybrid deep learning approach for skin lesion segmentation and classification. J Healthcare Eng

Thapar P, Rakhra M, Singh A (2022b) Comparing image feature extraction methods using dermoscopy noisy images. In: 2022b international mobile and embedded technology conference (MECON), pp 559–562. IEEE

Thapar P, Rakhra M, Singh A (2022c) The epidemiology of automatic skin cancer detection by comparative analysis of pre-processing and segmentation techniques. In: 2022c 3rd international conference on intelligent engineering and management (ICIEM), pp 894–899. IEEE

Zghal NS, Derbel N (2020) Melanoma skin cancer detection based on image processing. Current Med Imaging 16(1):50–58

Chapter 13
International Roughness Index Prediction Using Various Machine Learning Techniques on Flexible Pavements

Wasique Haleem Pandit, Krishna Pal Sharma, Nonita Sharma, Priyanka Tomar, and Shahnawaz Khan

13.1 Introduction

A pavement can be defined as the hard surface comprising of durable surface material that handles automobile and pedestrian activity carries. It also carries all the road traffic. It is defined as a construction done horizontally and supported by an onsite natural material. Existing documents must be analysed, as well as an investigation of the surface must be carried out before the construction of the pavement. The primary properties of the local stone and soil are considered, particularly in terms of toughness, rigidity, longevity, moisture susceptibility, shrinkage, and inflation capacity with time (Janani et al. 2019). To establish the required parameters, site testing, experimental calculations based on the type of soil, or laboratory-based studies and tests are conducted. The testing is carried out at its most vulnerable stage, which is generally when it has the maximum moisture content. Then it is determined what type of performance to anticipate when there is a high volume of traffic. The maximum value of embankment slopes and cuttings are calculated, as well as the density index that is to be achieved during construction. In a typical rural pavement, the wearing course is the upper most covering of the pavement. It is built

W. H. Pandit · K. P. Sharma (✉)
Dr. B. R. Ambedkar National Institute of Technology, Jalandhar, India
e-mail: sharmakp@nitj.ac.in

N. Sharma · P. Tomar
Indira Gandhi Delhi Technical University for Women, Delhi, India

S. Khan
Design and Information and Communications Technology, Bahrain Polytechnic, Isa Town, Bahrain

N. Sharma et al. (eds.), *Big Data Analytics in Intelligent IoT and Cyber-Physical Systems*, Transactions on Computer Systems and Networks,
https://doi.org/10.1007/978-981-99-4518-4_14

of compacted stone, asphalt, or concrete, and provides immediate vehicle support, a smooth and gripping surface, and water protection for the base course and natural formation. The base course adds essential strength, rigidity, and durability to the natural formation. It ranges in thickness from 4 to 40 inches depending on the type of traffic (moderate traffic or heavy traffic) (https://www.britannica.com/technology/road/The-modern-road#ref592118). The subbase acts as a layer of protection and a temporary working platform between the base course and the natural formation.

Depending on the flexural stiffness, there are two types of pavements: flexible and rigid. In flexible pavements, stone shards are crushed into place in the manner of the macadam or combined with bitumen to form asphalt. The diameter of the stones is between 1.5 and 1 inch to retain the workability. To get the bitumen fluid mixed with the stone, it must first be heated at temperatures between 300 and 400 °F. The hot mixture is applied to the road area in layers with the help of a paving machine. The thickness of the layers is twice that of the stone size. The layers are carefully rolled before the mixture cools and hardens. To save money on heating, bitumen emulsions or cutbacks are increasingly being used, in which either an emulsifier is used to treat the bitumen binder, or it is diluted with a lighter petroleum component that evaporates after rolling. Asphalts can now be mixed and laid at room temperature, thanks to these treatments.

Every nation necessitates to have well-defined measures to estimate the quality of roads (Janani et al. 2019). We must consider the lifelong expenses, riding quality, vehicle obstruction owing to repair costs, maintenance ease, and the influence of weather conditions while deciding whether to have flexible or rigid pavements. It is sometimes difficult to choose among them.

The qualities of the foundation material are determined in the lab, but site tests are also carried out to ensure that the building process follows the designer's goal. Designers usually take into account the likelihood of structural collapse due to a single overload as well as damage caused by the passage of multiple regular loads. Trucks are nearly exclusively responsible for both forms of failure.

13.1.1 Flexible Pavement

Inflexible pavements, small stone fragments and other materials are crushed into the pavements or combined with bitumen to create asphalt. The uppermost structural layer of a flexible pavement forms a shield that shields the supporting foundational course material from water and traffic while simultaneously generating enough resistance for tires, reducing the noise from vehicles in metropolitan areas, and reflecting enough light for nighttime driving. These types of surfaces are created by spray and chip seal which is a thin coating of the bituminous film along with stone or by a thin coating of asphalt (https://www.britannica.com/technology/road/The-modern-road#ref592118). For restoration of the existing asphalt surface or on the pavements carrying low to medium traffic loads the chip seal and spray are sprayed over the base course. It's inexpensive, efficient, and impenetrable, and it lasts around ten

Fig. 13.1 Shows an image of a flexible pavement (https://civiljungle.com/difference-between-flexible-pavement-rigid-pavement/)

years. Its biggest flaw is that it produces a lot of noise. For maintenance, additional spray coating with a bitumen surface treatment is usually needed. With heavier traffic loads or in urban regions, asphalt paving is used. Figure 13.1 is the image of flexible pavement.

13.1.2 Rigid Pavement

Rigid pavements are built from Portland cement concrete and have thickness of around 6–14 inches as shown in Fig. 13.2. This structure helps in seeping of water and other material into the surface through gaps which may weaken the structure (https://www.britannica.com/technology/road/The-modern-road#ref592118). When the local natural material is unsuitable for use as a foundation course, tiny amounts of lime, hydraulic cement, pozzuolana, or bitumen might be used to "stabilize" it which enhances the strength and stiffness of the pavement.

Over the period of time, traffic and environmental factors lead to degradation of the pavement. Over the course of their design life, performance is a broad word that describes how pavement conditions evolve. Road agencies have devised performance metrics to evaluate the quality of pavement during the past few decades (Bashar and Torres-Machi 2021; Abdelaziz et al. 2018; Chandra et al. 2013).

Authors in this work explore the possibility of using different models to evaluate structural conditions of the road. International Roughness Index (IRI) is predicted using different machine learning models using data sets collected from CRRII. Different models have been implemented to determine which model provides the best accuracy in terms of roughness index. Different models have been explored for the same kind of pavements with the same set of independent variables and dependent variables in accordance with the need of the site location and the prediction model to increase the accuracy. IRI using ANN (Artificial neural network) and XGB-Regressor

Fig. 13.2 Shows an image of a rigid pavement (https://civiljungle.com/difference-between-flexible-pavement-rigid-pavement/)

gave the lowest mean absolute percentage error in terms of data analysis as compared to the other models that were used.

How well a road structure is maintained and small repairs are done will impact how long it lasts. The kind of pavement, sides, drainage, traffic facilities, and privilege on the road are all factors in the process of maintaining the road. It involves the on-time repairing of cracks and potholes preventing water from seeping through the surface and the removal of rubbish thrown by the people on the road. It also involves the maintenance of road markings, indicators, and signs. In order to improve traction in tough winter climates, sand must be spread, salt must be distributed, and snow and ice must be cleared off the pavement.

Determination of the pavement condition and maintaining that condition of the pavement for a longer time depends on a number of variables and factors which need to be predicted with maximum accuracy to determine the optimum time at which intervention can be made to facilitate the serviceability of the pavement. The type and extent of current distresses, as well as the material qualities of the pavement structure, can all be used to assess the pavement's health. as well as estimating the quality of the construction Unfortunately, this is not the case. For both project and network level pavement study, a particular method of evaluation is neither practical nor cost-effective. Artificial neural networks (ANNs), a machine learning technique, is inspired by the human brain and hence replicates brain behaviour. ANNs generally have an input layer, an output layer, and multiple hidden layers. XGB-Regressor is also an accurate machine learning technique. XGBoost stands for "Extreme Gradient Boosting" and is a gradient boosting trees method implementation. The XGBoost is a well-known supervised machine learning model with features such as speed, parallelization, and performance. The goal of this research is determine the best machine learning model for IRI prediction.

13.1.3 Various Parameters Used in the Research

The following parameters that were taken into consideration while doing the research are briefly explained below.

13.1.3.1 Texture Depth

The total roughness of the surface is measured by texture depth, which facilitates vehicle braking and water dispersion from the contact areas between the road and the vehicle tires (www.nationaltesting.co.uk/road-surfaces/texture-depth-measurement). The macrotexture of a road surface is recognized to play a substantial influence in skid resistance at medium to high speeds. The estimation of texture depth is directly proportional to the mean profile depth (Fig. 13.3).

The test technique defined in BS EN13036-1:2010 is the simplest way to determine the depth of surface roughness (which replaced BS598-105:2000). This test pours a known quantity of glass spheres onto a surface allowing it to spread in a circular pattern. As the particles move across the surface, they fill in the low-spots. The diameter of the resulting surface is measured when the glass spheres cannot be spread any farther.

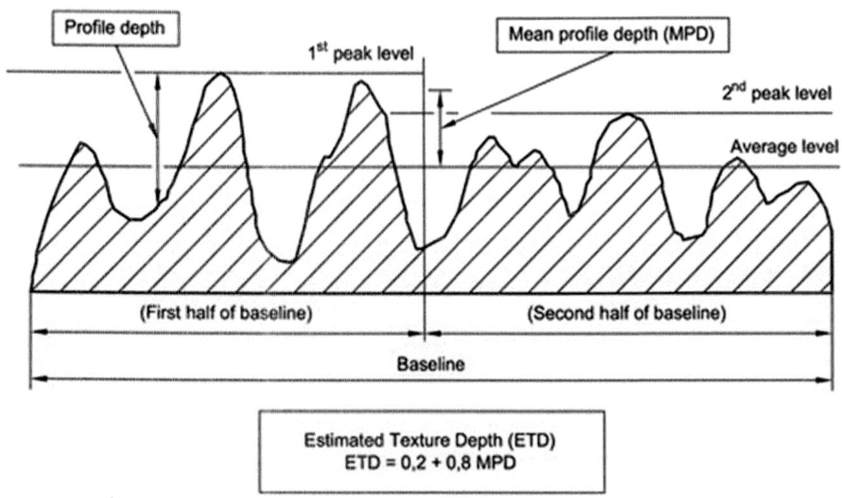

Fig. 13.3 Shows an image how texture depth is calculated (https://www.iso.org/obp/ui/#iso:std:iso:13473:-2:ed-1:v1:en)

13.1.3.2 Edge Break

A fractured or uneven edge of a road wearing surface is known as an Edge Break. Edge breaks commonly occur when the road shoulder is deteriorated, the pavement at the edge of the roadway is not strong enough, or water penetrates the pavement via the shoulders. Tyre damage can occur when there are significant edge breaks and edge drops. Figure 13.4 is the image of edge break on a flexible pavement.

13.1.3.3 Rutting

Rutting is the gradational emergence of conducted depressions in an asphalt face due to weight limits and poor base construction. Figure 13.5 is the image of rutting on flexible pavement. When large buses break the asphalt along a path over time, tyre routines occur. However, the base must be replaced or reconfigured to allow lesser vehicle business (Qi-sen et al. 2009), if an asphalt face develops rutting. A pattern is a long-term face depression formed by business moving over a flexible pavement's wheel pathways. routines develop in phases, with a slight, unrecoverable deformation being each time a heavy vehicle passes. As the lifetime of the pavement increases, face distortion may be accompanied by heave along each side of the pattern. The depth of rutting is measured as the perpendicular distance among top of the heave and the bottom of depression. There are three primary causes of rutting in flexible pavements

(A) asphalt subcaste problems,
(B) structural subcaste problems and
(C) weak sub-grade problems.

Fig. 13.4 Shows an image of an edge break (http://lgam.wikidot.com/edge-break)

Fig. 13.5 Shows an image of rutting (https://www.researchgate.net/figure/Rutting-distress-3-The-use-of-reinforcement-in-flexible-pavement-In-order-to-optimize_fig1_321983200)

13.1.3.4 Loss of Surface Material

When an asphalt surface loses its original shape, it creates material tension leading to difficulties (Singh et al. 2016). Figure 13.6 is the image of it. Pavement collapse requires adequate maintenance like crack sealing to prevent cracks from widening.

13.1.3.5 Pothole

Potholes form when water seeps over a lengthy period of time via being facefissures. Figure 13.7 is the image of a pothole on a flexible pavement. This will affect in large holes in the asphalt, which will spread and damage vehicles (Singh et al. 2016). When water seeps into the ground beneath the pavement, the same thing happens. It'll take up more space under the pavement if it freezes, and the pavement will expand, bend, and fracture, weakening the material pavement. The pavement also compresses as the

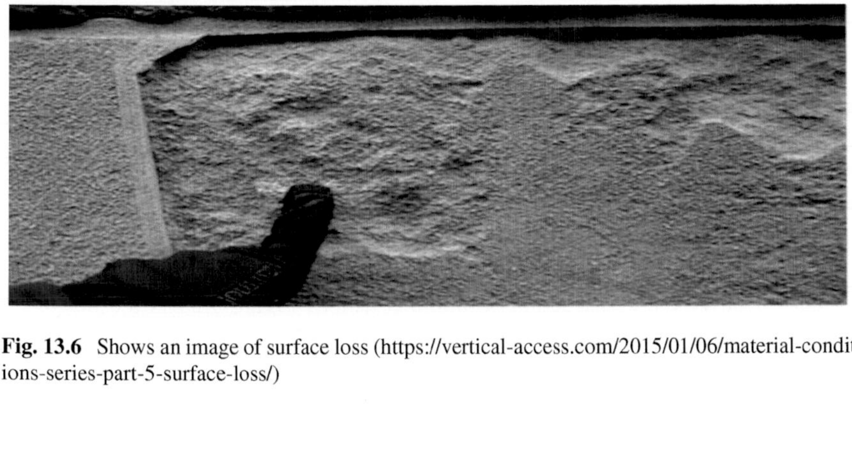

Fig. 13.6 Shows an image of surface loss (https://vertical-access.com/2015/01/06/material-conditions-series-part-5-surface-loss/)

Fig. 13.7 Shows an image of pothole (https://www.cityworks.com/blog/10-fascinating-facts-about-potholes/)

ice melts, leaving holes or spaces in the face beneath the pavement where water can get in and comestuck. However, the pavement will deteriorate and crack, If the water freezes and thaws constantly. Pieces of the thruway material weaken as motorcars and exchanges drive over the weak position in the road, causing the material to be displaced or broken down by the weight, performing in the pothole.

13.1.3.6 Ravelling

Water entrance and breakdown of the asphalt top coat lead to ravelling. As water and heat may harm the asphalt surface, link among asphalt bitumen and aggregate rock gets disintegrated. When an asphalt surface begins to ravel, it loses its impermeable characteristics and allows water to penetrate (Singh et al. 2016). When water seeps into an asphalt surface, it causes more cracks and eventually pavement breakdown. The deterioration of an asphalt road surface is known as ravelling. Figure 13.8 shows the image of ravelling. Raveling can occur due to age, traffic, dust, moisture, poor compaction and poor aggregate mix.

13.1.3.7 Cracking

The bulk of problems with pavement failure is caused by asphalt. Although soil settlement and exposure to the environment cause asphalt surface to naturally fracture over time, some less frequent cracking problems might be bought on by poor mix design for pavement construction (Colombier 2004). Figure 13.9 is the image of cracking on flexible pavement. There are several types of cracking:

Fig. 13.8 Shows an image of raveling (https://copavementsolutions.com/asphalt-raveling/)

Fig. 13.9 Shows an image of cracking (https://sableasphalt.com/bad-pavement-cracks-when-crack-sealing-just-wont-cut-it/)

- Alligator Cracking: Alligator cracking develops when the asphalt foundation and subgrade start to compress as a result of heavy automobiles. Alligator cracking frequently occurs at junctions when cars are stuck for a long time. The first cracks will start to show up and spread as a result of water penetration and increasing asphalt foundation tension.
- Edge Cracking: Substandard foundational material, Water, and heavy traffic are the prime causes of edge cracking along road side.
- Block Cracking: Block cracking results from the asphalt's seasonal expansion and contraction. If the mix design of an asphalt surface is too rigid, it will not allow for seasonal density variations, resulting in block cracking.
- Joint Cracking: Joint cracking occurs when an asphalt overlay procedure covers over a flexible concrete foundation.

- Transverse Cracking: Transverse fractures in asphalt roads are caused by settling or moving foundation material, inadequate paver performance, and high temperatures.
- Linear Cracking: The most common cause of linear cracking, which runs parallel to the road, is pavement fatigue.

13.1.3.8 Chainage

In surveying, a distance recorded in meters along an illustrative line, such as the middle line of a road or railroad, is referred to as "chainage". The phrase initially appeared in 1620, when 66-foot chains (also known as Gunter's chains), were employed to measure linear structures like roads and railroads. The chain usually had a hundred links, and one acre was equal to ten square chains. A mile is equal to 80 chains, and cricket fields are typically 1 chain length (www.Designingbuildings.co.uk). The 100-foot-long Ramsden chains were created in the eighteenth century.

Even though this method is no longer employed, the name has persisted, particularly in the context of railroads, where it can be used to specify the position of bridges and stations. Cumulative longitudinal lengths will be measured using a tool like an odometer and reported along the length of the railway from a datum that will be set as 0 at a certain location along the railroad. In most cases, this is enough to specifically identify elements like stations and bridges.

Authors in this work aim to apply the machine learning models to predict IRI. The performance of various models is compared so as to determine the most appropriate model. During the comparative evaluation, it is determined that XG-Boost yields optimal performance (Shwartz-Ziv and Armon 2022).

13.2 Literature Review

A literature review was conducted to have a thorough understanding of pavement performance and modeling approaches. A large number of research have been carried out all over the world to build pavement performance models.

Authors in Janani et al. (2019) suggested a new approach for prioritizing pavement maintenance sections based solely on the functional qualities of the pavement using ANNs. This can be used to avoid the number of expensive, time-consuming, and traffic-disrupting tests in order to assess the structural features of pavements. Furhter, authors in Nader Abdelaziz et al. (2018) analyzed the various independent variables in CRRII using Pearson correlation coefficient to construct an appropriate IRI prediction model. Authors in Abdelaziz et al. (2018) proposed a new method based on the precise integration method (PIM). The efficiency of the proposed model over traditional methods is also established in Abdelaziz et al. (2018).

Along the time, researchers employed machine learning models in this domain. For instance, authors in Panos et al. (Georgiou et al. 2018) used ANNs and support vector machines (SVMs) for predicitng IRI (Hamdi et al. 2017). The experiment was carried out to check the effectivness of these models and it was observed that ANN outperforms SVM. Similar work was also carried out by Mehran et al. (Mazari and Rodriguez 2016) to properly strategize maintenance, repair, and reconstruction of roads through IRI. Authors also pointed out that current correlation methods are not practical as it depends solely on IRI.

Similarly, Minu et al. (Li et al. 2018) stated that state highway agencies spend a huge amount and time every year to evaluage the structural and functional condition of existing, in-service pavements. Further, Muralikrishna and Veeraragavan in (Lucey et al. 2019) discussed pavement deterioration models that are built to anticipate performance and make appropriate maintenance decisions at the proper time. Statistical Packages for Social Sciences (SPSS) was used to create roughness and deflection progression equations. The ideal timing and option of maintenance procedures were calculated using road user cost models and life cycle cost analyses. Table 13.1 shown below, gives a brief summary of the studies that were conducted for our meta-analysis.

Now, it is evident from the table that ANN has proven its efficacy and hence authors in this chapter employ ANN in addition to other machine learning models.

Table 13.1 Comparison of various studies with machine learning models

Journal/year	Method	Model/s	Data source	Accuracy (%)
Journal of the Transportation Research Board 2021 (Bashar and Torres-Machi 2021)	Used a machine learning model and predicted the IRI of pavements	Random forest (RF), ANN, and SVM	LTPP	–
International Journal of Pavement Engineering 2020 (Abdelaziz et al. 2018)	A model that predicted road index for flexible pavements of international roads	ANN, MLR	LTPP	75
International Journal of Pavement Engineering 2019 (Janani et al. 2019)	Deployed a model focusing on the functional aspects of pavements. Pavement maintenance was mapped	ANN	Surveyed on roads of 14 villages of Tamil Nadu	65
Hindawi Publications 2018 (Georgiou et al. 2018)	A comparison based study that predicted pavement roughness using soft computing	ANN, SVM	Data collected annually from a motor way for seven years	88

13.3 Methodology

During literature review, it is evident that IRI prediction depends on various factors (Lucey et al. 2019). For the purpose of creating prediction models for structural and functional pavement issues, several researchers used the artificial neural network (ANN) approach (Plati et al. 2015). Additionally, it has been shown that an ANN model is accurate and reliable. This is relevant to the assessment of the functional state of pavements. The support vector machine (SVM) approach, a potential soft computing technology developed from statistical learning theory but rarely applied to pavement engineering issues, also emerged as a promising method for the prediction of roughness. The authors studied the working of various machine learning algorithms including ANN and SVM. The methodology employed is illustrated in Fig. 13.10. Along with ANN and SVM, the few other models that are going to be implemented on our dataset are XG-Boost Regressor, Gradient Boosting Regressor, Stochastic Gradient Descent (SGD) Regressor, Elastic net, Lasso-Lars, and Linear regression.

13.3.1 Input

CRRII dataset has 11 attributes and 1092 rows as shown in Fig. 13.11.

13.3.2 Data Analysis

The data is preprocessed by dropping null and missing values so as to infer meaningful prediction. In order to have enhanced understanding of the data, exploratory analysis is done and is depicted in Table 13.2.

13.3.3 Feature Selection

Further, correlation among independent and dependent variables is evaluated to determine the most dominating features. This is done through Pearson correlation and the corresponding results are demonstrated in Table 13.3.

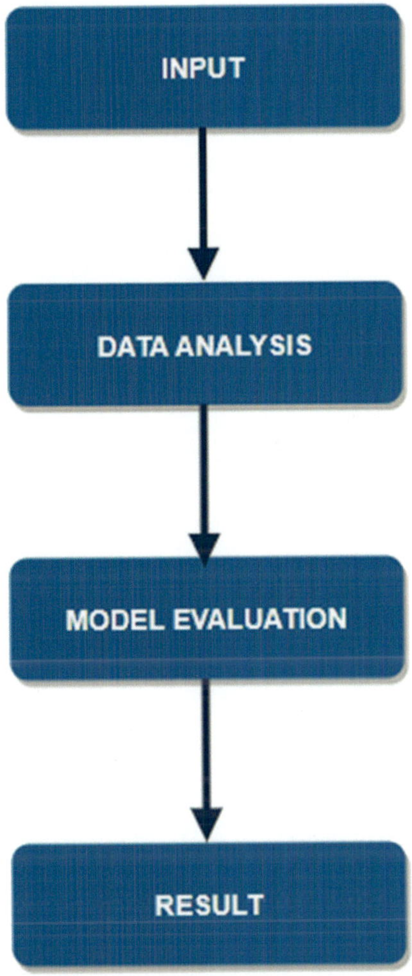

Fig. 13.10 Shows an image of methodology

13.3.4 Model Evaluation

As discussed earlier, here the authors aim to employ various Machine Learning models to predict the IRI. Here, most promising machine learning models have been discussed in this subsection.

13.3.4.1 Artificial Neural Network

As briefed earlier, ANN is inspired by functionng of the human brain and has interconnected layers of neurons (Sollazzo et al. 2017) as illustrated in Figure 13.12. Here, each layer transfers the information to a subsequent layer which processes it

	Road_Name	Chainage (Km)	International Roughness Index (m/km)	Cracking (sqm)
0	SR No 1 Vijayawada to Nuzvid Road	6.00	6.16	175.81
1	SR No 1 Vijayawada to Nuzvid Road	7.00	5.06	149.19
2	SR No 1 Vijayawada to Nuzvid Road	8.00	3.26	20.08
3	SR No 1 Vijayawada to Nuzvid Road	9.00	4.27	168.34
4	SR No 1 Vijayawada to Nuzvid Road	10.00	3.05	6.54
...	...	...	...	...
1087	SR No 262 Vellanki to [illegible]	1.00	5.40	14.50

Fig. 13.11 Shows a snapshot of dataset

using some activation function. In order to enhance the effectiveness of ANN, the number of neurons and layers (hidden) can be enhanced.

13.3.4.2 Support Vector Machine

SVM primarily aims to attempt supervised learning that mainly performs classification (Sultana et al. 2019) using a hyperplane as shown in Fig. 13.13. It works to find the hyperplane so as to have maximum margin. In addition to classification, SVM can also be used for regression.

13.3.4.3 Linear Regression

Linear Regression is used for continuous dependent variable when there is a linear correlation among independent and dependent variables as illustrated in Fig. 13.14. Here, x-axis represents the independent variable and y-axis indicates the dependent variable.

13.3.4.4 Ensemble Methods

Ensemble methods such as boosting employs various prediction models so as to integrate the results from multiple models (Sultana et al. 2020; Yadav and Sharma 2018). Two most popular variants of boosting are as follows.

Table 13.2 Descriptive statistics of data set

	Chainge (km)	IRI (m/km)	Cracking (sqm)	Wide structural crack (sqm)	Raveling (sqm)	Pothole (Nos)	LSM[a] (sqm)	Rutting (mm)	Edge break (sqm)	Texture depth (mm)
Count	1090.0	1090.00	1090.00	1090.00	1090.00	1090.0	1090.00	1090.0	1090.0	1090.0
Mean	11.570	5.545	54.463	17.196	29.545	1.431	20.682	6.170	98.69	0.486
Std	19.10	2.051	111.21	65.38	53.51	4.89	79.78	4.95	159.4	0.20
Min	0.720	1.930	0	0	0	0	0	0.97	0	0.23
25%	3.01	4.11	0	0	0	0	0	3.19	6.385	0.37
50%	7.010	5.155	9.180	0	10.59	0	0	4.65	32.48	0436
75%	14.01	6.51	54.75	5.297	30.42	1.01	4.815	7.52	121	0.52
Max	270.21	15.09	1238.9	977.3	600.11	50.1	1223.9	33.7	1046	2.70

[a] Loss of surface Material

Table 13.3 Pearson correlation

Parameters	IRI
Chainage (km)	0.034
Cracking (sqm)	0.329
Wide structural crack (sqm)	0.199
Raveling (sqm)	0.294
Pothole (Nos)	0.355
Loss of surface material (sqm)	0.318
Rutting (mm)	0.680
Edge break (sqm)	0.352
Texture depth (mm)	0.398

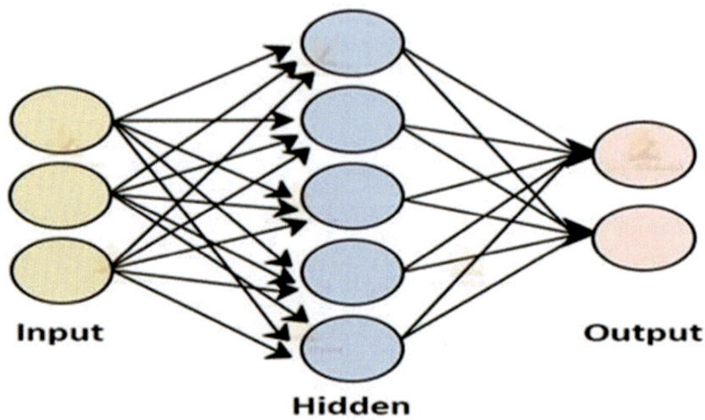

Fig. 13.12 Demonstration of artificial neural network (https://www.analyticsvidhya.com/blog/2021/07/understanding-the-basics-of-artificial-neural-network-ann/)

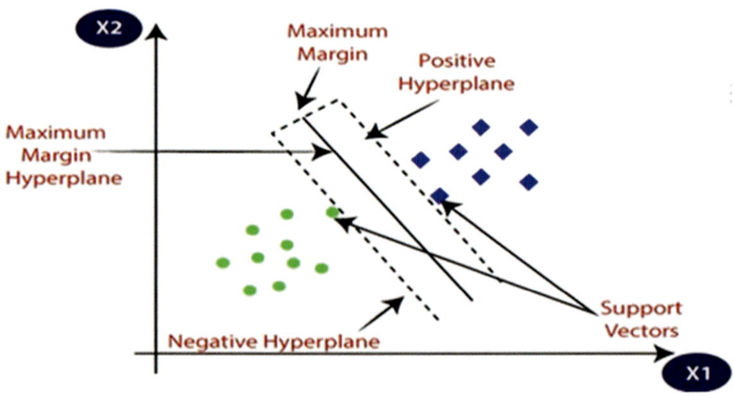

Fig. 13.13 Illustration of support vector machine (https://www.javatpoint.com/machine-learning-support-vector-machine-algorithm)

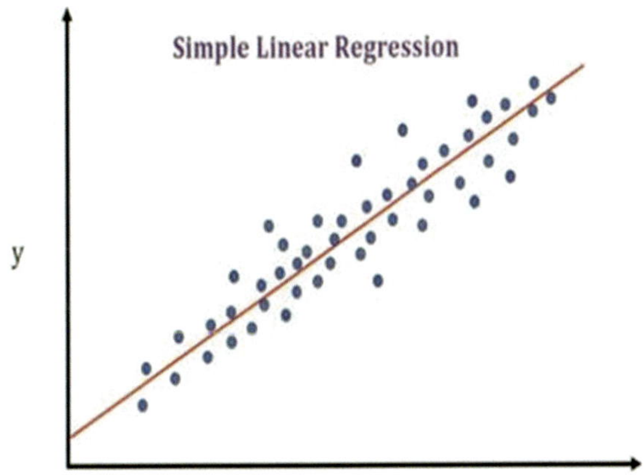

Fig. 13.14 Simple linear regression (https://medium.com/@dhaval.sony.504/everything-in-short-about-simple-linear-regression-633fc9f8dd65)

Gradient Boosting

Gradient boosting employs different prediction models in sequnce and thus error of one model impacts the subsequent model. It is named as gradient boosting as it minimizes the loss function by incorporating the algorithm for gradient descent (Sharma et al. 2021).

XGBoosting

XGBoost leverages decision trees and is designed to optimize the prediction. It has demonstrated remarkable efficiency for tabular and structured data (Sharma and Sikka 2021).

13.3.4.5 Stochastic Gradient Descent (SGD) Regressor

To quickly and easily fit linear regressors and classifiers to a loss function such as SVM and LR, stochastic gradient descent (SGD) is used. Although SGD has been around for a while in the machine learning community, large-scale learning has just lately been given a lot of attention. Large-scale, sparse machine learning problems that frequently arise in text classification and natural language processing have been resolved using SGD. Stochastic Gradient Descent provides the following benefits:

- Simplicity of application.
- Efficiency.

Stochastic Gradient Descent has a number of drawbacks, including:

- SGD involves a variety of hyperparameters, including the parameter of regularization and total number of iterations.
- It is sensitive to feature scaling.

13.3.4.6 Lasso Lars

LassoLars is a lasso model that, unlike the coordinate descent implementation, employs the LARS technique to obtain the precise solution. Least Angle Regression (LARS) is a regression approach for high-dimensional data (i.e., dataset and number of attributes are very large). Least Angle Regression is comparable to forward stepwise regression in certain ways. Because it is used with data that has several properties, LARS determines the attribute that is most significantly associated with the goal value at each stage. There might be several attributes with the same association. In this case, LARS averages the attributes and advances in the same direction as the attributes. This is why the technique is known as Least Angle Regression. LARS, in essence, takes jumps in the most ideally computed path while avoiding overfitting the model. Algorithm:

- Normalize all variables such that they have a value of 0 and a unit variance.
- Identify the variable with the highest correlation to the residual. Continue to move the regression line in this manner until we reach another variable with a similar or greater correlation.

Note: The discrepancy between the observed and predicted values is referred to as the residual.

- Adjust the regression line's inclination such that it is between and when two variables have the same correlation (i.e., the minimum angle between the two variables).
- Repeat doing this until you feel the model is large and "generic" enough, or until all of our data has been used.

13.3.4.7 ElasticNet

Variable selection and regularisation are both done concurrently using the regression method called Elastic Net. The primary idea underlying the elastic net is regularisation. When the model is overfitted, regularisation is considered. Overfitting is an issue that arises when a model performs well on a training dataset but poorly on a test dataset; In this case, regularisation is a method to lessen mistakes by properly fitting a function in the training dataset. Penalties can be applied to these operations.

Penalties come in two varieties: l1 and l2. The lasso regression model and the ridge regression model are two types of regularisation models that apply l1 and l2 penalties, respectively. As was previously mentioned, the lasso regression model includes the absolute size of the coefficient as a penalty term. As a penalty to the loss function, the ridge regression adds the squared magnitude of the coefficient.

Lasso is an acronym for least absolute shrinkage and operator. As the name implies, in lasso regression the coefficients are reduced to the absolute minimum and, if this is not feasible, the coefficient is removed from the models. The ridge regression includes all of the predictive factors in the model by offering an l2 penalty since it does not remove the coefficients from the model, which means it does not distinguish between significant and less important predictive variables in the model. By including them in the model together with their squared magnitude, it seeks to reduce the unbiased coefficient.

13.3.5 Data Visualization with Respect to Target Attribute

After data cleaning, the relationship between the different attributes with the target attribute scatter plot was used. It has been observed that the most influential parameter among these attributes is rutting. It was also observed that the relationship between other parameters and IRI in the scatter plot is almost unaffected with changing values. This means if the value of the parameter is changed the effect on the IRI is almost negligible. Therefore the effect on pavement distress is not much affected by parameters other than rutting.

For example Fig. 13.15 shows the relationship between Chainage vs IRI and from the scatter plot it was observed that increasing or decreasing the value of Chainage has the least effect on the pavement distress. The Figs. 13.15, 13.16, 13.17, 13.18, 13.19, 13.20, 13.21 and 13.22 shows absence of any relationship between the attributes (plotted on y axis) and IRI (plotted on x axis).

Figure 13.23 is plotted between rutting and IRI. It shows a positive linear kind of relationship.

13.3.6 Feature Importance

The significance of the relative parameter in accordance to XGB Regressor model is illustrated in Fig. 13.24.

From Fig. 13.24 it is evident that the most important feature is rutting while the least important is Chainage.

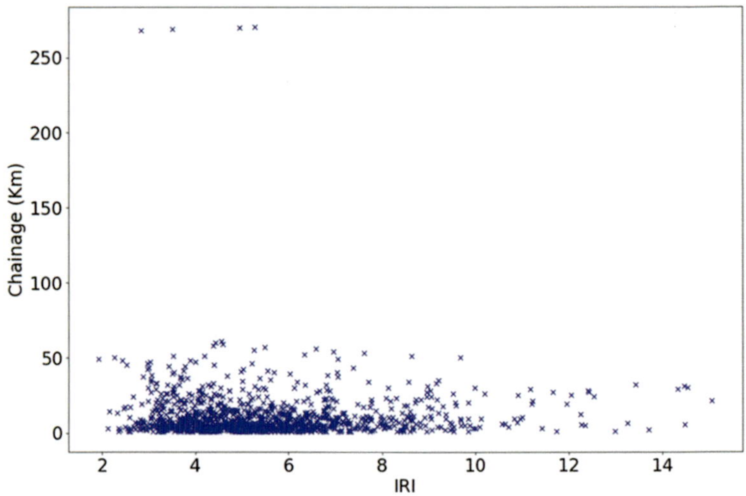

Fig. 13.15 Chainage versus IRI

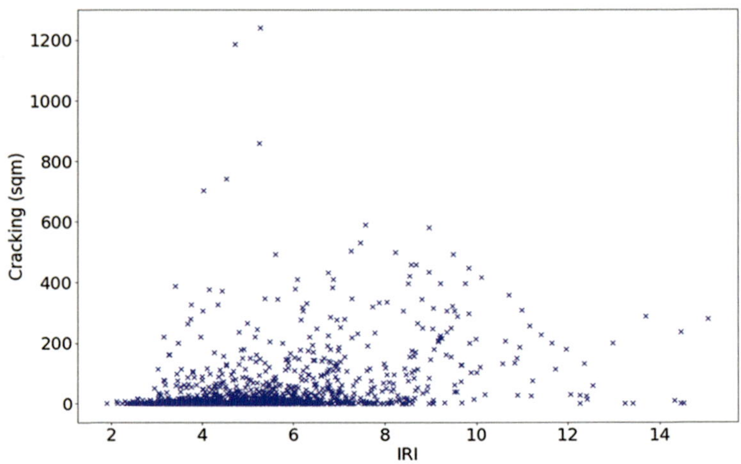

Fig. 13.16 Cracking versus IRI

13.3.7 Performance Evaluation Metrics

In order to measure the performance of various machine learning models, different metrics namely Mean Absolute Error (MAE), Mean Squared Error (MSE), and Root mean squared error (RMSE) are used. Lower value of MAE and RMSE indicates the effectiveness of the forecasting model.

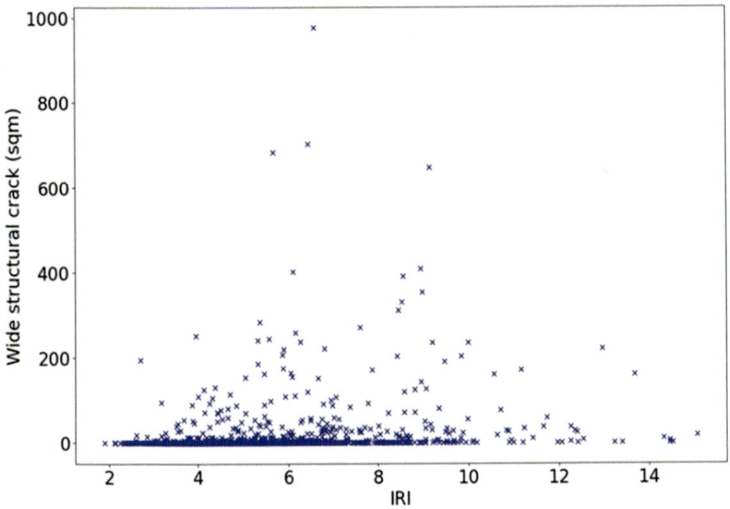

Fig. 13.17 Wide structural crack versus IRI

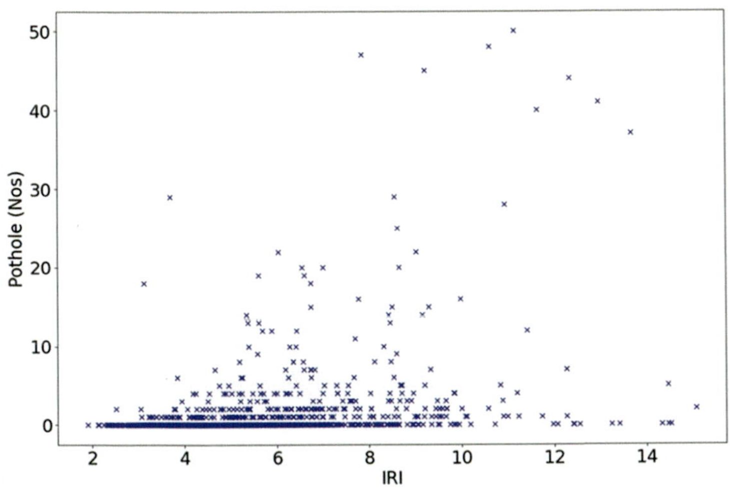

Fig. 13.18 Pothole versus IRI

13.4 Results and Discussion

In this chapter, the methodology used in this study is explained thoroughly. Further, this chapter gives an overview of the various machine learning algorithms used in this work. In the end, different performance evaluation matrices that are used to compare different algorithms are given.

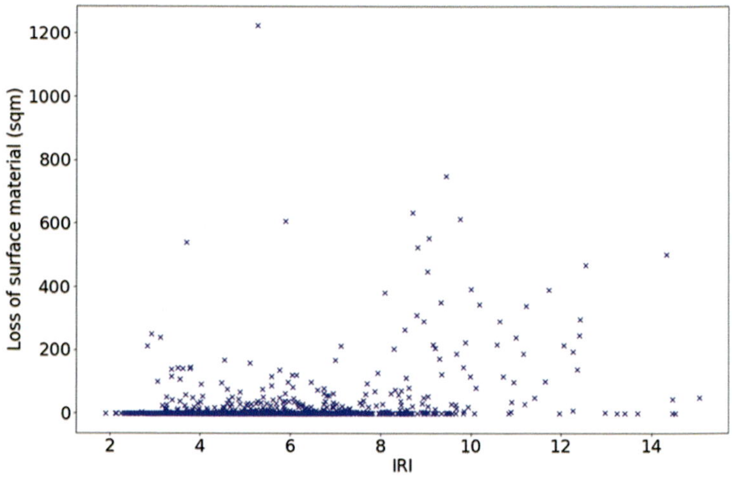

Fig. 13.19 Loss of surface material versus IRI

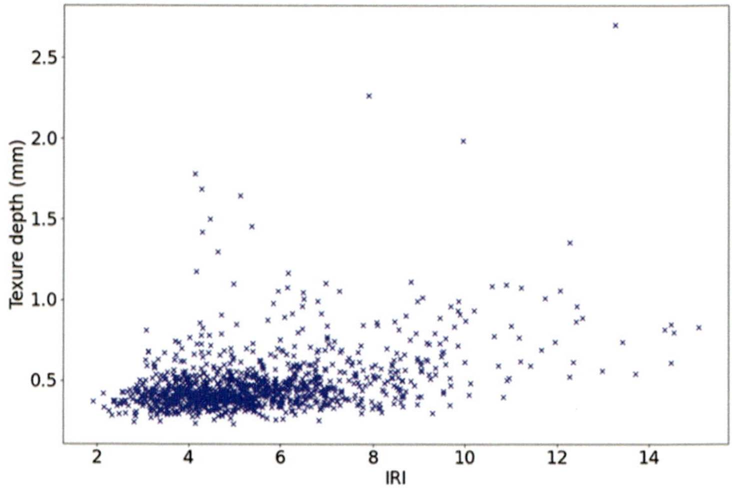

Fig. 13.20 Texture depth versus IRI

The performance metrics for various machine learning models is illustrated in Table 13.4.

Similar results are also graphically represented in Fig. 13.25.

From the comparative analysis it is evident that the ANN and XGB Regressor outperforms comparative models.

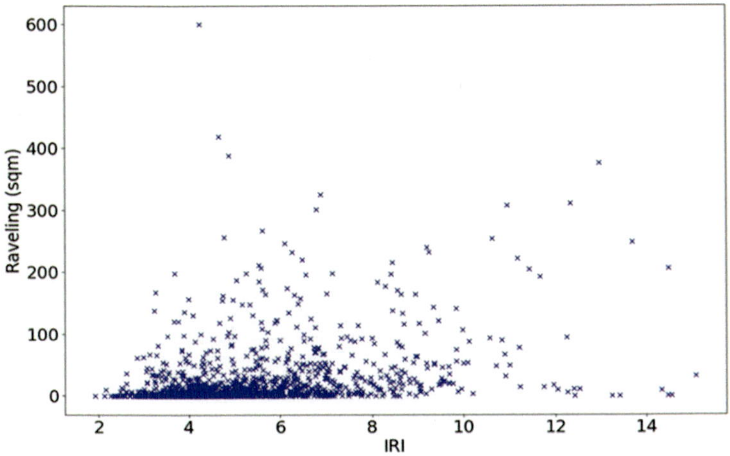

Fig. 13.21 Raveling versus IRI

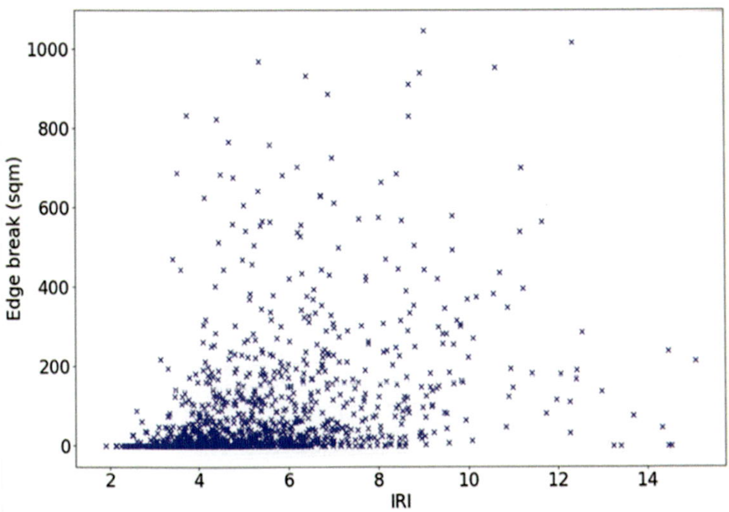

Fig. 13.22 Edge break versus IRI

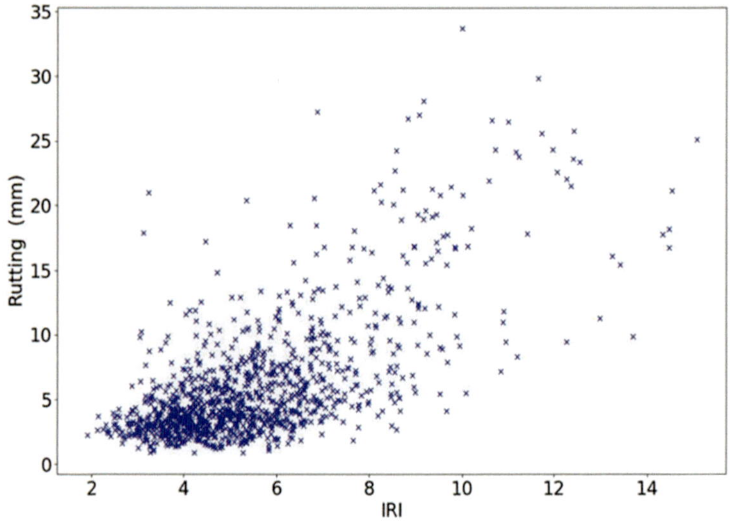

Fig. 13.23 Rutting versus IRI

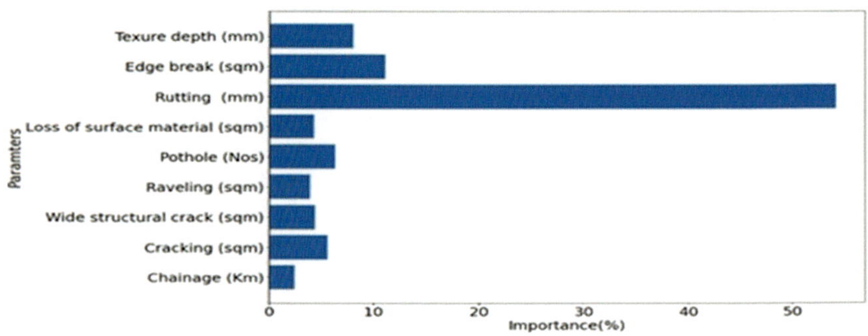

Fig. 13.24 Feature importance

Table 13.4 Comparative analysis

Model name	MAE (% age)	MSE (% age)	RMSE (% age)
ANN	18.13	2.08	1.40
SVR	18.62	1.90	1.44
Bayesian ridge	20.07	1.84	1.42
LassoLars	29.04	3.44	1.92
Linear regression	20.09	1.85	1.42
XGB regressor	17.69	1.54	1.30
ElasticNet	24.67	2.41	1.62
Gradient boosting regressor	17.99	1.52	1.29
SGD regressor	20.15	1.84	1.42

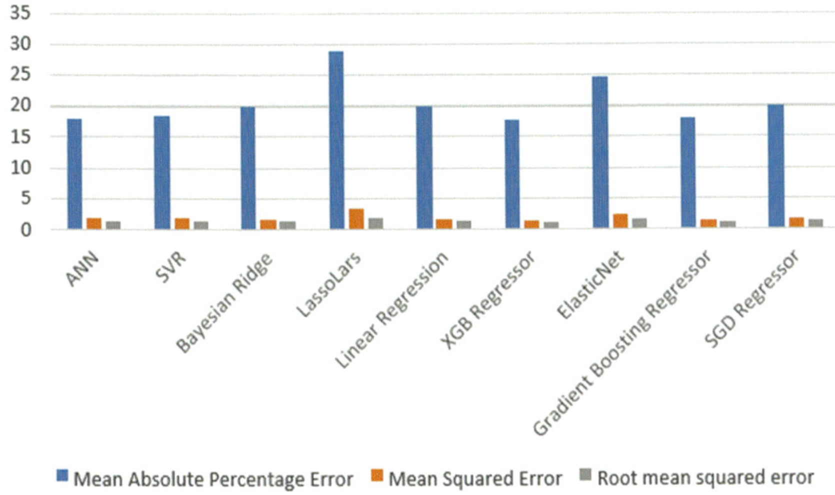

Fig. 13.25 Visualization of final result

13.5 Conclusion

This research work aimed to devise an efficient machine learning model to predict the IRI. In order to carry out the experiment, dataset from CRRII is used. During the experimental evaluation, ANN and XGB are observed to outperform the comparative models. This research can be helpful for maintenace of roads and thus the number of accidents can be drastically reduced. Further, timely maintenance of roads also leads to cost effectivenss.

This research can still be improved upon and a lot of automation can be done in terms of data feeding so that these models can be used in real time basis in both small and large scale projects. This research can also be modified and data from airports can also be fed to machine learning models and that will greatly reduce the inconvenience that is caused to the passengers in small airports due to closure of runways for maintenance since no real time monitoring is done in these small airports.

References

Abdelaziz N, Abd El-Hakim RT, El-Badawy SM (2018) International roughness index prediction model for flexible pavements. Int J Pavement Eng

Bashar MZ, Torres-Machi C (2021) Performance of machine learning algorithms in predicting the pavement international roughness index. Transp Res Rec 0361198120986171

Chandra S, Sekhar CR, Bharti AK, Kangadurai B (2013) Relationship between pavement roughness and distress parameters for Indian highways. J Transp Eng 139(5):467–475

Chopra T, Parida M, Kwatra N (2017) Development of pavement distress deterioration prediction models for urban road network using genetic programming. Hindawi

Colombier G (2004) Cracking in pavements: nature and origin of cracks. In: Prevention of reflective cracking in pavements. CRC Press, pp 14–29

Georgiou P, Plati C, Loizos A (2018) Soft computing models to predict pavement roughness: a comparative study. Hindawi

Hamdi H, Hadiwardoyo SP, Correia AG, Pereira P, Cortez P (2017, June) Prediction of surface distress using neural networks. In: AIP conference proceedings, vol 1855, No 1. AIP Publishing LLC, p 040006

http://lgam.wikidot.com/edge-break

https://civiljungle.com/difference-between-flexible-pavement-rigid-pavement/

https://copavementsolutions.com/asphalt-raveling/

https://medium.com/@dhaval.sony.504/everything-in-short-about-simple-linear-regression-633fc9f8dd65

https://sableasphalt.com/bad-pavement-cracks-when-crack-sealing-just-wont-cut-it/

https://vertical-access.com/2015/01/06/material-conditions-series-part-5-surface-loss/

https://www.analyticsvidhya.com/blog/2021/07/understanding-the-basics-of-artificial-neural-network-ann/

https://www.britannica.com/technology/road/The-modern-road#ref592118

https://www.cityworks.com/blog/10-fascinating-facts-about-potholes/

https://www.iso.org/obp/ui/#iso:std:iso:13473:-2:ed-1:v1:en

https://www.javatpoint.com/machine-learning-support-vector-machine-algorithm

https://www.researchgate.net/figure/Rutting-distress-3-The-use-of-reinforcement-in-flexible-pavement-In-order-to-optimize_fig1_321983200

Janani L, Dixit RK, Sunitha V (2019) Prioritisation of pavement maintenance sections deploying functional characteristics of pavements. Int J Pavement Eng

Li J, Zhang Z, Wang W (2018) International roughness index and a new solution for its calculation. American Society of Civil Engineers

Lucey J, Fathi A, Mazari M (2019) Predicting pavement roughness as a performance indicator using historical data and artificial intelligence. In: Airfield and highway pavements 2019

Mazari M, Rodriguez DD (2016) Prediction of pavement roughness using a hybrid gene expression programming-neural network technique. J Traffic Transp Eng (English Edition) 3(5):448–455

Minu PK, Sreedevi BG (2014) Development of pavement roughness model and maintenance priority index for Kerala state highway I. Int J Eng Res Technol 3:908–913

Muralikrishna P, Veeraragavan A (2011) Decision support system for performance based maintenance management of highway pavements. J Transp Res Board 22:05

Nguyen H-L, Pham BT, Son LH (2019) Adaptive network based fuzzy inference system with meta-heuristic optimizations for international roughness index prediction. Appl Sci

Plati C, Georgiou P, Papavasiliou V (2015) Simulating pavement structural condition using artificial neural networks. Struct Infrastruct Eng 12(9):1127–1129

Qi-sen Z, Yu C, Xue-lian L (2009) Rutting in asphalt pavement under heavy load and high temperature. In: Geo. Hunan international conference, pp 39–48

RoadBotics—make data-driven decisions

Sharma N, Mangla M, Mohanty SN, Pattanaik CR (2021) Employing stacked ensemble approach for time series forecasting. Int J Inf Technol 13(5):2075–2080

Sharma N, Sikka G (2021, May) Autoregressive techniques for forecasting applications. In: 2021 2nd international conference on secure cyber computing and communications (ICSCCC). IEEE, pp 551–554

Shwartz-Ziv R, Armon A (2022) Tabular data: deep learning is not all you need. Inf Fusion 81:84–90

Singh R, Ali F, Kumar D (2016) Assessment of damages caused to infrastructure due to geotechnical failures by the tropical cyclone winston in Viti Levu, Fiji

Sollazzo G, Fwa TF, Bosurgi G (2017) An ANN model to correlate roughness and structural performance in asphalt pavements. Elsevier

Sultana N, Sharma N, Sharma KP, Verma S (2020) A sequential ensemble model for communicable disease forecasting. Curr Bioinform 15(4):309–317

Sultana N, Sharma N, Sharma KP (2019, April) Ensemble model based on NNAR and SVR for predicting influenza incidences. In: Proceedings of the ınternational conference on advances in electronics, electrical and computational ıntelligence (ICAEEC)

www.Designingbuildings.co.uk

www.nationaltesting.co.uk/road-surfaces/texture-depth-measurement

Yadav S, Sharma N (2018, December) Homogenous ensemble of time-series models for indian stock market. In: International conference on big data analytics. Springer, Cham, pp 100–114

Chapter 14
A Deep Learning Model for Visual Sentiment Analysis of Social Media

Krishna Pal Singh Tiwari, Nonita Sharma, Preeti Vats, Manik Rakhra, and Divyansh Sharma

14.1 Introduction

14.1.1 Overview

Each day social media is getting preferred that's why people are posting their everyday tasks as well as their feeling and ideas in the microblogging system like Facebook, Instagram, Flickr and also twitter consequently containing much more details of the people. Our purpose in this research study is to immediately infer favorable, neutral, and unfavorable human perspectives from pictures submitted on Flickr as well as Instagram (Ortis et al. 2020). Based upon the feeling and sentiment inferred from their images, the authors may estimate a person's psychological wellness (Ortis et al. 2020). A regular technique is to look for components in an image that are related to human feelings, such as items (e.g., toys, birthday celebration

K. P. S. Tiwari (✉)
Cognizant Technology Solutions Pvt. Ltd, Howrah, India
e-mail: krishnapal.tiwari@cognizant.com

N. Sharma · P. Vats
Department of Information Technology, Indira Gandhi Delhi Technical University for Women, Delhi 110024, India
e-mail: nonitasharma@igdtuw.ac.in

M. Rakhra
Department of Computer Science and Engineering, Lovely Professional University, Phagwara, Punjab 144411, India
e-mail: manik.23538@lpu.co.in

D. Sharma
Bow Valley College, Calgary, AB, Canada
e-mail: d.sharma921@mybvc.ca

N. Sharma et al. (eds.), *Big Data Analytics in Intelligent IoT and Cyber-Physical Systems*, Transactions on Computer Systems and Networks,
https://doi.org/10.1007/978-981-99-4518-4_15

Fig. 14.1 Sobbing girl (Khorrami et al. 1510)

cakes, gun). Computer vision techniques such as face-scanning and also emotion acknowledgment can rapidly recognize the child in Fig. 14.1.

Facial emotions are important in nonverbal communication because they can reveal deeper human emotions as well as intent. The identification of facial expressions during social communication is a difficult process (Song et al. 2018). Deep Learning has proven to be superior to image processing methods for facial emotion detection. The six fundamental emotions that have been characterised historically include joy, sorrow, surprise, and fear. A FER (Facial Emotion Recognition) is a method that attempts to understand multiple facial muscles and classify them into the basic emotions mentioned above (Song et al. 2018).

This project aims to create a machine learning system that can be used in practice. Face expressions can be used to identify emotions and process them. It discusses three steps that are involved in emotion detection: facial, heart, and hand detection, features extraction and emotion classification (Kumar et al. 2020).

14.1.2 Visual Sentiment Analysis

Facial emotions are important in nonverbal communication because they can reveal deeper human emotions as well as intent. It is a challenging task to identify facial emotions in social communication. Deep Learning has proven to be superior to image processing methods for facial emotion detection. Happiness, sadness, surprise, and fear are the six basic emotions that have been described historically.

A FER (Facial Emotion Recognition) is a method that attempts to understand multiple facial muscles and classify them into the basic emotions mentioned above. FER is often implemented using many hand-crafted features, such as SIFT, LBP, and HoG. However, this approach is not able to simultaneously address multiple factors. Convolutional neural networks (CNN) have been shown to be promising when applied in FER research.

This project aims to create a machine learning system that can be used in practice. Face expressions can be used to identify emotions and process them.

Fig. 14.2 Picture showing women with emotion (Ding et al. 2017)

Convolutional networks (ConvNets) have been applied to a number of issues, including high-dimensional shallow feature encodings, visual identification problems, and large-scale picture classification systems. In a number of image classification challenges, CNN-based algorithms have been shown to attain the best levels of accuracy (Fig. 14.2).

14.1.3 Deep Learning

It is a sub-field of machine learning that depends only on ANNs (Artifi-cial Neural Networks). The human brain is portrayed as being mimicked by neural networks, and the same is true of deep learning. Deep learning has the advantage that not everything needs to be explicitly programmed. In deep learning, we must train a model on a dataset and then refine it until it makes nearly accurate predictions on both the testing and validation datasets (Sharma and Sharma 2021). Deep learning models are very helpful in resolving the dimensionality issue since they can focus on specific features on their own, with barely any input from the programmer. The DL model can be broadly divided into two components (Fig. 14.3).

Feature Extraction Phase

In this phase, authors train deep architectures on a large dataset by extracting a feature using the cascade of different layers. Authors simply input the images and then feed them to other layers.

Classification

At this stage, the photos are categorised into the relevant class. In Fig. 14.4, the relationships between AI, ML, and DL are mentioned, as seen in the diagram below.

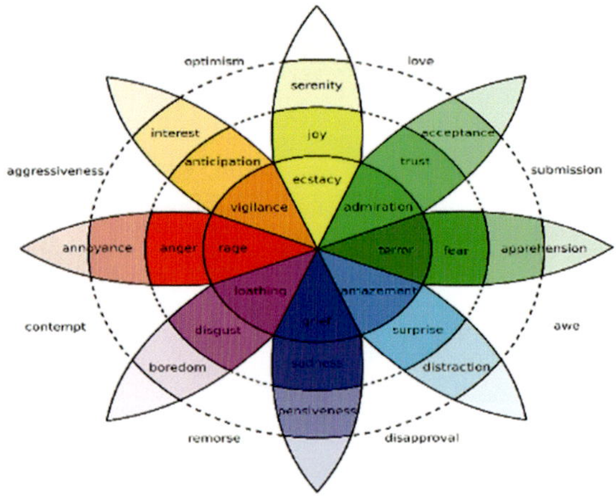

Fig. 14.3 Plutchik's wheel of emotions

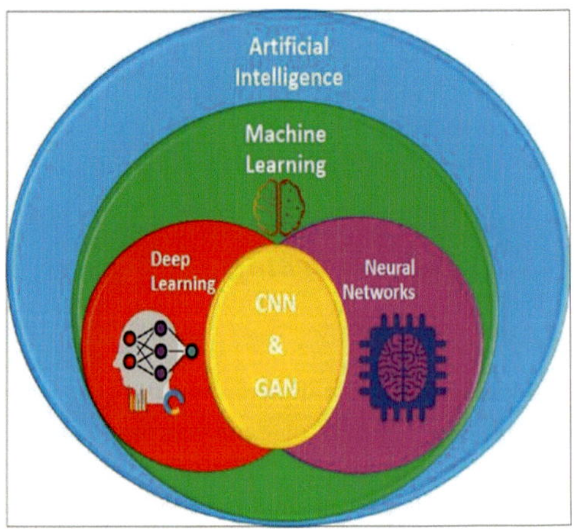

Fig. 14.4 Relationship between neural network, DL and CNN (Meng et al. 2017)

14.1.4 Convolutional Neural Networks

CNN is a Deep Learning calculation that can detect symbolism, assign esteem (intelligible and prejudicial instruments) to the image's various components/components, and distinguish one from the other. In comparison to previous stage calculations, ConvNet requires more preparation. While the old channels are handmade, with s, ConvNets can pre-use these channels' highlights with sufficient practice neurons that react to upgrades just in the limited area of the review field, also called as the Reception Field. The assortment of such fields penetrates to cover the whole obvious surface. CNN, which can, without much of a stretch, recognize and arrange objects with negligible preparation, prevails concerning breaking down visual pictures. It can, without much of a stretch, recognize the necessary highlights by its numerous lines structures. CNNs' indeed are the basic machine learning algorithms, for example, where a more powerful model increases artificial intelligence by demonstrating various sorts of human biological brain activity. A deep neural network has an input layer, various hidden layers, including completely connected layers, normative layers, multiple convolution layers, average pooling, and layers with multiple connections and an output layer. Few of these stripes are convolutional, which means they utilize a mathematical figure to pass information from one layer to the latter. This is same as some of the functions of the human visual brain.

The general CNN model is represented below in Fig. 14.5.

This stage involves moving each filter to every location on the image that is feasible. In a similar manner, drag the filter around the image to check how the feature lines up. Finally, for one feature, we will get the output shown in Fig. 14.6. After repeating the process for the other two filters, we obtain the convolution results for all of them.

Activation Layer

The actual work in convolutional neural networks provides a connection between the model's various layers. This layer contains a few functions such as ReLU, sigmoid, and SoftMax.

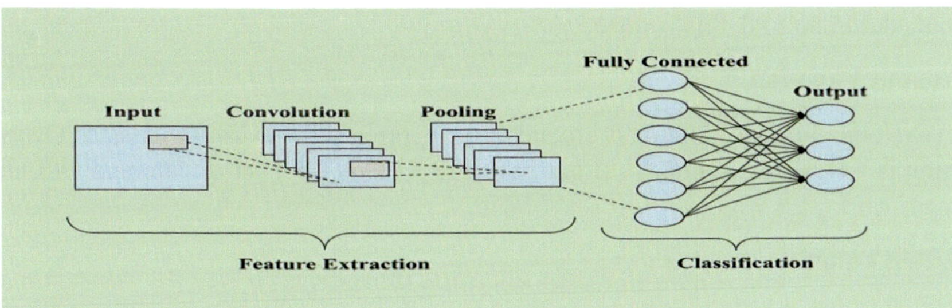

Fig. 14.5 General CNN model

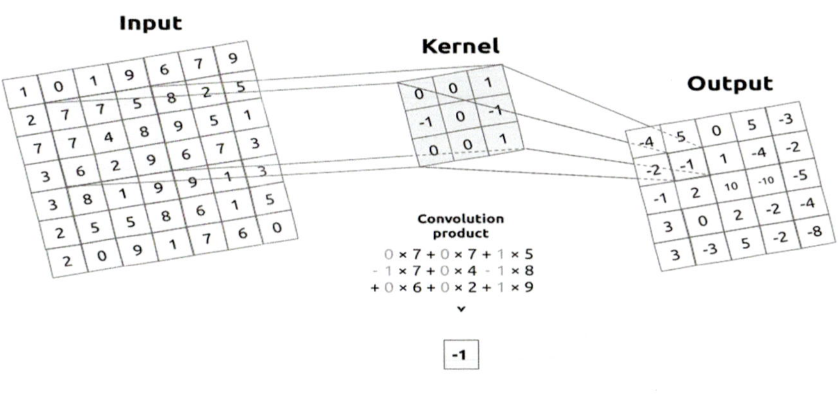

Fig. 14.6 Convolution (Meng et al. 2017)

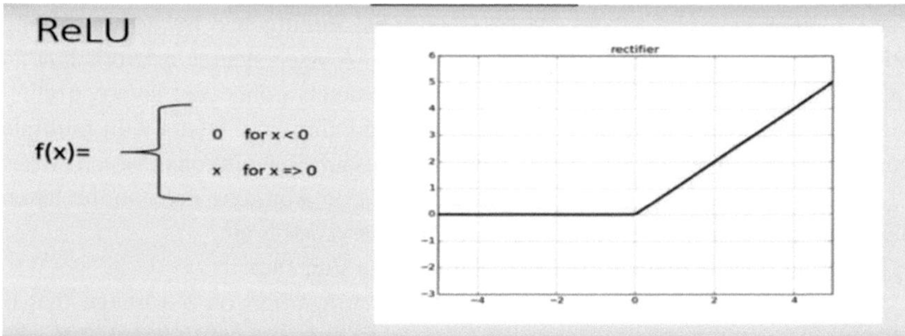

Fig. 14.7 ReLU function (Liu et al. 2017)

ReLU Layer

A node is activated by the Rectified Linear Unit (ReLU), a function that is used to determine when an input has crossed a predetermined threshold. As shown in Fig. 14.7, the output has a linear connection with the dependent variable when it reaches a particular threshold but is zero when the input is less than zero. All negative values in the convolutional results are changed to zeros in this layer. To stop the values from adding up to 0, this is done.

Dropout Function

In CNN, the dropout function is crucial. It helps prevent the over-fitting issue. Over-fitting is a sign that a model has gotten too dependent on a set of data and will no longer work well with different sets of data.

Flatten Function

Using this function, the pooled feature map is transformed into a single-dimensional array and then forwarded to the following layer.

SoftMax

SoftMax builds on this idea by developing an universe with multiple classes. SoftMax gives each class in a multi-class problem a decimal probability. These decimal probabilities must add up to 1.0. This extra restriction makes training converge more quickly than it would otherwise.

Pooling Layer

This layer reduces the image to a smaller frame. In this instance, the max-pooling function is utilised. Following are the steps: Choose a stride, a window size (often 2 or 3), walk the matrix over the ReLU results, and choose the highest value from each window. Let's examine the outcomes of pooling the first filtered image with either a matrix size of 2 or a stride of 2. The largest or maximum value is 1, so we keep track of it and advance the matrix by two steps. The pooling of a single 2*2 matrix

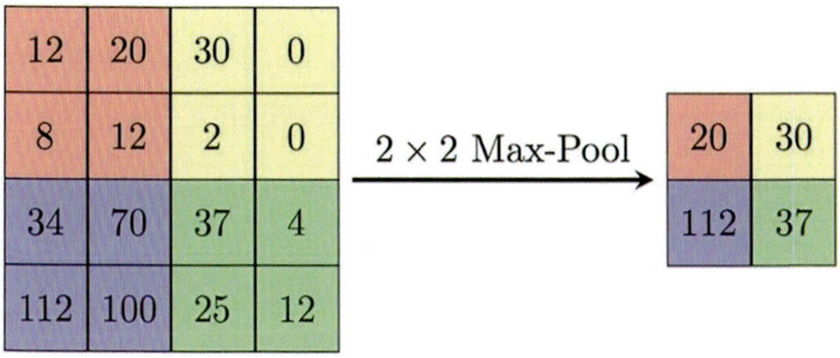

Fig. 14.8 Pooling layer (Yang et al. 2018)

is shown in the figure. After pooling the entire image, we obtain the result shown in Fig. 14.8.

Batch Normalization

A deep neural network training technique called batch normalisation normalises the two. The final layer 2*2 matrix 2 input to a layer for each mini batch is this pooling. This approach will decrease the number of epochs required to train the model.

Fully Connected Layer

In essence, the convolutional network's fully connected layer learns a (possibly non-linear) function in the pertinent low-dimensional, somewhat invariant feature space. The classification occurs in this layer, which is the bottom layer. Our filtered and condensed photos are combined into a single list or vector in this step.

14.1.5 Transfer Learning

Transfer Learning (TL) is an machine Learning optimization technique that emphasizes transferring knowledge gathered while solving one task to another but a similar task (Priyavrat and Sikka 2021). Using transfer learning, the training cost is minimized, accuracy is improved, and low generalization error is obtained.

To obtain the best performance in comparison to existing CNN models in different implementations, building and training a CNN architecture from scratch is an arduous procedure. Therefore, different models may be retrained according to the applications and is also used for feature engineering. Machine Learning and Knowledge Discovery in Data (KDD) have made huge colossalress in numerous knowledge engineering areas, including classification, regression, and clustering. The pre-trained model or wanted model segment can be incorporated straightforwardly into another CNN model (Aggarwal et al. 2023). The weight of the pre-trained models might be freezing in some applications; during the development of the new models, and weight can be updated with slower learning rates, which allows the pre-trained model to behave like weight initialization when the new model is learned. The pre-trained model may also be utilized as a weight initialization, classifier, and extractor. Firstly, we use the transfer learning approach, but it doesn't provide better results. After that we fine-tuned hyper parameters of our DL models and subsequently fine tuning provides us better results.

14.1.6 Fine Tuned CNN Models

Building a deep learning model from scratch is no easy task. Here some changes in the architecture of the deep learning model can be done as per the problem to be solved. Fine-tuning a deep learning model is a required step to improve the precision of anticipated outcomes (Pall et al. 2022). We gradually update the weights beginning with the most minimal level layers and working our way to the top. It learns a lot from pre-trained weights while training fine-tuned models. In the wake of training and testing, we can contrast our networks.

14.1.7 Traditional Methods

Different machine learning algorithms were employed in conventional approaches for picture identification and categorization (Pall et al. 2022).

Workflow of Traditional Methods

- Pre-Processing: This step is mainly done to remove noise from the image. This step mainly resizes the images. Various filters can be used in this step. This step can also help in increasing the accuracy.
- Object segmentation: This step mainly focuses on decreasing the search time by segmenting the image. This step helps in the detection of an object or region of interest.
- Feature Extraction: This step helps in the extraction of the desired features. Various machine learning can be used in the identification of desired features.
- Classification: This step helps in classification i.e. this step determines to which category the determined object belongs.

In machine learning a huge dataset can't be processed. At one value in the machine learning the model stops giving better accuracy and remains confined to one.

14.1.8 Research Objectives

- Development of various models coupled with Transfer Learning which will help in the Visual Sentiment Analysis.
- To use Facial Emotion images for the detection and prediction Various Emotion.
- To increase the accuracy of various pre-trained models in different datasets.
- To give a comparative analysis of different models in the detection process.
- The presented pre-trained DL models have an end-to-end structure that is totally autonomous and does not require custom feature extraction techniques.

14.1.9 Thesis Organization

The work has been organized into six chapters: Sect. 14.1 gives a brief introduction along with motivation and objectives. Section 14.2 discusses literature surveys of the work. Section 14.3 gives the problem definition and model architecture. Section 14.4 describes Dataset and Evaluation metrices and discusses the experimentation and results. Section 14.5 discusses the conclusion and future scope.

14.2 Literature Survey

An overview on visual sentiment analysis is represented in this section. VSA is a very recent area various sections are defined to show the previous work. At the end of the Literature review there are summary of all the related work.

Ortis et al. (2021) work used FER-2013 as a origin of design new CNN models. They developed these two CNN architectures. Because they were simple and used fewer training parameters, the architectures created expressly for the FER-2013 dataset were adequate. Both models are proposed on the FER-2013 dataset, their accuracy was better than 65%. Their models were able to predict the model and had the lowest number of parameters to train than other models, but were still capable of performing give human-like accuracy (65%). It was the best model for the job dataset.

Studies in the subject of visual sentiment detection are mainly concerned with issues like modelling, detecting, and utilising emotions communicated through facial or physical gestures (Kaya et al. 2012).

One new area of research is facial expression and nonverbal sentiment analysis. Researchers have used multimodal sentiment analysis on videos; however, more work needs to be done in relation to visual sentiment.

Researchers investigated the sentiment of adjectives over 100 images annotated by 42 subjects. They found it was possible to predict whether a pair of adjective–adjective words were present in an image by using aspects of light, saturation and sharpness to improve sentiment prediction.

Gonçalves et al. (2013) introduced an AI that used machine learning and linear SVM outputs to create visual sentiment analysis. ANPs, a semantic construction used in the SentiBank approach, paired "adjectives" and "nouns" for visual detectability, producing pairs like "cute bear," "beautiful sunrise," "tasty meal," and "dreadful accident."

The ANP model, proposed in Hasan et al. (2018) for constructing a visual sentiment ontology, inspired the SentiBank detector bank. The process involved the use of Flickr and YouTube APIs to identify images with emotive qualities. By part-of-speech tagging and extracting candidate pairs based on sentiment strength, named entities, and popularity, individuals were then able to produce a pool of adjective-noun pair candidates to search.

In the paper (Al-Halah et al. 2019) they have presented "Sentiment Networks with Visual Attention (SentiNet-A) is an layered network that investigates visual attention in order to improve picture sentiment analysis. A multi-layer neural network has been developed and implemented into a CNN-based image recognition system. Extensive trials on two benchmarks back up our assertion and proposition" (Corchs et al. 2019).

In the paper (Vadicamo et al. 2017) they have presented a hybrid approach for real-time sentiment analysis using a combination of text and picture modalities was proposed in research (info-graphic). For text, picture, and mixed data, the model exceeded the baselines, with sentiment classification accuracy of around 88, 76, and 91%, respectively (Gonçalves et al. 2013). The decision module might also aid in the detection of neutral ambiguities that were indicative of sarcasm.

The issue of visual sentiment based on convolutional neural networks is addressed in this study, where the sentiments are predicted utilising a variety of affective cues (Hasan et al. 2018). They have described a two-branch, end-to-end, weakly supervised deep architecture for learning discriminative representations.

This study (Al-Halah et al. 2019) proposes a novel cross-modal technique for semantic content correlation that links captions to images. An image-text combination is identified by this model using a joint attention network, and the relationship between the words in the caption is then determined using that information as the query (Al-Halah et al. 2019). The prediction of picture sentiment is based on the correlation of two modalities, textual meaning and visual content.

In this paper (Corchs et al. 2019), MASAD has 38k samples with image-text pairs and 57 aspects with seven domains. Extensive tests on the MASAD dataset have revealed that models trained on this dataset have a strong generality potential. The following research directions, we feel, are worth investigating: More expressive model architectures for learning inter multimodal knowledge are being developed. We will try to figure out how to extract the context and forecast the sentiment at the same time.

Recently, researchers have been creating machine learning models specific to sentiment analysis for visual media, such as photographs or videos. They were able to achieve higher accuracies of above 20% on sentences involving adjectives and nouns (Vadicamo et al. 2017). In contrast to prior models that employed convolutional neural networks for image classification and only classifiers for sentiment analysis with SVMs, others constructed a model on ImageNet before training it particularly

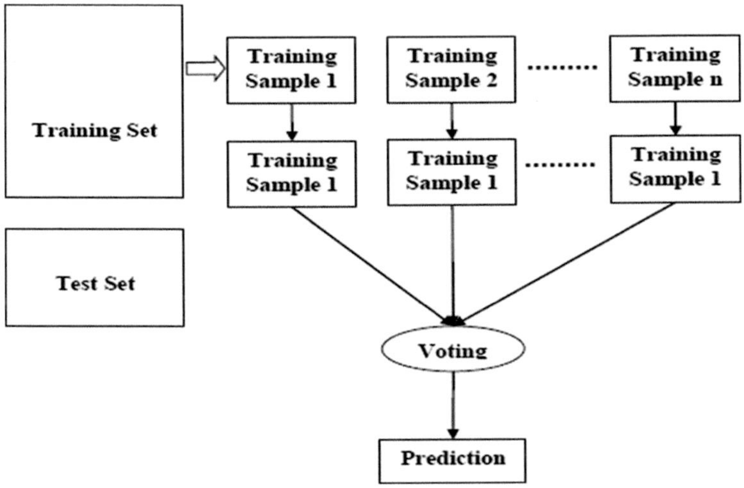

Fig. 14.9 General training set categorization

for sentiment detection with SVMs (Machajdik and Hanbury 2010). The researchers also considered the inclusion of LSTM (long short-term memory) in order to create more naturalistic sentence summaries by incorporating both object grounded nouns and affective descriptors.

Visual sentiment analysis can increase accuracy, covering a wider variety of languages and media types. It is not always accurate, missing the context of a situation or an entity (Fig. 14.9).

14.3 Methodology and Implementation

CNNs will be used for this project. The dataset will then be selected. Then, we will do the appropriate pre-processing. Then we'll be Divide the data into validation, testing, and training. Next, we create the model from the data and train it is using the dataset. This will produce the desired result. This proposed method aims to predict 6 facial expressions viz. You can do this by using CNN's Sequential Model. This model will be further enhanced with the layers of different algorithms that can be used. These algorithms will be used for activation functions, learning rates etc.

14.3.1 Sentiment Classification

Various machine learning approaches that can be used for this purpose are discussed below in detail.

CNN

CNN is a DL approach which is used to make the machines learn from their past experiences. This is a type of supervised technique in which data is first fed to the model and then it predicts the unseen corpus.

Support Vector Machine

Support Vector Machine (SVM) is a machine-learning technique used for classification of sentiment in this study. Each sample is labelled corresponding to one of the classes from the given set of training samples. An SVM training model assigns new samples to each class. Multiclass SVM makes labels to instances which are assigned from finite set of elements.

K-Nearest Neighbor

KNN is a supervised ML approach. It follows a majority voting method for similarity to K-nearest neighbor which is calculated using any of the distance methods. Euclidean, Manhattan and Minkowski distance are the commonly used distance methods (Fig. 14.10).

14.3.2 Various Steps Involved in Visual Sentiment Analysis

Data Acquisition

In the data acquisition process, a trained data scientist finds datasets and Machine Learning models to train. Various methods are used to generate data from different sources are as follows:

- Data Generation
- Data Discovery
- Data Augmentation.

The dataset consists of a very huge number of pictures to get more accurate results. The data in this model consists of grayscale pictures of faces 48 × 48 pixels. 35,887 facial images of different emotions were collected to categorize into 7 different classes. These faces were captured using a smartphone and blurry images will not give proper results. Capturing the faces in the middle of the image was important to more reliably categorize their expressions." The test set has 3,589 cases, whereas the training set has 28,709 examples. And the validation set has the same 3,589 items as the test set.

Data Pre-processing

There are a few changes necessary to any image before it is run through the machine. You can change the number of pixels or shrink the size if not necessary and remove any additional noise. This will give you much more accurate results and higher accuracy. Additionally, data augmentation is carried out in this methodological stage. The

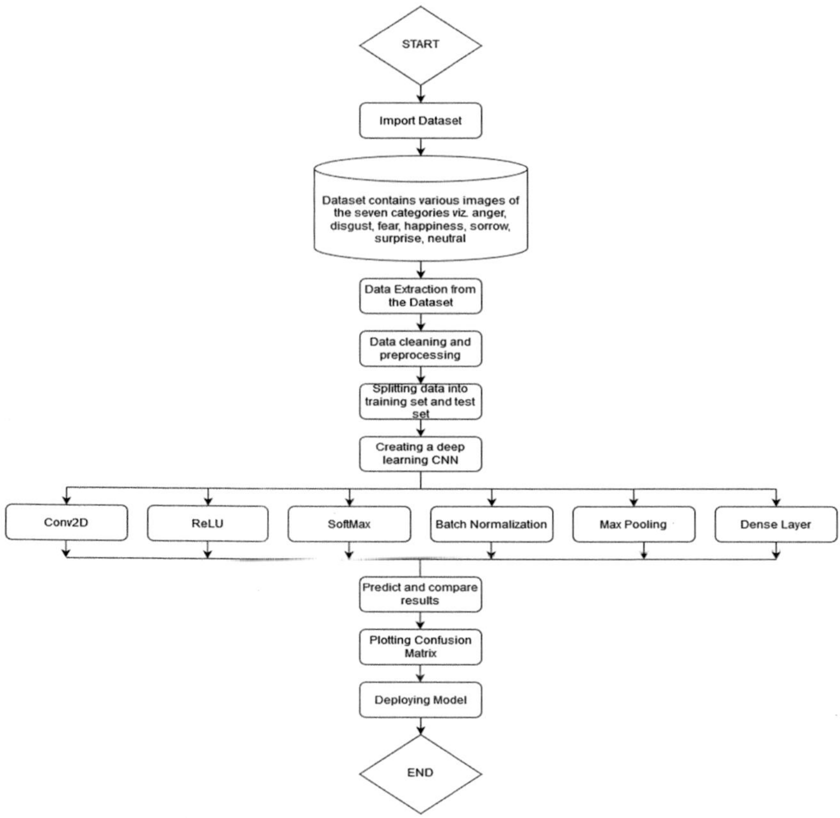

Fig. 14.10 Working of machine learning model

term "data augmentation" describes the process of greatly expanding the amount of photographs without actually taking new ones.

Feature Extraction

The various features that must be extracted for this project are:

Color-Histogram

The simplest visual feature taken from RGB colour channels is the Color Histogram. A 256-dimensional histogram displaying the distribution of pixel values may be obtained for each channel. The feature vector is then normalised to unit length and referred to as a feature vector (Fig. 14.11).

GIST Descriptor

For expressing real-world situations in a computer model, the GIST descriptor is proposed. The picture is initially filtered using numerous Gabor filters of varying sizes and orientations. Using the filters, you may capture the texture information

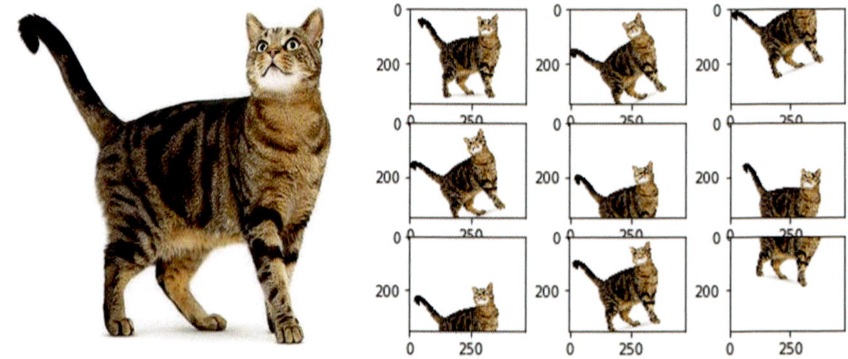

Fig. 14.11 Image augmentation

in the image. In this thesis, we investigate using three scales, with the number of orientations under each scale fixed at eight, eight, and four, respectively. We construct 20 feature maps in total, one for each of the 20 Gabor filters. We divided each map into four grids. Each grid's energy is defined as its average response value. All the energies are eventually merged into a 320-dimensional GIST feature vector.

Bag-of-Visual-Words (BoVW)

Bag-of-Visual-Words is similar to bag-of-words (BoW) in that the words are descriptors calculated on picture patches surrounding key points. The difference of Gaussian functions (DoG) applied in the scale space is used to locate the position of key points. The Hessian-Affine area detector is another popular approach for key points discovery. The dominant direction and size are determined by examining the gradient values around the identified key point. As a result, the extracted feature is insensitive to translation, scaling, rotation, and lighting changes. Instead of extracting local features around key points, it has been discovered that the approach employing intensively sampled local areas may obtain equivalent results for visual identification.

14.3.3 Building and Designing the Model

The various methods used for classification of this thesis are CNN, SVM and KNN. So, the working of all three of them is discussed below.

Convolutional Neural Networks

CNN is a deep-learning algorithm which takes an input image, assigns various weights and biases to the image so that it can be differentiable from one another. It mainly consists of four layers. The explanation of the various layers is given below (Fig. 14.12).

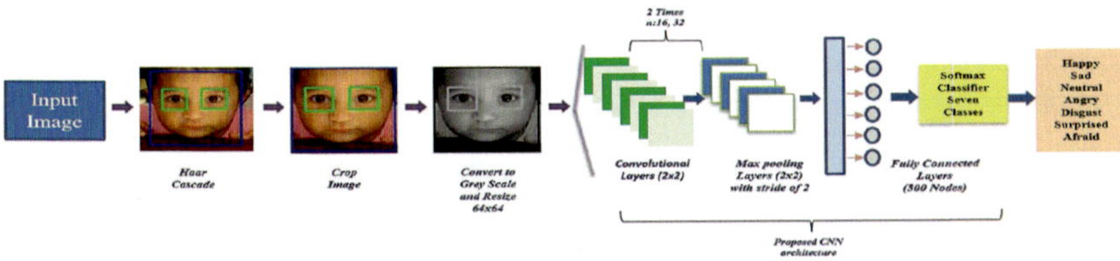

Fig. 14.12 CNN model overview (Hamester et al. 2015)

Convolution Layer

This layer of convolution neural network is used to perform mathematical functions in the model.

Pooling Layer

To summarize the images in the input, convolutional layers apply learned filters to the image. The dimensions of the filter being applied to the image is smaller than the size of the pixels setting up thc image.

Activation Layer

In CNN, the actual work gives a connection b/w the different layers of the model. A few functions like ReLU and sigmoid are present in this layer.

Fully Connected Layer

A fully Connected layer is a feed-forward NN. It's the last few layers in the network and creates output for the entire input. Recognition and classification on the image are performed in this layer.

The CNN model was chosen for this task because it requires less human oversight and pre-processing of the data. It is a self-learning algorithm that identifies key components in images that have been divided into test and train set in an 20:80 ratio.

With this model, the picture size is initially set at 48*48. The picture is then sent into the first convolution layer. The picture is then sent through a convolutional with 64 kernels of 15*15 size. The process is repeated up until the input size reaches 1*11. The input images are processed in this layer before the mathematical operations are carried out.

The max-pooling layer is an additional crucial layer. The max-pooling layer's primary goal is to reduce the size of the feature map. The main information source for this layer is the feature map. The Dense Layer is the most interconnected layer in the model, concluding that every neuron from earlier layers is connected to the layer below it. The output data is returned by this layer. The proposed model's model summary is shown below.

```
Layer (type)                 Output Shape              Param #
=================================================================
conv2d (Conv2D)              (None, 48, 48, 32)        320
_________________________________________________________________
conv2d_1 (Conv2D)            (None, 48, 48, 64)        18496
_________________________________________________________________
batch_normalization (BatchNo) (None, 48, 48, 64)       256
_________________________________________________________________
max_pooling2d (MaxPooling2D) (None, 24, 24, 64)        0
_________________________________________________________________
dropout (Dropout)            (None, 24, 24, 64)        0
_________________________________________________________________
conv2d_2 (Conv2D)            (None, 24, 24, 128)       73856
_________________________________________________________________
conv2d_3 (Conv2D)            (None, 22, 22, 256)       295168
_________________________________________________________________
batch_normalization_1 (Batch (None, 22, 22, 256)       1024
_________________________________________________________________
max_pooling2d_1 (MaxPooling2) (None, 11, 11, 256)      0
_________________________________________________________________
dropout_1 (Dropout)          (None, 11, 11, 256)       0
_________________________________________________________________
flatten (Flatten)            (None, 30976)             0
_________________________________________________________________
dense (Dense)                (None, 1024)              31720448
_________________________________________________________________
dropout_2 (Dropout)          (None, 1024)              0
_________________________________________________________________
dense_1 (Dense)              (None, 7)                 7175
=================================================================
Total params: 32,116,743
Trainable params: 32,116,103
Non-trainable params: 640
```

Model: "sequential"

Support Vector Machine

Support Vector machine is a supervised machine-learning model implemented in this thesis. It is a very effective and accurate model so it is chosen to compare with CNN. SVM divides the linear and non-linear data into classes. In SVM, we need to maximize the margin between two classes. The classification that helps finding the maximum optimal margin is called as hyper-plane (Fig. 14.13).

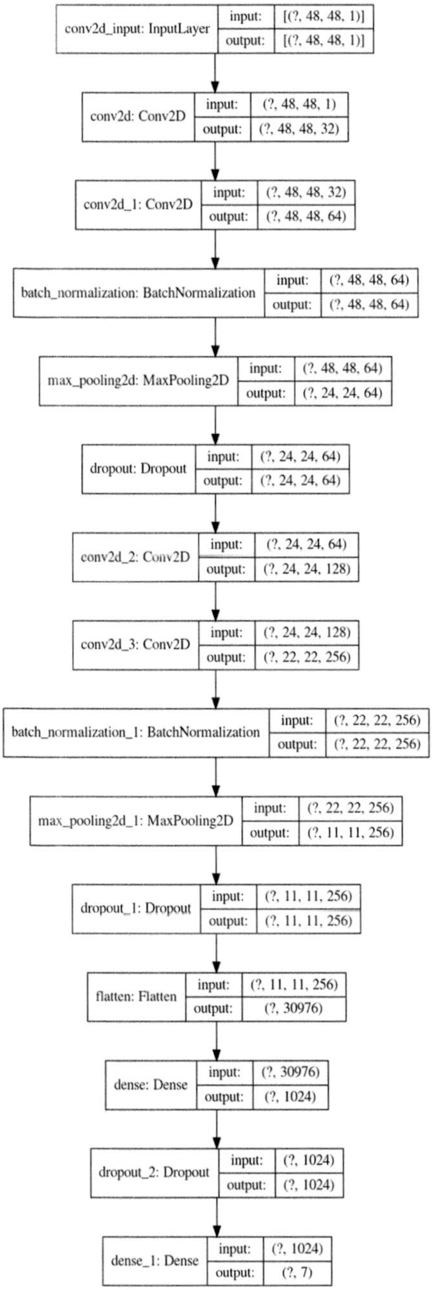
conv2d_input: InputLayer
input: [(?, 48, 48, 1)]
output: [(?, 48, 48, 1)]
conv2d: Conv2D
input: (?, 48, 48, 1)
output: (?, 48, 48, 32)
conv2d_1: Conv2D
input: (?, 48, 48, 32)
output: (?, 48, 48, 64)
batch_normalization: BatchNormalization
input: (?, 48, 48, 64)
output: (?, 48, 48, 64)
max_pooling2d: MaxPooling2D
input: (?, 48, 48, 64)
output: (?, 24, 24, 64)
dropout: Dropout
input: (?, 24, 24, 64)
output: (?, 24, 24, 64)
conv2d_2: Conv2D
input: (?, 24, 24, 64)
output: (?, 24, 24, 128)
conv2d_3: Conv2D
input: (?, 24, 24, 128)
output: (?, 22, 22, 256)
batch_normalization_1: BatchNormalization
input: (?, 22, 22, 256)
output: (?, 22, 22, 256)
max_pooling2d_1: MaxPooling2D
input: (?, 22, 22, 256)
output: (?, 11, 11, 256)
dropout_1: Dropout
input: (?, 11, 11, 256)
output: (?, 11, 11, 256)
flatten: Flatten
input: (?, 11, 11, 256)
output: (?, 30976)
dense: Dense
input: (?, 30976)
output: (?, 1024)
dropout_2: Dropout
input: (?, 1024)
output: (?, 1024)
dense_1: Dense
input: (?, 1024)
output: (?, 7)

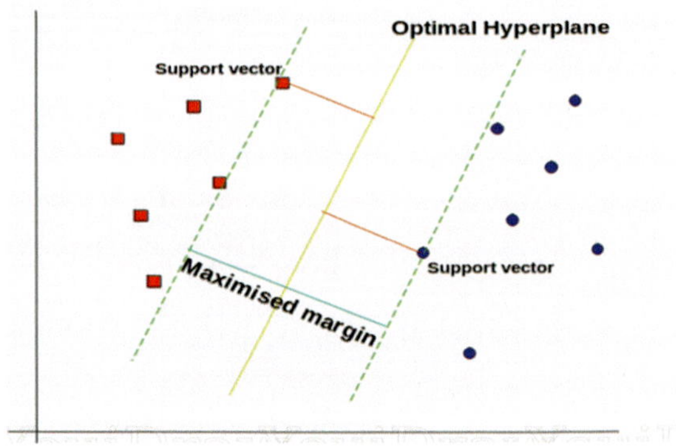

Fig. 14.13 SVM hyperplane

14.3.4 Model Flow

SVM Kernels are used to add more dimensions to the low dimensional space to make it easier to divide the data. So, the various kernels that can be used in SVM are as follows.

Linear Kernel

Linear kernel is the type of kernel which is used when the data can be separated linearly which means using a single line. It is one of the most commonly used kernels of the SVM as it can train the large data models.

Polynomial Kernel

This kernel function represents the similarity of vectors in the set of data in a feature space over the polynomials.

Radial Basis Function Kernel

This kernel is used for performing various kernelized learning algorithms. It is generally used for classification purpose.

In this project, the data is first divided for SVM classification into various classes depending upon the corresponding diseases in plants. Then, the linear function kernel is used as it is a fast learning algorithm and is helpful for large data-sets.

The size of the iterations in our model is kept as 500 which means the data is passed through the model 500 times to train the model (Fig. 14.14).

K-Nearest Neighbor

KNN is an algorithm that memorizes all the available classes and detects the new data on that basis. K is the number of nearest neighbors in KNN. We have to define the value of K. For e.g. If we take the value of K as 3 that means the 3 points from the object will be selected and then their distance will be evaluated. The point with the closest distance will be chosen. The distance measures in choosing the K-nearest can be any of the following. Some most commonly used distances are explained below.

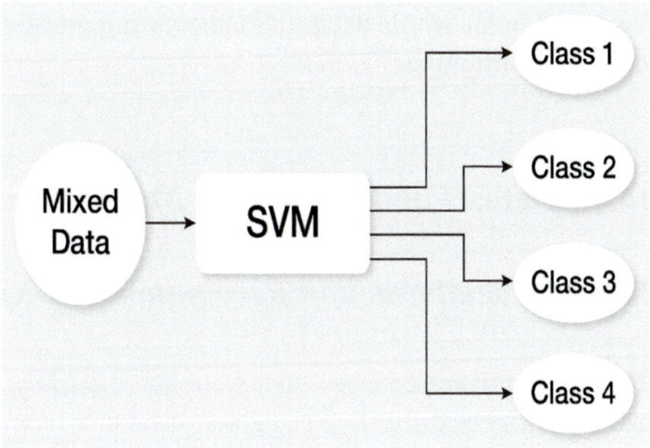

Fig. 14.14 Classification of data in SVM (Sun et al. 2016)

- Euclidean Distance

This distance is the most commonly used method in KNN. It is the measure of a line between 2 points (×1 and ×2) on a straight line.

- Manhattan Distance

This distance is the sum of difference between the x-coordinate and y-coordinate (Fig. 14.15).

$$d = |x1-x2| + |y1-y2|$$

Prediction is slow in the KNN algorithm and is called as a lazy learner because it does not have a classifier in the learning phase. This algorithm simply memorizes the whole dataset and there is no training in the KNN algorithm. Each time the item that is to be predicted is given to the KNN model, it searches for its nearest

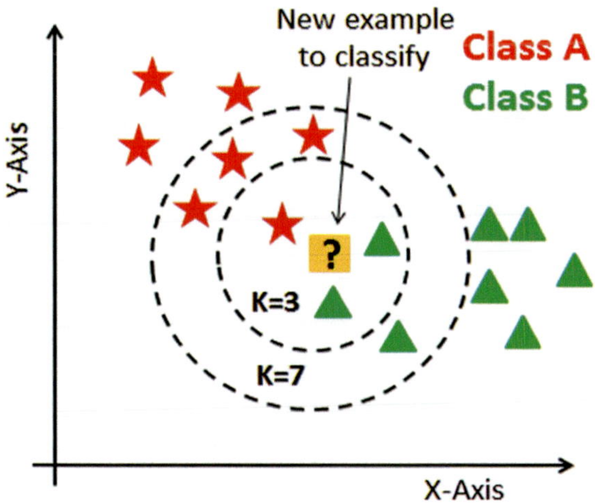

Fig. 14.15 KNN model overview (Liu et al. 2019)

neighbors in the whole dataset. So, the testing process in this model is quite lengthy and time-consuming.

14.4 Dataset and Evaluation Metrics

14.4.1 Facial Emotion Recognition (FER) Dataset

'Grayscale images of faces' that were automatically registered make up the information. It can be difficult to categorise each face based on the emotion that is shown in their expression. "emotion" and "pixels" are the two columns in "train.csv." The "feeling" column contains a number code that spans the entire range of emotions visible in the image, from 0 to 6, inclusive. A string of values encased in "quotes for each image" serves as the "pixels" column's description of row-major pixel values. Based on the pixel values in the "train.csv," the model's objective is to identify the emotion that is present (Fig. 14.16).

Dataset Details

Sr. no.	Class name	# Training image	# Validation image	# Testing image
1	Angry	3995	958	958
2	Disgust	1987	497	497
3	Fear	1760	440	440
4	Happy	2008	502	502
5	Neutral	1816	454	454
6	Sad	1826	456	456
7	Surprise	1683	421	421

Fig. 14.16 Sample image from FER dataset

14.4.2 Evaluation Metrics

With the aim of analyzing the performances of DL models on the basis of various hyper parameters like Validation Accuracy, precision, accuracy, recall and F1 score. Confusion Matrix, and AUC curve have been calculated in this study. In this chapter, definitions of various parameters used in this study are given and then in the next chapter these parameters have been drawn out for different models.

- **Accuracy**

Use Evaluation Metrics to make the right Decision and run Agile. It is the outcome of the true positives and true negatives., It is also defined as performance measure for model. It can be calculated as

$$\mathrm{Acc} = [(\mathrm{TP} + \mathrm{TN})/(\mathrm{TP} + \mathrm{FN} + \mathrm{TN} + \mathrm{FP})] * 100\%$$

- **Training and Testing Accuracy**

The term "training accuracy" refers to the utilization of identical pictures for both training and testing. The test accuracy denotes the trained model's ability to recognize images that were not utilized during training.

- **Validation Accuracy**

It is the type of accuracy which is measured on the sample of data held back while training the model. This kind of accuracy gives an estimate of the model skill.

- **Precision**

It is defined as the fraction of the TP labelled by sum of the true positive and false positives.

$$\mathrm{Precision} = \mathrm{TP}/(\mathrm{TP} + \mathrm{FP})$$

- **Recall**

The recall is defined as fraction of the sum of positive by the number of positive accurately classified as Positive. It is the capability of the model to detect the correct positive and is defined as

$$\mathrm{Recall} = \mathrm{TP}/(\mathrm{TP} + \mathrm{FN})$$

- **F1-Score**

It is known as the "harmonic mean" of the model's recall and accuracy. In other words, it is the weighted average of Precision and Recall. Typically, increasing accuracy causes recollection to decline, and vice versa. The authors occasionally want to consider both precision and recall at once. F1 score can be calculated as

$$\text{F1-Score} = 2 * (\text{Precision} * \text{Recall})/(\text{Precision} + \text{Recall})$$

The authors can compute the F1 Score for each class in a multi-class classification since the authors know the Precision and Recall for each class.

14.4.3 Confusion Matrix

A confusion matrix is used to gauge how well the classifier or model is doing. An error matrix is a list of all correct and incorrect predictions. Both the actual values and the values that the classifier had predicted are disclosed. Its main function is to rate the effectiveness of classification in machine learning systems, especially supervised learning algorithms. As shown below, the authors would have a 2 × 2 matrix with four values for a binary classification problem. Below is an illustration of a confusion matrix.

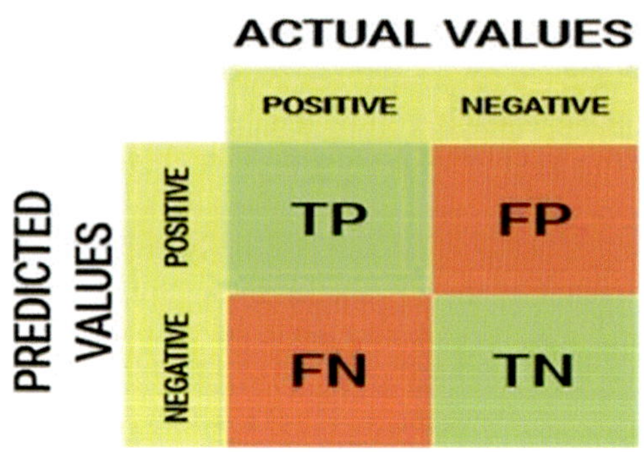

True Positive (TP)

- When the predicted and observed values agree.
- When the value matched what the model predicted and was favourable.

True Negative (TN)

- When the actual value and model predictive was negative and same.

False Positive (FP)—Type 1 error

- When the model create a positive result but the actual value was negative. Also referred to as the Type 1 mistake.

False Negative (FN)—Type 2 error

- When the model predicted a negative value but the actual value was positive. Also known as the Type 2 error.

14.5 Experimentation and Results

The authors now begin training our model with the FER-2013 dataset. As discussed earlier, the training set would consist of 28,821 images. There are already labels provided so that tells the model the kind of expression each image input has. For this example, the authors make it go through with 100 epochs as the authors think this would be more convenient. While the training is processing, the authors can observe that the accuracy of the model keeps on increasing. The final accuracy that the authors achieve by training the model is ≈89%. The authors can now go ahead and use the test set.

Method	Accuracy rate (%)
GoogleNet	65.20
(VGG) + (SVM)	66.31
Convolution + inception layer	66.40
BW (bags of words)	67.40
Attentional ConvNet	70.02
ARM (ResNet-18)	71.38
ResNet	72.40
VGG	72.70
CNN (this work)	89.23

We now begin testing our model with the FER-2013 test set. Our test set consists of 7067 images. For testing the syntax is model.predict (test_set). After running the code, we compare the output with the test labels. With this comparison we can tell how accurate our model turned out to be. The resulting accuracy we got after testing our model came out to be ≈90%

The results of plotting the confusion matrix and classification report are given in the table below. How the confusion matrix looks like can be roughly seen. The predicted labels outline the results of the model's predictions of the expressions while the actual labels outline the actual expression of the image. There is also a heatmap that depicts the difference in the range of accuracies of each predicted label and the actual label (Fig. 14.17).

Classification report of all the scores is given below. This points out the precision of the various labels and shows the recall, F-1 score and support (Figs. 14.18 and 14.19).

	Predicted happy	Predicted angry	Predicted disgust	Predicted fear	Predicted surprise	Predicted sad	Predicted neutral
Actual happy	543	9	87	55	121	128	18
Actual angry	20	70	8	1	4	5	3
Actual disgust	153	3	444	36	123	193	66
Actual fear	41	1	34	1540	103	71	35
Actual surprise	98	2	58	94	753	188	23
Actual sad	119	5	111	45	210	632	17
Actual neutral	18	3	62	47	27	17	623

[47] #CLASSIFICATION REPORT PLOTTING

Fig. 14.17 The output snapshot of the confusion matrix table

	Predicted happy	**Predicted angry**	**Predicted disgust**	**Predicted fear**	**Predicted surprise**	**Predicted sad**	**Predicted neutral**
Actual happy	543	9	87	55	121	128	18
Actual angry	20	70	8	1	4	5	3
Actual disgust	153	3	444	36	123	193	66
Actual fear	41	1	34	1540	103	71	35
Actual surprise	98	2	58	94	753	188	23
Actual sad	119	5	111	45	210	632	17
Actual neutral	18	3	62	47	27	17	623

Fig. 14.18 Confusion matrix

Classification Report

	Precision	Recall	F1-score	Support
Happy	0.49	0.63	0.55	961
Angry	0.70	0.60	0.65	111
Disgust	0.60	0.39	0.47	1018
Fear	0.84	0.84	0.84	1825
Surprise	0.55	0.63	0.59	1216
Sad	0.51	0.47	0.49	1139
Neutral	0.79	0.77	0.78	797

Based on the results which we got in our research it indicates that application deep learning can be useful to predict Facial emotions. Our system can be utilized in many ways to benefit efficiency. In the future, we may investigate additional deep learning

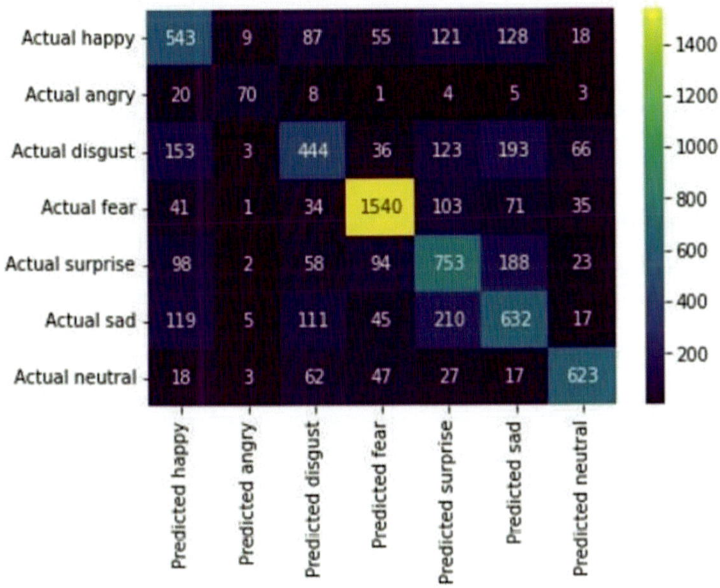

Fig. 14.19 Heatmap of confusion matrix

techniques to see more efficient results. This can be beneficial in developing AI that can read human facial expressions which will significantly help human–machine interaction.

14.6 Conclusion and Future Scope

In this study, the authors have done fine-tuning on deep learning models and proposed a small-sized CNN for the classification of Emotions in Facial Emotion Recognition dataset.

This work achieves state of the art accuracy of 90% and used the Dataset which consist of approx. 34 thousand of images with the proposed model which is based on Convolution. Different types of deep-learning models were used but the highest accuracy was achieved by CNN.

This proposed study has given better results than most Emotion Classification methods used in the literature survey. In future work, I would like to extend our work on different datasets and will try to improve the performance in the classification of Sentiment.

References

Aggarwal T, Sharma N, Aggarwal N (2023) Gunshot detection and classification using a convolution-GRU based approach. In: Noor A, Saroha K, Pricop E, Sen A, Trivedi G (eds) Proceedings of emerging trends and technologies on intelligent systems. Advances in intelligent systems and computing, vol 1414. Springer, Singapore. https://doi.org/10.1007/978-981-19-4182-5_8

Al-Halah Z, Aitken A, Shi W, Caballero J (2019) Smile, be happy:) emoji embedding for visual sentiment analysis. In: Proceedings of the IEEE/CVF international conference on computer vision workshops, pp 0–0

Corchs S, Fersini E, Gasparini F (2019) Ensemble learning on visual and textual data for social image emotion classification. Int J Mach Learn Cybern 10(8):2057–2070. Springer Science and Business Media LLC

Ding H, Zhou SK, Chellappa R (2017) Facenet2expnet: regularizing a deep face recognition net for expression recognition. In: 2017 12th IEEE international conference on automatic face and gesture recognition (FG 2017). IEEE, pp 118–126

Gonçalves P, Araújo M, Benevenuto F, Cha M (2013) Comparing and combining sentiment analysis methods. In: Proceedings of the first ACM conference on online social networks—COSN '13. ACM Press

Hamester D, Barros P, Wermter S (2015) Face expression recognition with a 2-channel convolutional neural network. In: 2015 International joint conference on neural networks (IJCNN). IEEE, pp 1–8

Hasan A, Moin S, Karim A, Shamshirband S (2018) Machine learning-based sentiment analysis for twitter accounts. Math Comput Appl 23(1):11. MDPI AG

Kaya M, Fidan G, Toroslu IH (2012) Sentiment analysis of Turkish political news. In: 2012 IEEE/WIC/ACM international conferences on web intelligence and intelligent agent technology. IEEE

Khorrami P, Paine T, Huang T (2015) Do deep neural networks learn facial action units when doing expression recognition? arXiv preprint arXiv:1510.02969v3

Kumar A, Srinivasan K, Cheng W-H, Zomaya AY (2020) Hybrid context enriched deep learning model for fine-grained sentiment analysis in textual and visual semiotic modality social data. Inf Process Manag 57(1):102141. Elsevier BV

Liu X, Kumar BV, Jia P, You J (2019) Hard negative generation for identity-disentangled facial expression recognition. Pattern Recognit 88:1–12

Liu X, Kumar B, You J, Jia P (2017) Adaptive deep metric learning for identity-aware facial expression recognition. In: Proc. IEEE Conf. Comput. Vis. Pattern Recognit. Workshops (CVPRW), pp 522–531

Machajdik J, Hanbury A (2010) Affective image classification using features inspired by psychology and art theory. In: Proceedings of the international conference on multimedia—MM '10. ACM Press

Meng Z, Liu P, Cai J, Han S, Tong Y (2017) Identity-aware convolutional neural network for facial expression recognition. In: 2017 12th IEEE international conference on automatic face and gesture recognition (FG 2017). IEEE, pp 558–565

Ortis A, Farinella GM, Battiato S (2020) Survey on visual sentiment analysis. IET Image Process 14(8):1440–1456. Institution of Engineering and Technology (IET)

Ortis A, Farinella GM, Torrisi G, Battiato S (2021) Exploiting objective text description of images for visual sentiment analysis. Multimed Tools Appl 80(15):22323–22346. Springer Science and Business Media LLC

Pall A, Sharma N, Sharma K, Wadhwa V (2022) A systematic review of deep learning techniques for semantic image segmentation: methods, future directions, and challenges. In: Handbook of research on machine learning

Priyavrat SN, Sikka G (2021) Multimodal sentiment analysis of social media data: a review. In: Singh PK, Singh Y, Kolekar MH, Kar AK, Chhabra JK, Sen A (eds) Recent innovations in computing. ICRIC 2020. Lecture notes in electrical engineering, vol 701. Springer, Singapore. https://doi.org/10.1007/978-981-15-8297-4_44

Sharma R, Sharma N (2021) Application of machine learning in precision agriculture. In: Mangla M, Satpathy S, Nayak B, Mohanty SN (eds) Integration of cloud computing with internet of things. https://doi.org/10.1002/9781119769323.ch8

Song K, Yao T, Ling Q, Mei T (2018) Boosting image sentiment analysis with visual attention. Neurocomputing 312:218–228. Elsevier BV

Sun M, Yang J, Wang K, Shen H (2016) Discovering affective regions in deep convolutional neural networks for visual sentiment prediction. In: 2016 IEEE international conference on multimedia and expo (ICME). IEEE

Vadicamo L, Carrara F, Cimino A, Cresci S, Dell'Orletta F, Falchi F, Tesconi M (2017) Cross-media learning for image sentiment analysis in the wild. In: Proceedings of the IEEE international conference on computer vision workshops, pp 308–317

Yang H, Ciftci U, Yin L (2018) Facial expression recognition by de-expression residue learning. In: Proceedings of the IEEE conference on computer vision and pattern recognition, pp 2168–2177

Chapter 15
A Study of Deep Learning Methods for Automatic Cancer Detection and Classification in Histopathological Whole-Slide Images

Javaid Ahmad Wani, Nonita Sharma, Manik Rakhra, Arun Singh, and Reena

15.1 Introduction

Histopathology is essential in medical imaging. Consequently, automatic histopathological computer vision has a significant influence on the overall affordability, reliability, and accessibility of healthcare. In the first stages when a person experiences symptoms of any disease or cancer he/she undergoes the examination under CT-Scans MRIs or X-ray. Then the detection of suspicious things under that scan is still doubtful to radiologists and examiners. The samples of tissues are taken from the organ and sent to histopathological labs. Thus, Histopathology or Biopsy test is done for the final confirmation of diseases or cancer especially in ML, specifically DL-based models. DL models have outperformed in a variety of disciplines, along with clinical applications and profound features in healthcare.

J. A. Wani (✉)
Qualcomm India Private Limited, New Delhi, India
e-mail: Javaidcse14@gmail.com

N. Sharma
Indira Gandhi Delhi Technical University for Women, Delhi, India
e-mail: Nonitasharma@igdtuw.ac.in

M. Rakhra · A. Singh
School of Computer Science and Engineering, Lovely Professional University, Phagwara, Punjab, India

Reena
Computer Science, Edge Hill University, Ormskirk, Lancashire, England
e-mail: reenae@edgehill.ac.uk

N. Sharma et al. (eds.), *Big Data Analytics in Intelligent IoT and Cyber-Physical Systems*, Transactions on Computer Systems and Networks,
https://doi.org/10.1007/978-981-99-4518-4_16

15.1.1 Whole Slide Imaging Technology in Histopathology

Under the heading of "digital pathology," a number of technologies have been developed, such as whole slide imaging (WSI), which involves scanning whole histologic sections using a digital slide scanner to create "digital slides". With WSI, wide-field and high-resolution microscopy pictures may be quickly collected, providing incredibly precise information regarding tissue morphology. Multiple histological parameters can be quantified using WSI and computational technologies, which offers insights into disease pathophysiology and tissue/organ biology.

WSI scanner is a robotic microscope that can digitize an entire glass slide by combining or stitching together individual pictures using software. The collected image of a slide can be seen, zoomed, and moved around spatially on a computer screen after the digital file has been retrieved, much like a traditional light microscope. The two processes that make up WSI are, in general, the digitalization of glass slides using specialized hardware (a scanner made up of an optical microscope and a computer-connected digital camera), which produces a digital image, and the viewing and/or analysis of that image through software that is in charge of image creation and management. The images obtained using this technology are frequently referred to as whole slide images, WSIs, whole slide scans, or digital slides. Each virtual image represents a complete glass slide. The last ten years have seen steady advancements in WSI technology, and today there are a number of commercially accessible scanners that can capture digital images. To create digital photographs, there are primarily two methods. While some models employ a line-scanning approach that produces linear scans of tissue sections, most models use a tiling system, in which the original slide is obtained as tiles. In order to make a single digital image of the histologic slice using either method, the tiles or line scans must be stitched together and smoothed down using a specialist software.

The field of transplantation medicine is another one where the use of WSI as a tool for histological assessment is growing. The liver is the second most transplanted organ in the US, just after the kidney.

Histopathological examination is the diagnostic and research tool for tissue disorders. An integrated suite of histopathological samples has greatly aided doctors and researchers in the world of clinical research. The identification of cancerous tissues is a critical issue for clinicians in providing appropriate oncology treatments. A whole slide image (WSI) is a digitized scanning of the tissues on the glass slide that allows the samples taken to be stored digitally on the computer system in the form of a digital picture. India is witnessing more than one million cases of breast cancer per year. WSI processing and storage have greatly aided professionals while also encouraging researchers to develop more reliable and efficient fully automated analyses, and cancer diagnosing models.

15.1.2 Deep Learning

It is a Machine Learning discipline that is entirely dependent on ANNs (Artificial Neural Networks). The neural networks are depicted to mimic the human brain; likewise, deep learning is also the human brain mimic (Sharma et al. 2021). The beauty of deep learning is that we do not have to program everything in deep learning explicitly. In deep learning, we have to string a model on the training dataset and also improve it till it predicts nearly correctly on both the testing and validation datasets. Deep learning models can focus on precise features on their own, with hardly a small input from the programmer, and are quite beneficial in sorting out the dimensionality issue (Kaur et al. 2022). The DL model can be broadly divided into two components.

Feature extraction phase: In this phase, we train deep architectures on a large dataset by extracting a feature using the cascade of different layers. We simply input the images and then feed them to different layers (Sharma et al. 2018).

Classification: In this phase, the images are classified into the respective class. Machine learning is a subset of Artificial Intelligence, while Deep Learning is a subset of Machine Learning, as seen in the Fig. 15.1.

15.2 Literature Survey

In (Kather et al. 2019), authors evaluated various DL models, fine-tuned them and trained on the NCT-HE-100K, which was composed of 100,000 HE patches, and tested on the CRC-VAL-HE-7K, which contains about 7000 HE patches. VGG19 has shown outstanding performance 94.3% in classifying the 9 different classes of cancer. Authors in Mishra et al. (2018) have carried out a comparative study of different deep learning models and a self-build model for the classification of

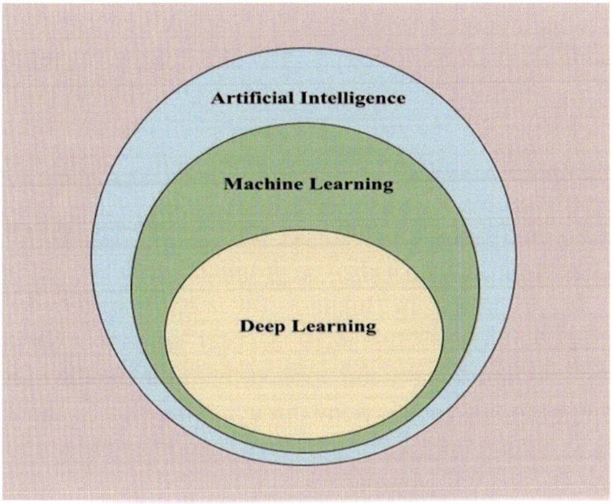

Fig. 15.1 Relation between AI, ML and DL

the Osteosarcoma tumor images. The main aim was to come up with an efficient and accurate classification model. The self-build model has shown a magnificent performance than the fine-tune models. The dataset utilized contains about 64k image samples, which were annotated manually with the help of experienced pathologists. The accuracies of VGGNet, LeNet, AlexNet, and the self-build model were recorded as 67%, 67%, 73%, and 92% respectively.

Further, authors in Babaie et al. (2017) have introduced a dataset under the title of Kimia path24, which has 24 different types of tissue classes that were selected on the bases of their texture pattern. Three different methods employed for the retrieval and classification of the patches. The Bag-of-words method did not perform well, the local binary pattern (LBP) and CNN-based method have shown relatively good accuracy of 41.33%, and 41.80% respectively. Irum et al. performed patch-based DL that was implemented for the detection and classification of breast cancer and the relatively small dataset consists of about 300 images of four different classes of breast cancer, which were taken from the publicly available were utilized, and 70k patches were created from the same dataset for training and validating the methods and achieved an accuracy of 86% (Modeling 2021). Similarly authors in Kumar et al. (2017) introduced a dataset that is freely accessible from the KIMIA Lab official site. The dataset is composed of about 960 histopathological image samples of 20 different types of tissues. The LBP method showed a slightly good performance of 90.62%, BoVW achieved better accuracy followed by CNN of 96.50%, and 94.72%, respectively. Authors in Tsai and Tao (2020) collected Colorectal cancer (CRC) histopathological samples and utilized them as the exploratory dataset to validate optimized parameters, and the potential of the 5 most widely DL models were used to accurately classify colorectal cancer tissues evaluated by comparing performance on CRC-VAL-HE-7K, and CRC-VAL-HE-100K datasets and achieved an accuracy of 77 and 79%.

Sobhan et al. used the Kimia-Path 24C dataset which contains 24 WSIs from various tissue classes (Riasatian et al. 2021). The whole dataset is designed to resemble retrieval work activities in clinical practice. Color is a vital feature in histopathology and Color, was completely disregarded in the Kimia-Path24 dataset, with all patches stored as gray-scale since retrieved from Colored WSIs. In the Kimia-Path24C dataset, the color feature has to give great significance. To extract the interesting patches, K-means clustering and the Gaussian Mixture Model (GMM) segmentation algorithms were employed. VGG16, Inception, and DenseNet models were used as feature extraction to provide further initial findings for setting a benchmark and have achieved accuracies of 92%, 92.45%, and 95.92% respectively. Computer Science has given promising results over the decades in science and technology. Authors in Bukhari et al. (2020) acknowledged the use of computers in medical diagnosis first in 1995. Later, authors in Khvostikov et al. (2021) developed digital chest X-rays and applied them for the diagnosis of lung cancer. The cancer diagnosis was mostly done through X-rays throughout the 70 and 80 s. Many studies have been proposed for the classification and identification of various cancers using the DL and ML approaches.

Further research work done by Borkowski et al. (2019) compared models for the classification of colon cancer from the LC25000 dataset and achieved the accuracy of 99.67% for MobileNetV2 and a loss of 1.24%. Sun et al. carried out their research on the LIDC lung cancer database and used CNN for the examination of lung nodules and attained an accuracy of 89.9% in their study. Here (Howard et al. 2017), the authors used six different datasets in their study and compared CNN with DFCNet for classifying lung cancer and attained accuracies of 77.6 and 84.6%.

15.2.1 Research Gaps

There are some problems in the detection process:

1. Small nodules and less contrasted nodules are difficult to observe in CT-Scans.
2. Some nodules/cancer-causing cells can be missed by the models.
3. Some tissues of the lungs and some arteries can be wrongly detected as lung nodules.
4. Using the models directly on the data can give false positives often.
5. The key challenge in this is Whole Slide Imaging (WSI) that develops gigapixel files with the resolution of 100,000 × 100,000 causing morphological variance which causes difficulty in the visual understanding and learning from images.
6. The above-surveyed studies and algorithms have been performed on smaller datasets.
7. The quality of image enhancement in WSI and its application with CNN is required.
8. Various models like Alex Net, and VGG16 face vanishing gradient problems.

15.3 Methodology and Implementation

15.3.1 Problem Definition

WSI processing and storage have greatly aided professionals while also encouraging researchers to develop more reliable and efficient fully automated analysis diagnosing car models. ML, specifically DL-based models, strengthen medical imaging analysis software solutions. DL with a CNN is a rapidly growing area of histopathologic image analysis (Filho et al. 2018).

The continuous technological advancements in digital scanners, image visualization methods, as well as the incorporation of AI-powered methodologies offer opportunities for new and emerging technologies (Masood et al. 2018a). Their advantages are numerous, including easy accessibility via the web, exclusion of physical storage, and no risk of smudging depletion or slide breakdown, to name a few. Several hurdles, including heavy price, technical difficulties, and specialist reluctance to adopt an innovation, have hampered it being used in pathology (Masood et al. 2018b).

As per reports in cancer statistics, millions of people die of lung cancer and colon cancer as both have low survival chances. The reasons behind this are smoke, dysfunction of the lungs, and alcohol. It is almost impossible to find a way of healing cancer in its terminal stages. Early diagnosis of cancer can increase the survival rate. After years of advancement in the field of technology computed tomography was able to develop high resolution in the images and in the CT scans needed to be observed and scanned by radiologists. Screening of X-rays and CT scans only takes plenty of time. Substantial technological advancements have resulted in the implementation of novel digital imaging potential solutions in Histopathology WSI is gaining popularity among many pathologists for diagnosing, academic, and scientific research (Bejnordi et al. 2018). Automatic analysis of histopathology image data has greatly aided doctors and researchers in medicine, primarily because of the abundance of labeled data and technology that really can be used for analysis purposes, and specialists from various fields of computer. Despite the advancements in the field of healthcare, there is a scope for further research to increase the survival rate and early diagnosis of cancer (Zoph et al. 2017).

15.3.2 Proposed Work Methodology

The overall methodology of our proposed work is discussed below.

Step 1: Data Pre-processing

Pre-processing refers to the removal of extra data after giving it to the classifier. The motive of preprocessing is to enhance data by suppressing unwanted distortions or enhancing particular visual properties that are important for any further analysis and processing. Some main forms of data preprocessing involve outlier identification, missing value treatments, and removing undesired or noisy data. It is the filtering of data done to improve the accuracy of the model.

Step 2: Data Augmentation

Data augmentation refers to significantly increasing the number of images without taking the new images. We augment the image dataset by flipping the images from left to right or from right to left, rotating, zoom-in or zoom-out, and cropping. These are some data augmentation techniques that I have used in our proposed work. This technique is also used to reduce overfitting.

Step 3: Data Splitting

Data splitting is the process of dividing the data into two or more subsets. It is the most important feature of data science, especially for constructing data-driven

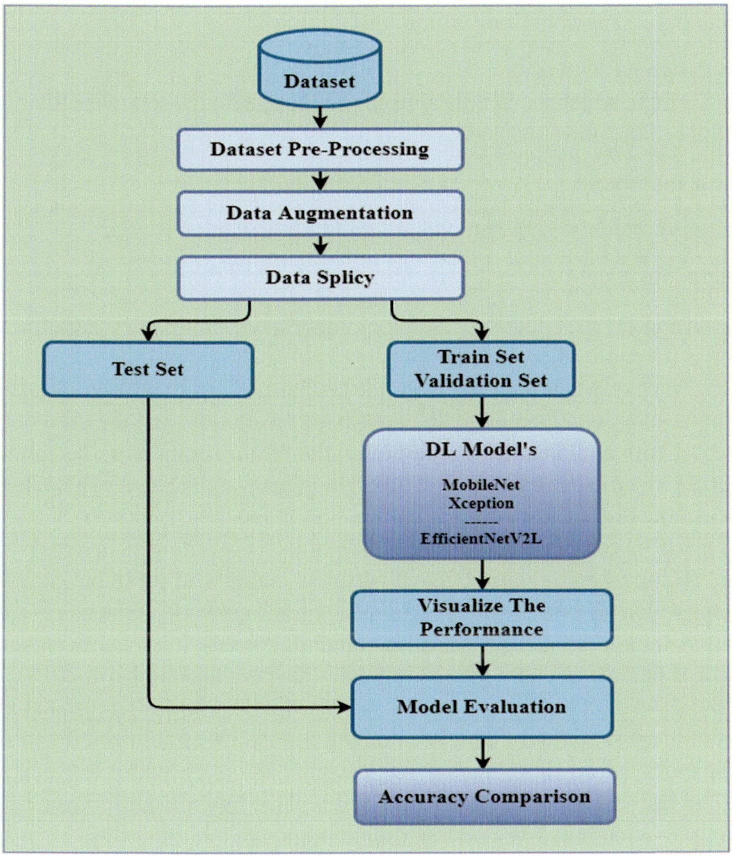

Fig. 15.2 Proposed methodology

models. This method aims to provide help in the design of data models and data-driven processes. We have divided the dataset into three subsets: training, validation, and test sets.

The proposed methodology is visually depicted in Fig. 15.2.

15.3.3 Models Used in Our Study (DL Models)

MobileNetV2

The Mobile NetV1 was introduced to decrease complexities, and it was developed so light that it was able to be used on mobile devices. It reduces the required memory. Mobile NetV1 uses depth wise separable convolution. MobileNetV2 uses three structures in it.

(1) Depth wise Separable Convolution

This operation has two parts.

Depth wise in which the filter is applied per channel, and a pointwise filter is used asan output of the previous phase.

(2) Linear Bottleneck

Pointwise is combined with bottleneck, and linear activation is used

(3) Inverted Residual

The expansion layer is added at the block input's beginning, and output is added together as output for the whole block.

The two blocks in it have three layers: the first layer is in the convolution layer with ReLu proceeding with the next with depthwise convolution, and the third one does convolution with nonlinearity. We have used the Adam compiler in this model and the categorical_crossentropy loss function. The general architecture of MobileNetV2 is shown in the Fig. 15.3.

This model is similar to the Mobile Net except it is based on an inverted residual structure (Hatuwal and Thapa 2020). It performs depth-wise separable convolution operations which are building blocks for many neural network architectures (Kieffer et al. 2017). It improves the performance of mobile models. It is a lightweight model and has less parameter count than MobileNet (Komura and Ishikawa 2018). There are 2,264,389 parameters in this model out of which 2,238,597 are trainable. The weights of the first 20 layers have been frozen and then we fine-tuned the model by using the cross-entropy loss function and Adagrad algorithm with a learning rate of 0.001. The last layer has been truncated and replaced with a dense layer having a SoftMax activation function to classify the images into five classes.

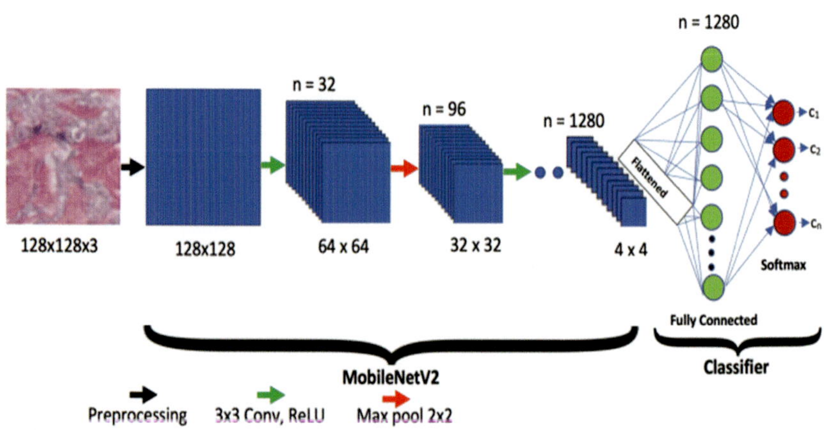

Fig. 15.3 Architecture of MobileNetV2

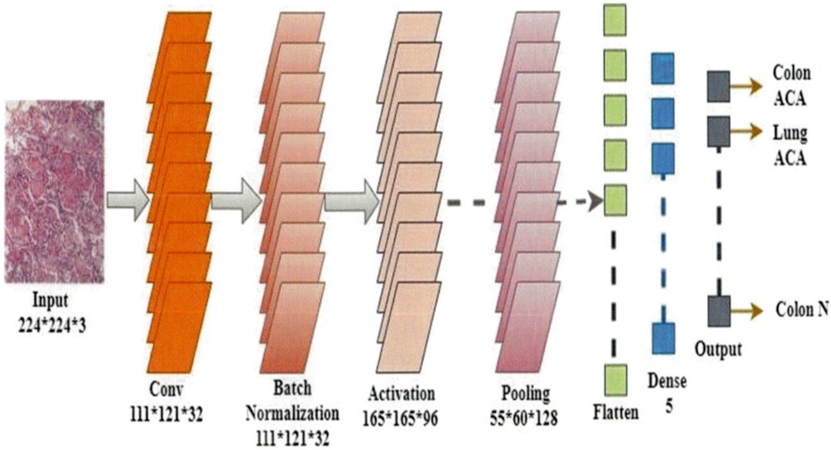

Fig. 15.4 Architecture of Xception model

Xception Model

It was developed by Google scholars and is inspired by the inception model and ResNet. The convolution operations are depthwise and pointwise in it. They perform shortcut operations like ResNet. We have used the activation function as softmax in this model. We have used categorical_crossentropy as a loss function in it. In this model there are 84,936,983 total parameters out of which 84,740,315 are trainable. The architecture of Xception is shown below in Fig. 15.4.

This model is inspired by Inception V3 and stands for 'Extreme Inception' (Farahani et al. 2015). Both the models have the same parameters but the performance measure is dependent upon the use of parameters (Chollet 2017). It consists of 36 linear stacks of depth-wise convolutional layers and it has residual connections. There are 20,871,725 parameters in this model out of which 20,187,197 are trainable. In this model the weights of the first 16 layers have been frozen and then finetuned the model by using the cross-entropy loss function and Adam algorithm with a learning rate of 0.001 and the last layer has been truncated and replaced with a dense layer having a SoftMax activation function to classify the images into five classes.

EfficientNetV2L

This family of CNNs has higher speed in training than previous ones. EfficientNetV2 uses a 3 × 3 kernel size. In this model, we have used the global average pooling 2D layer and activation function as SOFTMAX and compiler as SGD. The architecture of EfficientNetV2L is given in Fig. 15.5.

This model is inspired by Inception V3 and stands for 'Extreme Inception' (Chollet 2017). Both the models have the same parameters but the performance measure is dependent upon the use of parameters (Urban et al. 2018). It consists of 36 linear stacks of depth-wise convolutional layers and it has residual connections. There are 20,871,725 parameters in this model out of which 20,187,197 are trainable. In this

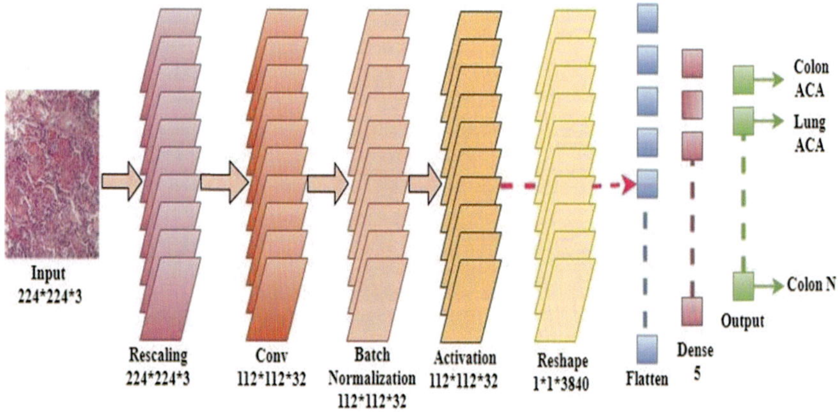

Fig. 15.5 Architecture of EfficientNetV2L

model the weights of the first 16 layers have been frozen and then finetuned the model by using the cross-entropy loss function and Adam algorithm with a learning rate of 0.001 and the last layer has been truncated and replaced with a dense layer having a SoftMax activation function to classify the images into five classes.

NasNet Large

This model is one of the finest models of the CNN family, which has been trained on millions of images. The model is acquainted with classifying thousands of images of large datasets, and it reuses this approach in the classification of different problems. The input shape is (331, 331, 3) in this model by default. In this model, we have used SOFTMAX as an activation function. There are 84,936,983 total parameters out of which 84,740,315 are trainable.

We have popped the last three-layer and added a new dense layer that contains 38 neurons representing the number of classes. The last layer we added is the soft-max layer to classify images. We utilized the Adam optimizer and categorical cross-entropy. The learning rate parameter was set to 0.0001, as it is always set between 0.01 and 0.0001. The minimum the learning rate, the minimum will be the loss. The model trained 30 epochs were used to update the internal model parameters., and the batch size was set to 128 as the size of the dataset is very large.

A pre-trained version of this network has learned that over a million images from the ImageNet dataset may be loaded. The pre-trained model can categorize pictures into 1000 classes. The network accepts an image with a resolution of 224 by 224.

The general architecture of NasNet Large is shown in Fig. 15.6.

This model is inspired by the Neural Architecture Search (NAS) framework. This convolution neural network was designed by stacking copies of convolutional layers on the ImageNet dataset. NasNet achieved the accuracy of 82.7 and 96.2% on the ImageNet dataset. Different versions of NASNets can be created by varying the convolution layers and the size of filters that can outperform the accuracies of

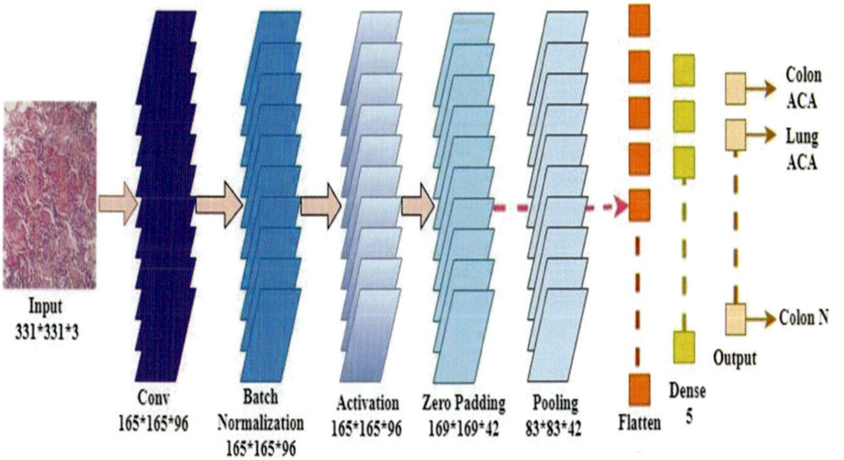

Fig. 15.6 Architecture of NasNet Large

human effort models (Modeling 2021). In this model there are 84,936,983 total parameters out of which 84,740,315 are trainable; This model has been fine-tuned by truncating the last two layers and replacing them with the dense layer having a SoftMax activation function for the classification of images into 5 classes. Adam algorithm has been used in this model with a learning rate = 0.0001 and loss function as categorical_crossentropy in this model.

CNN (Proposed Model)

In this study, there is a small-sized proposed CNN model of just 15 layers which is very small in comparison to the fine-tuned models used in this study but shows good results in classifying cancers. The dimensions of the input image are (224, 224, 3). There are a total of five convolution layers in this model. The output of every convolution layer is activated with the activation function. In this study ReLU activation function is used except for the last dense layer where the SoftMax activation function is used. There are a total of 42,756,997 parameters in this model out of which 42,754,117 are trainable. The first layer in our model is the convolution layer which analyzes the input image using 32 kernels of size 3 × 3. The convolution layers are used to detect patterns from the given input. The output from the previous layer is activated using the activation ReLU function. It is used as the default activation function in various types of neural networks due to its better performance. The output from the previous layer is normalized by batch normalization which helps in the regularization of the models. It eliminates the problem of overfitting in the model. This normalization is often placed after the convolution layer. The next layer is a max-pooling layer which has a pool size of 3, 3 which reduces the dimensions of the image thereby selecting the maximum element from the feature map. The next layer consists of dropouts, and they prevent overfitting and enhance the learning mechanism of the model. But switching off neurons above 50% can cause poor learning of the model. The second and third convolution layers perform convolution operations using 64 kernels of size 3×3 followed by the activation function, batch normalization, and dropouts. The fourth and fifth convolution layers use 128 kernels of size 3×3 followed by the abovementioned layers. The last two dense layers are stacked which contain 1024 neurons. In the last layer, we have used the SoftMax activation function to predict the multinomial probability distribution of 5 classes.

15.3.4 Model Evaluation

Model assessment is the method of analyzing the deep learning model's performance, as well as its strengths and limitations, using various evaluation criteria. Model evaluation is critical for determining a model's performance during the early stages of research, as well as for model monitoring. The model performance of the proposed models is listed in Tables 15.1 and 15.2.

15.3.5 Accuracy Comparison

We have implemented five different fine-tuned CNN models namely MobileNetV2, EfficientNetV2L, CNN, Xception, NasNet Large and achieved the validation accuracy of 98%, 96%, 94%, 99%, 99% respectively.

15.4 Experimentation and Results

Extensive experiments are carried out in this chapter on two publicly available datasets to assess the efficacy of the proposed automated cancer detection and classification systems developed in this thesis. The Experimentation and results are discussed in two parts, firstly on the KimiaPath24C dataset followed by the LC25000 dataset.

15.5 Conclusion and Future Scope

In this study, I have done fine-tuning on deep learning models and proposed a small-sized CNN for the classification of cancer images in the LC25000 dataset and KimiaPath24 dataset. Comparison of Five deep learning models have been done on both datasets. These models have achieved accuracies ranging from 93 to 99%. The highest accuracy of 99.69% was achieved in NasNet Large in the LC25000 dataset which has the size of 343 MB and has 88.9 M parameters and is 533 layers deep and on

Table 15.1 Validation accuracy of Kimiapath24C

Epochs	Accuracy									
	CNN		MobileNet		NasNet Large		Xception		EfficientNetV2L	
	Train	Validation	Train	Validation	Train	Validation	Train	Validation	Train	Validation
30	91.17	79.16	99.65	91.35	0.98	0.96	99.65	94.45	98.65	96.96
25	89.17	80.17	99.21	90.08	0.97	0.98	99.73	92.95	98.29	98.46
20	85.50	47.60	99.53	89.99	0.97	0.96	99.68	95.07	97.78	97.67
15	82.26	66.36	99.26	91.08	0.96	0.94	99.64	93.44	96.48	94.39
10	74.00	58.43	98.91	90.81	0.94	0.96	99.41	93.99	95.17	91.30

Table 15.2 Observed results of proposed models (LC25000)

Epochs	Accuracy									
	CNN		NasNetLarge		MobileNetV2		Xception		EfficientNetV2L	
	Train	Validation	Train	Validation	Train	Validation	Train	Validation	Train	Validation
30	0.9904	0.9435	0.9987	0.9969	0.9964	0.9837	0.9991	0.9999	0.9865	0.9696
25	0.9745	0.9629	0.9999	0.9982	0.9953	0.9629	0.9990	0.9996	0.9829	0.9846
20	0.9814	0.9136	0.9990	0.9741	0.9935	0.9642	0.9995	0.9989	0.9778	0.9767
15	0.9811	0.9167	0.9979	0.8295	0.9904	0.8823	1.0000	0.9999	0.9648	0.9439
10	0.9738	0.7915	0.9988	0.5897	0.9838	0.8236	0.9986	0.9950	0.9517	0.9130

Kimia Path 24 that has the highest accuracy of 96% achieved in Efficient Net V2L which is the latest model of the Efficient Net family i.e. Efficient NetV2L which was developed in 2021 has been used in this study. I have saved h5 files of the architecture and the weights of the models used in this study. This proposed study has given better results in terms of accuracy than most cancer classification methods used in the literature survey. The computer vision-based techniques and deep learning models can assist pathologists to diagnose the different types of cancer at low cost and time. In future work, I would like to extend our work on different datasets of histopathological images and will try to improve the performance in the classification of cancers.

References

Babaie M, Kalra S, Sriram A, Mitcheltree C, Zhu S, Khatami A, et al (2017) Classification and retrieval of digital pathology scans: a new dataset. In: 2017 IEEE conference on computer vision and pattern recognition workshops (CVPRW). IEEE, Honolulu, HI, USA, pp 760–768

Bejnordi BE, Mullooly M, Pfeiffer RM, Fan S, Vacek PM, Weaver DL, et al (2018) Using deep convolutional neural networks to identify and classify tumor-associated stroma in diagnostic breast biopsies. Free PMC article

Borkowski AA, Bui MM, Thomas LB, Wilson CP, DeLand LA, Mastorides SM (2019) Lung and colon cancer histopathological image dataset (LC25000). arxiv logo (Electrical Engineering and Systems Science > Image and Video Processing), 2

Bukhari SUK, Syed A, Bokhari SKA, Hussain SS, Armaghan SU, Shah SSH (2020) The histological diagnosis of colonic adenocarcinoma by applying partial self supervised learning. Medrxiv (the preprint server for health sciences)

Chollet F (2017) Xception: deep learning with depthwise separable convolutions. In: 2017 IEEE conference on computer vision and pattern recognition (CVPR). Honolulu, HI, USA

Chollet F (2017) Xception: deep learning with depthwise separable convolutions. In: Proceedings of the IEEE conference on computer vision and pattern recognition (CVPR)

Farahani N, Parwani A, Pantanowitz L (2015) Whole slide imaging in pathology: advantages, limitations, and emerging perspectives. Pathol Lab Med Int 23–33

Filho AOC, Silva AC, Paiva AC, Nunes RA, Gattass M (2018) Classification of patterns of benignity and malignancy based on CT using topology-based phylogenetic diversity index and convolutional neural network. Sci Direct (Elsevier) 81:200–212

Hatuwal BK, Thapa HC (2020) Lung cancer detection using convolutional neural network on histopathological images. Int J Comput Trends Technol 68(10):21–24

Hirra I, Ahmad M, Hussain A, Ashraf MU, Saeed IA, Qadri SF, et al (2021) Breast cancer classification from histopathological images using patch-based deep learning modeling. IEEE Access 9

Howard AG, Zhu M, Chen B, Kalenichenko D, Wang W, Weyand T, et al (2017) MobileNets: efficient convolutional neural networks for mobile vision applications. arxiv (Cornell University)

Kather JN, Krisam J, Charoentong P, Luedde T, Herpel E, Weis C-A, et al (2019) Predicting survival from colorectal cancer histology slides using deep learning: a retrospective multicenter study. PLOS Med 16(1):e1002730

Kaur S, Rani R, Garg R, Sharma N (2022) State-of-the-art techniques for passive image forgery detection: a brief review. Int J Electron Secur Digit Forensics 14(5):456–473

Khvostikov A, Krylov A, Mikhailov I, Malkov P, Danilova N (2021) Tissue type recognition in whole slide histological images. In: 31st international conference on computer graphics and vision, September 27–30, 2021, Nizhny

Kieffer B, Babaie M, Kalra S, Tizhoosh HR (2017) Convolutional neural networks for histopathology image classification: training versus using pre-trained networks. arxiv (Cornell University)

Komura D, Ishikawa S (2018) Machine learning methods for histopathological image analysis. Comput Struct Biotechnol J 16:34–42

Kumar MD, Babaie M, Zhu S, Kalra S, Tizhoosh HR (2017) A comparative study of CNN, BoVW and LBP for classification of histopathological images. In: 2017 IEEE symposium series on computational intelligence (SSCI). IEEE, Honolulu, HI, USA

Masood A, Sheng B, Li P, Hou X, Wei X, Qin J, Feng D (2018a) Computer-assisted decision support system in pulmonary cancer detection and stage classification on CT images. J Biomed Inform

Masood A, Sheng B, Li P, Hou X, Wei X, Qin J, Feng D (2018b) Computer-assisted decision support system in pulmonary cancer detection and stage classification on CT images. J Biomed Inform 79:117–128

Mishra R, Daescu O, Leavey P, Rakheja D, Sengupta A (2018) Convolutional neural network for histopathological analysis of osteosarcoma. J Comput Biol 25(3):313–325

Riasatian A, Babaie M, Maleki D, Kalra S, Valipour M, Hemati S, et al (2021) Fine-tuning and training of densenet for histopathology image representation using TCGA diagnostic slides. Sci Direct

Sharma S, Juneja A, Sharma N (2018) Using deep convolutional neural network in computer vision for real-world scene classification. In: 2018 IEEE 8th international advance computing conference (IACC). IEEE, pp 284–289

Sharma N, Mangla M, Mohanty SN, Gupta D, Tiwari P, Shorfuzzaman M, Rawashdeh M (2021) A smart ontology-based IoT framework for remote patient monitoring. Biomed Signal Process Control 68:102717

Tsai M-J, Tao Y-H (2020) Deep learning techniques for colorectal cancer tissue classification. In: 2020 14th international conference on signal processing and communication systems (ICSPCS). IEEE, Adelaide, SA, Australia

Urban G, Tripathi P, Alkayali T, Mittal M, Jalali F, Karnes W, Baldi P (2018) Deep learning localizes and identifies polyps in real time with 96% accuracy in screening colonoscopy. Gastroenterology 155(4):1069–1078

Zoph B, Vasudevan V, Shlens J, Le QV (2017) Learning transferable architectures for scalable image recognition. arxiv.org (Cornell University)

Chapter 16
Analysis of Video Summarization Techniques for Resource Optimization in Multimedia Applications

Rakhi Akhare, Subhash K. Shinde, and Monika Mangla

16.1 Introduction

Nowadays, both academics and industry have given Computer Vision a lot of attention (Murugan et al. 2018). The fields of computer science of image processing and computer vision are both quite fascinating. With the aim of automating processes, computers or machines are built to interpret at a high level from the input digital photos or videos. One of the numerous approaches it employs is video processing. Operations like video analysis and video summarization are carried out during video processing. Because of the enormous amount of data that today's media produces, real-time computer vision is becoming more and more common in multimedia applications. One of the often used method for computer vision is video summarization (Murugan et al. 2018).

Although the video was previously used only for TVs and theatres, it is now a commonly used way for the entertainment that are used by many different user groups. The amount of video footage produced has increased significantly as a result of the development of more sophisticated video recording and rendering technology. Based on a recent survey most of the internet traffic includes the video data which require high bandwidth and speed. Also storage and computing of this big data is also a very tedious job (Bora and Sharma 2018a).

R. Akhare (✉)
CSED, Lokmanya Tilak College of Engineering, Navi Mumbai, India
e-mail: rakhiakhare2508@gmail.com

Subhash K. Shinde
Professor and Vice Principal, Lokmanya Tilak College of Engineering, Navi Mumbai, India

M. Mangla
Dwarkadas J Sanghvi College of Engineering, Mumbai, India

N. Sharma et al. (eds.), *Big Data Analytics in Intelligent IoT and Cyber-Physical Systems*, Transactions on Computer Systems and Networks,
https://doi.org/10.1007/978-981-99-4518-4_17

Multimedia information indexing and retrieval has grown in importance over the past few years as a result of the explosive growth in multimedia information, the development of internet communication, and the development of digital video technologies. Numerous studies have been conducted on the retrieval and analysis of videos using different features present in the videos like visual, textual, audio etc. This investigation demonstrates that there is a significant amount of rich heterogeneous information related to video content when designing retrieval applications. The extraction of high-level semantic information from low-level audio or visual data makes it extremely difficult to retrieve information from audio or visual data. The act of summarizing videos is crucial because it makes it possible to browse through vast video collections more quickly and to index and access content with greater efficiency (Basavarajaiah and Sharma 2019).

Analyzing and comprehending the video content is a laborious task due to the enormous size of the video file. Additionally, browsing the video database is time consuming task, which makes the network slow to transmit video. To maximize the usability of these big recordings, the vast video data needs to be managed correctly and efficiently. Any computer vision technique that is used on a huge video becomes extremely computationally and storage-intensive. Since video contains duplicate information in consecutive frames, while doing the video processing in computer vision, all the frames are not needed. Sometimes many of the frames are irrelevant for particular types of applications. In these situations, a video summary can be useful to quickly understand the topic of the video. Therefore, a summary video is useful when a user just has time to briefly scan the video's information (Elkhattabi et al. 2015).

16.1.1 Video Summarization

Video summary is the technique of creating and presenting an insightful abstract or summarized perspective of the entire video. It can also serve as a highlight of the original video routines that assist users with making decisions about deletion, selection, and consumption of content. A collection or sequence of significant keyframes that provides a comprehensive overview of the original video can also be referred to as a video summary.

A video is made up of various components, such as a series of consecutive frames, shots, and scenes. Scenes have a certain amount of shots, and those shots have a number of frames that are gathered together to construct a specific kind of video.

A video summary includes following major components:

- Frames: Frames are temporally ordered images that make up a video and which contain important information are called key frames.
- Video Clips: Video clips are known as video segments which are formed by arranging frames in temporal order.

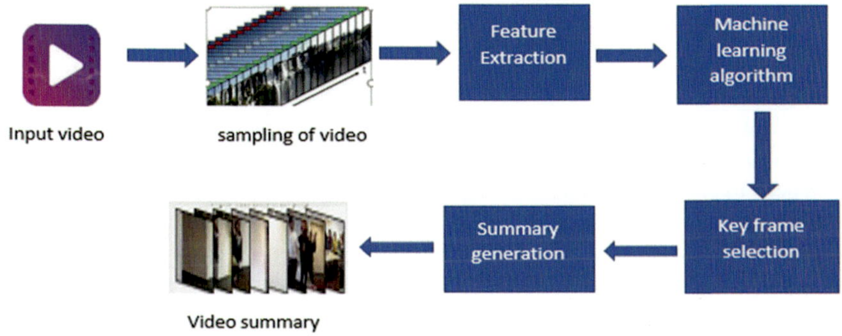

Fig. 16.1 Video summarization steps

- Features: There are basically two types of features present in the video summary, visual and textual. Visual features are the characteristics or patterns that are taken out of a video and used to create a video summary. They provide information about the video's content.
- Textual descriptors: Textual descriptions are summaries of the video's content written in text form.

Video summarization involves preprocessing the incoming video, which is subsequently sampled into a number of frames, as seen in Fig. 16.1. Following sampling, features are taken from individual frames in accordance with the needs of the output summary. Different machine algorithms can be used to identify the semantic relationship between frames after feature extraction. Depending on the dataset's accessibility, supervised machine learning techniques are used when labelled data is available; otherwise, unsupervised techniques are used. Keyframes for the final output can be created from the frames that carry the most crucial information. After choosing your keyframes, organize them all in a logical sequence to create your video summary. Remove all redundant keyframes from the summary to increase its quality.

The crucial phase of the summarizing process is feature extraction. Features represent information present in the videos. Based on features there are mainly two types of features that are considered for producing the summary. These include the created features like textual, HOG, SURF, and learnt features extracted using ANN. Previously different image processing techniques were used by user to extract the features from the images. Deep features are extracted by using different deep learning models which are pretrained models using large datasets. Convolutional Neural Networks (CNNs), RNN are most popular models in the computer vision domain.

The chapter is arranged as Sect. 16.2 discusses the related work in the video summarization domain. Section 16.3 presents different categories of video summarization techniques. Different applications of video summarization are discussed in Sect. 16.4. Section 16.5 talks about the challenges in the area of video summarization. Section 16.6 explores the future work in the field of video summarization, lastly the conclusion and references.

16.2 Related Work

During the past few decades, video summarization has grabbed significant attention among the community of researchers. Resultantly, several algorithms and methodologies have been proposed by researchers. The goal of these proposed methods have been to produce a video summary given the contents of the film. The proposed methods may be primarily categorized into domain specific summarizer or generic summarizers as per authors in Haq et al. (n.d.), Pei et al. (2011). Here, content specific refers to the movies related to domains such as education, news, or sports etc. The approaches defined for domain specific summarization are difficult to apply to regular videos.

Authors in Cheng and Hsu (2006) presented an event-based video summarization that determines the energy of the frame by adding the energy of ground truth frame to that of the present frame. It is followed by identification of frames containing an event. Further, the video is summarized to produce frames containing events. For the same, an algorithm is proposed by authors in Damnjanovic et al. (2008) that combines different frames. Authors in Damnjanovic et al. (2008) also discuss methods for detecting events. The accuracy of the same was enhanced by authors in Kalaivani and Roomi (2017) by employing bootstrap aggregating method.

With the advancement in deep learning, it its application in video summarization is also observed. Authors in Kumar et al. (2017) used Histograms of Oriented Gradients (HOG) and Temporal Difference Maps (TDMap) for video summarization. Carrying the work further, authors in Elharrouss et al. (2020) proposed utilization of the Hierarchical Hidden Markov Model for video summarization. This proposal can be employed for rushed videos for objects and events.

The method for personal video summarization is also carried out by authors in Wang and Ngo (2012). The proposed method is a 2-step process where the first step detects a scene for structuring the movie and the second step creates a video summary using a subset of video scenes. Authors in Miniakhmetova and Zymbler (2015) also proposed a video summarization method that uses perceptual quality and content balance. Authors in Nahian et al. (1708) claimed that video can be clarified using distinctive or common (among multiple frames) actions. Clustering and key frame extraction is also suggested to be employed for video summarization by authors in Pritch et al. (2009), Jadon and Jasim 2019).

Authors in Raut and Gunjan (2020) discussed a static approach to summarize video that uses subsets of the frame for summarization of the input films. Here, the input video is mapped through HOG and thereafter k-mean clustering is used for video summarization. The approach for detecting human face and spectral clustering is proposed by authors in Mohan and Nair (2019). In order to detect these, it used size, quantity, and placement of face to detect the human face.

Chasanis et al. (2009) used the method of shot boundaries detection for removing repetitive frames and Stroboscopic effect. In order to remove the noise and anomalous data, authors in Peker and Bashir (2009) proposed the usage of collaborative representation of frames. The authors suggest employing minimal sparse reconstruction to remove anomalous data. Furthermore, it also uses sparse boundary and average percentage of reconstruction (APOR) to optimize the model.

Similarly, authors in Varghese and Nair (2015) suggested a modality correlation method to summarize videos. It has basic 3 operations namely determination of correlation, fusion of 2 correlations and finally determining high-scoring shots for summarization. Employment of Convolutional Neural Networks (CNN) was also suggested by authors in Wang et al. (2020) as an efficient choice for summarizing surveillance videos. The suggested approach uses shot segmentation to learn features of deep learning. The approach also uses memorability and entropy for summarization. The shot is used as a keyframe since it has a good memorability and entropy score.

Extending the work further, trajectory-based video summarization is used for a dynamic environment. As per this work, fixed cameras can efficiently detect things when deployed in a static background. This object can be tracked for its location using the trajectory method (Muhammad et al. 2020). Using a clustering algorithm, optical flow, and background-subtracted, Lai et al. (Bora and Sharma 2018b) presented a technique for object detection. In order to create a spatio-temporal trajectory after object detection, a sliding window is utilized to integrate the detected items in subsequent frames. Create a video summary by combining all the spatiotemporal trajectories. Where the camera is moving, these strategies are not very helpful.

The sparse dictionary learning method is an unsupervised learning methodology that aids in the creation of data logs and the observation of changes over time. Although this method still requires a lot of time, it will probably be useful for creating summaries that use dynamic summarizing in the future. For uncompelled movies, Lai et al. (2016) described an approach that builds a videography lexicon to display every video as a sequence of words. For dictionary creation, however, shot boundary and clustering algorithms are employed. Authors in Li et al. (2017) presented a sparse representation-based method for categorizing the frame and video's attributes. The video is first split into a number of patches using features and traditional sparse representation, then after acquiring unique frames, a video summary is constructed using the simultaneous block version of the block-based Orthogonal Matching Pursuit (SBOMP) algorithm.

16.3 Analysis of Video Summarization Methodologies

Previously, video summaries were written down. A textual summary used to be confusing in the majority of instances because there was no criteria by which the level of summary needed can be determined. Later, due to its simplicity of interpretation, summarizing the video in video format gained popularity. Figure 16.2 displays the classification of video summarization methods based on several parameters.

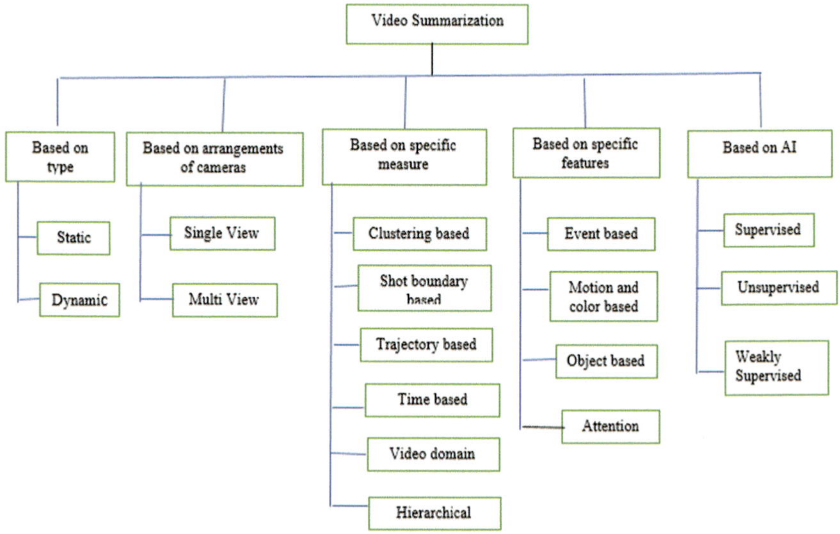

Fig. 16.2 Categorization of video summarization methods

16.3.1 Summarization Based on Types of Summary

Depending upon output summary that is produced, there are two ways to summarize videos:

Static storyboard: The important frames of the video are arranged in a static storyboard in a chronological order. Using a particular technique, the storyboard is created by choosing the important frames in relation to a parameter or the user's interest.

Dynamic summary: It is also called as video skimming. IT is the technique of extracting the most interesting or important video snippets from a film and using them to create a summary of the entire video. Since dynamic summary is only a portion of the original video itself, it can be understood quickly.

16.3.2 Summarization Based on Arrangement of Cameras

Single View Video Summarization: The behaviors of objects in a specific area are recorded or monitored by a single camera, and the recording of that single view is analyzed to produce a summary of the video.

Multiview video summarization: Number of cameras capture the image from a distinct viewpoint, and aim on the specific area or region with various angles when monitoring the unusual activity or object present in a specific zone e.g. CCTV cameras.

The summaries of all the videos taken by various cameras are used to create the final video. This kind of summary can be useful in a variety of settings, including families, businesses, banks, and other institutions. In multi-view videos, many times the contents of the videos are overlaid, therefore producing a summary from many points of view is a particularly difficult task.

16.3.3 Summarization Based on Some Specific Measures

There are several methods for creating video summaries, including general video summaries, domain-specific video summaries, and customized video summaries. Before, generic video summarizing was the most popular type, but now days, domain-specific video summary for things like sports, news, and movies is gaining popularity. As summarization is a subjective task, different users need different summary, so personalized summary can also be produced using interesting clues available in the video.

With the relevant and interesting contents in a video, viewers can create custom video summaries. Generic video summary refers to this method of emphasizing the video without taking into account any of the objects or events depicted in it. By simulating the way that motion, objects, sounds present while playing a video can be used to produce generic summary without knowing the semantic contents of the video.

Videos are frequently summarized using specific items and uncommon occurrences. Videos that are egocentric in nature, in particular, can be summed up by the characters and items they contain. Videos are sometimes represented as graphs with the video's events represented as nodes. Edges serve as a link between the occurrences. The same is true for summarizing events and objects using human faces and human activity. The user would choose a group of key frames, and by simply matching the frames, the summary with related key frames could be produced.

- **Clustering based Video summarization**: Finding related behaviors or attributes inside a frame is done using the clustering method. However, getting rid of erratic frames is also quite helpful. Compared to other methods mentioned earlier, the clustering method efficiently summarized the videos. There are different methods like K-means clustering, spectral clustering, clustering based on similar activities etc.
- **Shot boundaries-based video summarization**: A shot is typically regarded as the primary element of the video and is frequently utilized to depict various scenes and periods of time. These shot boundaries are identified to generate the summary.
- **Trajectory based Video Summarization**: When a video is playing continuously, trajectory-based video summarization can be used to follow moving objects, asthe surveillance footage was taken at a mall or while driving.
- **Time based video summarization**: Both live and previously recorded videos can be summarized. Depending on the time of the summary generator, it can be

either a fixed summary that uses recorded videos or a real-time summary that uses real-time video. The demand for speedy production makes live video summary difficult. A delayed output is an inaccurate output in real-time systems.

- **Video domain-based summarization**: Both the compressed and uncompressed versions of the original video can be used to summarize videos. As it is quicker than decompressed video, the majority of programmers employ a compressed format for videos. To retrieve pixel-level information from compressed video, more time and space are required during decoding. Comparatively, compressed video can only be. The summarizing of compressed domain videos is useful for further resource minimization.
- **Ordered summarization**: A video summary can occasionally be completed utilizing sequential order. Initially, a local frame-by-frame summary of the video's content was provided. Following that, a shot-level summary can be created utilizing this frame-level data. This type of summary can also be called as hierarchical summary

16.3.4 Summarization Based on Recent AI Technique

Now a days Computer Vision using AI is a popular among the researchers to analyze the massive visual data. Machine Learning algorithms and Deep Learning methods play a vital role in the video summarization domain.

Video summarization can be categorized as supervised or unsupervised. Supervised methods train a machine to generate video summaries using labelled data. There are no labelled data accessible when summarizing in an unsupervised manner. In accordance with the demands of the application, the frames are either categorized or clustered. While unsupervised approaches are utilized for clustering, supervised methods are often employed for classification. To create multiple clusters, similar types of frames are grouped together, and then keyframes frames are selected from each group to provide a video summary.

The summary of the videos heavily relies on machine learning. Since decades, a variety of solutions have been put out to perform the video summarization employing handcrafted features. Videos are formed by arrangements of frames in spatiotemporal sequence so, obtaining visual information from video is more challenging than from photos. Therefore, for object detection, event detection, action recognition, and other purposes, the visual attributes are helpful in interpreting the content of the video. These handcrafted features, however, are inefficient and inaccurate.

Deep learning performs well in the field of artificial intelligence (AI), where there is a vast amount of data and a significant need for processing space and capacity. Many deep features in DNN are automatically learned from provided large-scale input. Video input frames are used to select deep features. Many computer vision algorithms prefer deep features to features that were manually produced.

The following categories can be used to roughly group the deep-learning-based video summarizing techniques currently in use:

- **Supervised Learning technique**: In supervised learning techniques, the network is trained using labelled datasets, such as the TVSum and SumMe dataset, which contains frame-level annotations. Given how expensive and time-consuming video annotations are, the main problem with this is the necessity for a highly annotated dataset.
- **Unsupervised Learning technique**: A ground truth annotated dataset is not required for unsupervised learning techniques. It derived some knowledge from it using the internal data found in the provided films. The majority of Reinforcement Learning techniques use adversarial networks in unsupervised learning. An enormous number of training videos are required for training.
- **Weakly supervised Learning technique**: Weakly supervised methods that, like unsupervised methods, strive to reduce the requirement for big quantities of hand-labeled data. Despite their shortcomings in comparison to a complete set of human annotations, less expensive weak labels can nevertheless be employed to build powerful predictive models.

16.3.5 Summarization Based on Properties of the Video

Numerous details are added while a video is being recorded. This information can be used to create a video summary. Internal methods employ the information already present in the video, whereas external methods use additional information in the form of metadata that is automatically supplied when the video is being recorded. Internal and external approaches can occasionally be combined.

For instance, frame-level information i.e. image data and textual metadata can be integrated to improve the overall performance of video summarization. Another crucial aspect of a video that may be used with the visual content for an efficient summary is the audio. In comparison to visual features alone, audio visual features can produce greater results.

There are some other Video summarization techniques which are classified based on features.

- **Event based video summarization**: look for odd or suspicious characteristics. In order to create a video storyboard, the video summarization algorithm is used to combine all the frames with typical situations.
- **Motion and Color Based video Summarization**: Motion and color features are commonly used features to combine similar activities in video together to produce summary.
- **Object Based video summarization**: The frames containing desired objects are combined together from the video and find the most relevant frames among those to make summary.
- **Attention based video summarization**: The user's area of interest is determined using the attention-based models.

A single summary method may fit into several of the aforementioned categories. For instance, a method might involve leveraging internal feature information and an unsupervised machine-learning methodology to create video synopsis based on certain events, objects present in the real time videos. Depending on the method used to create the summary, a summarizing methodology may fit into more than one category.

16.4 Applications of Video Summarization

Wherever there are a lot of videos being recorded, automatic video summarizing is a must. The following is a list of uses for automatic video summarizing methods:

Surveillance Videos

Surveillance videos are running continuously, to extract the important activity from such long recording, such as thefts in shopping centers, traffic accidents, unusual behavior of individuals in polling places or during exams. You may also pick the sections of a film where a specific person or object first appears to summarize that section of the video.

Egocentric Videos

People having dementia, constantly record their activities by using some wearable camera equipment benefit from egocentric videos. However, it is hard to watch the entire days' worth of footage in order to identify a segment that is relevant to short-term memory loss syndrome people on that specific day. The people with these health issues would greatly benefit from a brief interpretation of these video recordings.

Medical Videos

A wide range of possibilities for automatic video summary are presented for medical activities like endoscopy, operations and surgery, diagnostic hysteroscopy etc. The length of the medical video varies greatly depending on how long the treatments are that are being shown. Long medical movies can be summarized to help medical professionals conduct in-depth examinations of the processes and to help medical students learn the procedures. The medical professionals will be able to immediately review the details of an earlier case or look up comparable cases, thanks to this.

Domain Specific Videos

Automatically summarizing news videos enables us to swiftly scan for the key trends reported in the News. One of the fascinating uses of video summary is the automatic creation of movie trailers and Sports video highlights. By representing the videos with summaries, it may also assist individuals in managing the movies that were captured on their mobile devices and facilitating simple access to the videos.

Internet videos are another kind of video that should be condensed before being shown to users. To choose which movie to watch, a summary or other preview of the video would be very beneficial.

Agricultural videos using drones and robots

Recent innovations like drones and robots are equipped with cameras, allowing them to record footage in numerous locations that are off-limits to humans. By summarizing these videos, the interpretation of this new class of videos will be made simpler (Table 16.1).

Table 16.1 Comparative analysis of video summarization techniques

Author and Year	Dataset	Feature used	Strength	Weaknesses
Meiet al. (2020)	TVSum50, YouTube videos	2DCNN for Spatial and 3DCNN for temporal features	Produce summary analogous to manual summary	Requires good-quality labeled dataset
Sridevi and Kharde (2020)	Visiocity	Continuity, Intent and Diversity	Different evaluations measures are considered to check quality of video	Execution time is more as size of dataset is very large
Kaushal et al. (2020)	A Gold standard dataset	Features which capture domain importance along with diversity models	Considers characteristics like representativeness, coverage, diversity for summary	Complexity is high
Kaushal et al. (2019)	SumMe, TVSum	Keyframe selection	Easy to collect unpaired dataset than labelled one	The quality of summary depends upon available data
Rochan and Wang (2019)	SumMe, TVSum	Attentive and distribution model	Short-term contextual attention insufficiency and distribution inconsistency issues are handled	It requires large training data
Ji et al. (2020)	Custom dataset	Color, texture, wavelet features	I-frames features are used to enhance speed	Redundancy is not handled
Panda and Roy-Chowdhury (2017), Feng Wang and Chong Wah Ngo (2016)	Tennis matches from 2013 and 2014 tournament	Frame selection using constraint satisfaction programming	Based on user's interest, distinct summaries can be created	Depends upon user expertise

(continued)

Table 16.1 (continued)

Author and Year	Dataset	Feature used	Strength	Weaknesses
Boukadida et al. (2018)	Open video project	Scoring and elimination is done using a rank-based method	It is applicable for all kinds of videos	Large number of calculations required at frame block level
Srinivas et al. (2016)	TVSum50	CNN based prediction of frame level shot importance	Enhance the performance of feature based methods	Lack of semantics information
Nahian et al. (2019)	OrangeVille, CoSum, SumMe, TVSum	Unsupervised object-level video summarization	Frame-level video summarization	High time and space complexity
Zhang et al. (2016)	Combined dataset and VTW dataset	Hierarchical recurrent neural network	Exploits long temporal dependency with Lesser computation operations	Not relevant to user's concern
Tsunoo et al. (2017)	TRECVID BBC rushes videos	Motion features	Multimodality is used for video synopsis	Complexity is high due to number of calculations

16.5 Challenges and Future Directions

Challenges

The following are a few difficulties or issues that this study highlights:

- Availability of annotated datasets.
- Accessibility of training video.
- Selection of interested frames or shots
- Duplication of data present in videos
- Variety of information
- Amalgamation of applications and real time models
- Complexity is high
- Time and space complexity.

Future Directions

The future research in this area should concentrate on:

- There is need to explore more on compressed video summarization domain.
- To focus on personalized video summaries to explore through large amount of visual data.

- Visualization of generated summary using proper tool or applications.
- Utilization of modern deep learning techniques in the field of computer vision to search through the vast amounts of available visual data.
- To improve video processing techniques to optimize resource utilization.

16.6 Conclusion

We deal with a tremendous amount of video data every day since it surrounds us. We spend a lot of time on social media, which also has a lot of videos. Due to the overabundance of digital video content, it is now necessary to distribute, classify, store, browse, and retrieve content systematically to optimize the available resources.

We conducted a thorough analysis of the state of video summarization in this chapter.

This analysis gave us the chance to talk about how summarization technology has changed over the past few years and what the future may hold, as well as to inform the relevant community about promising new developments and unresolved problems.

References

Basavarajaiah M, Sharma P (2019) Survey of compressed domain video summarization techniques. ACM Comput Surv (CSUR) 52(6):1–29

Bora A, Sharma S (2018a) A review on video summarization approcahes: recent advances and directions. In: 2018a international conference on advances in computing, communication control and networking (ICACCCN) ICACCCN), p 601–606

Bora A, Sharma S (2018b) A review on video summarization approcahes: recent advances and directions. In: 2018b International conference on advances in computing, communication control and networking (ICACCCN), pp 601–606

Boukadida H, Berrani SA, Gros P (2018) Automatically creating adaptive video summaries using constraint satisfaction programming: application to sport content. IEEE Trans Circuits Syst Video Technol 27(4):920–934

Chasanis VT, Likas CL, Galatsanos NP (2009) Scene detection in videos using shot clustering and sequence alignment. IEEE Trans Multimed 11(1):89–100

Cheng C-C, Hsu C-T (2006) Fusion of audio and motion information on HMM-based highlight extraction for baseball games. IEEE Trans Multimedia 8(3):585–599

Damnjanovic U, Fernandez V, Izquierdo E (2008) Event detection and clustering for surveillance video summarization. In: Proceedings of the ninth international workshop on image analysis for multimedia interactive services. IEEE Computer Society, Washington, USA

Elharrouss O, Almaadeed N, Al-Maadeed S, Bouridane A, Beghdadi A (2020) A combined multiple action recognition and summarization for surveillance video sequences. Appl Intell, 1–23

Elkhattabi Z, Tabii Y, Benkaddour A (2015) Video summarization: techniques and applications. Int J Comput Inf Eng 9(4):928–933

Haq HBU, Asif M, Ahmad MB, Video summarization techniques: a review

Jadon S, Jasim M (2019) Video summarization using keyframe extraction and video skimming, 2019. arXiv preprint arXiv:1910.04792

Ji Z, Jiao F, Pang Y, Shao L (2020) Deep attentive and semantic preserving video summarization. Neurocomputing 405:200–207

Kalaivani P, Roomi SMM (2017) Towards comprehensive understanding of event detection and video summarization approaches. In: Recent trends and challenges in computational models (ICRTCCM), 2017 second international conference, IEEE, pp 61–66

Kaushal V, Subramanian S, Kothawade S, Iyer R, Ramakrishnan G (2019) A framework towards domain specific video summarization. In: 2019 IEEE winter conference on applications of computer vision (WACV), pp 666–675

Kaushal V, Kothawade S, Iyer R, Ramakrishnan G (2020) Realistic video summarization through VISIOCITY: a new benchmark and evaluation framework. In: Proceedings of the 2nd International workshop on AI for Smart TV content production, access and delivery, pp 37–44

Kumar K, Shrimankar DD, Singh N (2017) Event BAGGING: A novelevent summarization approach in multiview surveillance videos. In: Innovations in electronics, signal processing and communication (IESC), 2017 international conference, pp 106–111

Lai PK, Décombas M, Moutet K, Laganière R (2016) Video summarization of surveillance cameras. In: 2016 13th IEEE International conference on advanced video and signal based surveillance (AVSS), pp 286–294

Li K, Li S, Oh S, Fu Y (2017) Videography-based unconstrained video analysis. IEEE Trans Image Process 26(5):2261–2273

Mei S, Ma M, Wan S, Hou J, Wang Z, Feng DD (2020) Patch based video summarization with block sparse representation. IEEE Trans Multimedia

Miniakhmetova M, Zymbler M (2015) An approach to personalized video summarization based on user preferences analysis. In: Application of information and communication technologies (AICT), 2015 9th international conference, pp 153–155

Mohan J, Nair MS (2019) Domain independent static video summarization using sparse autoencoders and K-means clustering. J Intell Fuzzy Syst 36(3):1945–1955

Muhammad K, Hussain T, Baik SW (2020) Efficient CNN based summarization of surveillance videos for resource-constrained devices. Pattern Recogn Lett 130:370–375

Murugan AS, Devi KS, Sivaranjani A, Srinivasan P (2018) A study on various methods used for video summarization and moving object detect ion for video surveillance applications. Multimedia Tools Appl 77(18):23273–23290

Nahian MA, Iftekhar ASM, Islam MT, Rahman SM, Hatzinakos D (2017) CNN-based prediction of frame-level shot importance for video summarization. arXiv preprint arXiv:1708.07023

Nahian MA, Iftekhar ASM, Islam MT, Rahman SM, Hatzinakos D (2019) CNN-based prediction of frame-level shot importance for video summarization. arXiv preprint arXiv:1708.07023

Panda R, Roy-Chowdhury AK (2017) Sparse modeling for topic-oriented video summarization. In: Proceedings of IEEE international conference on acoustics, speech and signal processing (ICASSP'17), pp 1388–1392. https://doi.org/10.1109/ICASSP.2017.7952384

Pei M, Jia Y, Zhu S-C (2011) Parsing video events with goal inference and intent prediction. In: Proceedings of international conference computer visual, pp 487–494

Peker KA, Bashir FI (2009) Content-based video summarization using spectral clustering. Mitsubishi electric research laboratories Cambridge, MA. University of Illinois at Chicago, Chicago, IL

Pritch Y, Ratovitch S, Hendel A, Peleg S (2019) Clustered synopsis of surveillance video. In: 6th IEEE International conference on advance video and signal base selection (AVSS 2009), Genoa, Italy, pp 2–4

Raut V, Gunjan R (2020) Video summarization approaches in wireless capsule endoscopy: a review. In: E3S Web of conferences, Vol 170, pp 03005

Rochan M, Wang Y (2019) Video summarization by learning from unpaired data. In: Proceedings of the IEEE/CVF conference on computer vision and pattern recognition, pp 7902–7911

Sridevi M, Kharde M (2020) Video summarization using highlight detection and pairwise deep ranking model. Procedia Comput Sci 167:1839–1848

Srinivas M, Pai MM, Pai RM (2016) An improved algorithm for video summarization–a rank based approach. Procedia Comput Sci 89:812–819

Tsunoo E, Bell P, Renals S (2017) Hierarchical recurrent neural network for story segmentation. In: Proceedings interspeech (INTERSPEECH'17), pp 2919–2923

Varghese, Nair KR (2015) An algorithmic approach for general video summarization. In: 2015 Fifth International conference on advances in computing and communications (ICACC), pp 7–11

Wang F, Ngo CW (2012) Summarizing rushes videos by motion, object and event understanding. IEEE Trans Multimedia 14

Wang F, Ngo CW (2016) Summarizing rushes videos by motion, object, and event understanding. IEEE Trans Multimedia 14(1):76–87

Wang X, Nie X, Liu X, Wang B, Yin Y (2020) Modality correlation-based video summarization. Multimedia Tools Appl, 1–16

Zawbaa HM, El-Bendary N, Hassanien AE, Kim TH (2018) Event detection based approach for soccer video summarization using machine learning. Int J Multimedia Ubiquitous Eng 7(2):63–80

Zhang K, Chao W-L, Sha F, Grauman K (2016) Video summarization with long short-term memory. In: Proceedings of European conference on computer vision (ICCV'16), pp 1–17. https://doi.org/10.1007/978-3-319-46478-7_47

Chapter 17
A Survey on Security Threats and Network Vulnerabilities in Internet of Things

Harish Kumar Saini, Monika Poriye, and Nitin Goyal

17.1 Introduction

Today the Internet has become omnipresent, and is affecting our daily life in inconceivable ways. We are moving into an era of more ubiquitous connectivity where a number of different devices will be connected over the network. We are getting into an age of the "Internet of Things" (IoT) (Khullar et al. 2022). Working of IoT includes the process of sensing the surroundings to collect some relevant information which is being forwarded over the communication network to some specified application for further action. IoT devices are designed to work autonomously without any human assistance. The merits of IoT and its applications are opening new possibilities of innovation and development. IoT system consists of devices with processing unit, memory, sensors, and actuators etc. to sense and preprocess the data and communication channel to transfer data over the network. The increased number of applications increases the amount and diversity of data being processed, stored and transmitted in different environments.

As IoT primarily uses internet as foundation to operate therefore all the threats that lie inside the internet are applicable on IoT also. The amount of potential threats is large because of the increased number of exposures, increased impact of attacks, every compromised IoT increases the probability of attacks and new and complex technology stack may increase the possibilities of new attacks (Scully 2017).

H. K. Saini (✉) · M. Poriye
Department of Computer Science and Applications, Kurukshetra University, Kurukshetra 136119, Haryana, India
e-mail: harishsaini5479@gmail.com

N. Goyal
Department of Computer Science and Engineering, Central University of Haryana, Mahendragarh 123031, Haryana, India

N. Sharma et al. (eds.), *Big Data Analytics in Intelligent IoT and Cyber-Physical Systems*, Transactions on Computer Systems and Networks,
https://doi.org/10.1007/978-981-99-4518-4_18

$$\text{Security Risk} = \frac{\text{Threat Level} \times \text{Probability of Attack} \times \text{Point of Exposure}}{\text{Security Measures Implemented}}$$

17.2 IoT Security Challenges

While designing efficient protection methods against security threats the following unique characteristics and constraints of IoT must be taken care of while designing an efficient security mechanism (Kranenburg and Bassi 2012):

- Manifold Technologies: IoT chains multiple technologies and each technology has its own vulnerabilities towards various attacks.
- Multiple Application areas: The IoT application areas include Healthcare, smart homes and smart farming etc. and each application has different security requirements.
- Scalability: The increasing number of smart devices is an important constraint in designing an effective protective mechanism.
- Availability: Availability means uninterrupted operation of the IoT system to achieve a standard availability of 99.999%. As protective mechanisms may take down the system or some IoT devices, administrators often hesitate to use these mechanisms.
- Data size: It is difficult to retain the privacy of massive data coming from various sensors over the network.
- Device Limitations: The limited computational, storage and battery capacity of the smart devices make them an easy target for Denial of Service attack.
- Remote Locations: Many IoT devices, specifically sensors, need to be installed at remote locations where an attacker can easily compromise these without being noticed.
- Mobility: It's very difficult to design a proficient security mechanism for dynamic environment where devices change their positions so frequently.
- Unpredictable behavior: the number of devices and their technologies makes it difficult to predict their behavior with others under different situations.
- Device Similarity: the uniformity in the design and technologies of devices makes them vulnerable to same type of threats.
- Device longevity: The device's long life may outlive their support and it's very difficult to protect these devices from attacks of new technology.
- No upgrade support: Most IoT devices are not designed to support updates and the abandoned software may be attacked by attackers.

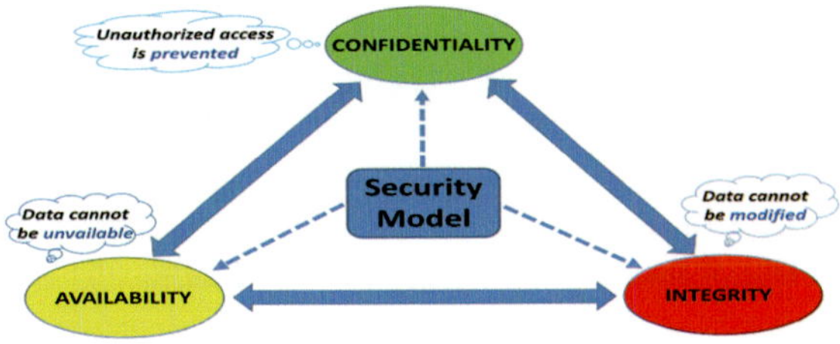

Fig. 17.1 CIA security model (Enthusiast and Security n.d.)

17.3 Security Goals in the IoT

The primary goal of security in the IoT system is to provide a security triad to maintain Confidentiality, Integrity and Availability (CIA) of information (Leloglu 2017; Nguyen et al. 2015; Mahmoud et al. 2015; Scully 2017) as shown in Fig. 17.1.

- Confidentiality: Data confidentiality is the ability to preserve the privacy of the user by ensuring the authorized access of the sensitive data. Confidentiality in IoT ensures that the devices will never transfer data to any unauthorized entity. Confidentiality includes privacy, separation and key management.
- Integrity: It's the ability to protect sensitive information from being modified by cyber criminals or by some external interference during communication. Integrity includes data integrity, Boot process, Authentication, Authorization and Accounting.
- Availability: Availability implies that the IoT devices, data and the communication link must be available whenever required. Availability includes counter measures, whitelisting, intrusion protection and management.

17.4 IoT Security Vulnerabilities

The vulnerabilities of the IoT system are flaws that increase the risk of hardware or software breaches to perform malicious activities. The most harmful IoT vulnerabilities are the following (Lopez et al. 2018; Rana et al. 2022b; Sharma et al. 2021; Zhang and Wang 2009; Dorai and Kannan 2011; Tait 2017; Paul 2019):

- Use of weak password.
- Unable to lock system after multiple unsuccessful attempts.
- Insecure network services.
- Insecure update mechanisms.
- Use of obsolete and insecure components.
- Inadequate confidentiality protection.
- Missing multiple authentications for sensitive programs.
- Use of poor encryption techniques for transport.
- Insecure information storage.
- Poor device management.

- Physical insecurity.
- Futile authentication and authorization.
- Insecurities in default settings.
- Lack of security in mobile, web and cloud interfaces.
- Lack of security features in software/firmware.
- Use of insecure third party components.

The hackers can use these vulnerabilities to realize their nefarious objectives.

17.5 Three Layer Generic IoT Architecture

Architecture is the blue print for any implementation. Since IoT consists of a numerous range of technology therefore there is no single universally agreed architecture but more than one reference architecture model can coexist. The essential feature of architecture includes availability, maintainability and functionality. Since the working of an IoT system includes numerous responsibilities ranging from sensing, processing, communication and finally to generate required output, we divide the architecture into different working layers on the basis of domain of responsibilities. Each device within the same layer may use heterogeneous technology to operate. The most generic and fundamental architecture is a three layer architecture as shown in Fig. 17.2.

17.5.1 Perception Layer

The perception layer is the physical accountable for sensing and collecting data about the surrounding. The primary tasks of this layer are:

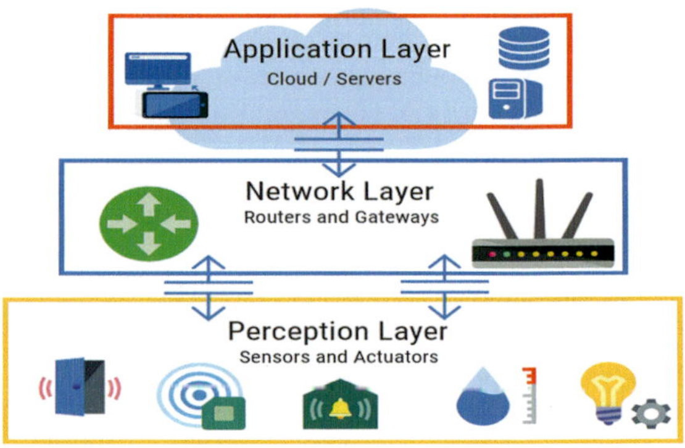

Fig. 17.2 Three layer architecture of IoT (Calihman 2019)

- To sense and record data.
- To preprocess the data.
- To establish communication link over the network.

This layer is also called as the sensor domain because it is responsible for recognizing things and collects the data from them. A number of sensing devices like barcodes, RFID and sensors etc. are connected to IoT devices to gather information. The sensing devices are selected as per the need of the application. The data sensed and gathered by these devices includes temperature, pressure, motion, vibration, location etc. This layer can be viewed as a collection of perception nodes having controllers and sensors etc. and perception network to communicate with the above layer (Jing et al. 2014). The common security threats of this layer fall in two categories: Environmental threats like fires, floods and earthquakes etc. which could cause substantial harms to the physical setup of IoT network and specially designed physical threats like tampering, fake node injection, node jamming and many more (Panagiotis et al. 2018).

17.5.2 Network Layer

The IoT devices of the perception layer or sensing domain need to be connected to a networking device called gateway. The network layer is responsible for providing a number of functionalities which mainly includes the collection and transfer of data from perception layer the cloud domain/Application layer (Calihman 2019). This layer is also known as fog domain and the devices in this domain are called fog devices. Every fog device is linked with a set of sensing devices of lower layer which transmits the collected information to this device. These fog devices further aggregate and preprocess the received information and transfer it to one or more servers of the cloud domain/Application layer. Every device in the fog layer is linked with each other to manage and coordinate all the available devices of sensing layer. The network layer protocols are responsible to maintain security and availability of data whenever data is traversed through the network. The various security threats to this layer include: Denial of Service, Man in the middle, Traffic Analysis, RFID Spoofing, RFID unauthorized access, Sinkhole attack etc.

17.5.3 Application Layer

This layer is also called as the cloud domain and is used to provide services to various custom applications that actually use data coming from the things. The cloud domain is a collection of various servers that hold the applications and provide the necessary computing facilities (Wu et al. 2010). This layer/domain is prone to attacks because of the vulnerabilities of applications. The common security threats of this layer are: code injection, buffer overflow, viruses, worms, Trojans, Spyware and adware, etc.

17.6 Classification of IoT Security Attacks

According to Kumar et al. (2021), Andrea et al. (2015), Chang and Li (2019) we can classify the attacks on theblIoT system in four different categories:

- Physical Attacks that cover all the attacks related to the physical layer.
- Network Attacks that cover all the attacks on the network layer.
- Software Attacks that includes all the attacks faced by an application layer.
- Encryption Attacks related to IoT protocols.

We now elaborate the various attacks presented in the above table.

17.6.1 Physical Attacks

Devices are the prime means through which attacks are instigated. Physical attacks are the type of attacks that result from breaches to the IoT devices with a target to affect the lifespan or functionality of the device. These attacks will work only if the attacker accesses the system through close vicinity. The various physical attacks are as follows:

- **Node Tampering**

Node tampering is where an attacker attempts to physically modify a node to gain access over the various components of the node in an IoT system (Perrig et al. 2004; Saibabu et al. 2020). The primary purpose of this attack is to get all privileges over a node to access sensitive information and use this compromised node to initiate different types of attack like (Butun et al. 2019):

- Attack against confidentiality by accessing confidential information.
- Attack against Integrity by sending fabricated data from compromised node to authentic users.
- Attack against availability by exhausting the services by generation frequent queries.

- **RF Interference of RFID's**

IoT systems usually operate in a noisy environment and RFID tags are susceptible to the surrounding disturbance, which may result in temporary loss of services. This attack can be instigated by producing electromagnetic interference signals with frequency similar to the RFID communication system. These signals will affect the RFID signals hampering communication (Li 2012). The main two types of interference are (Ullah 2018):

- The interference that hampers transmission of genuine information.
- The interpretation of counterfeit signal from one system as an authentic signal by other systems.

- **Node Jamming**

This attack is analogous to the RF Interference attack for the RFIDs but this attack operates on the WSNs. The attacker can jam the signal by interfering with the radio frequencies of the WSN nodes. A successful attack jams the sensor nodes resulting in denial of service (Mpitziopoulos et al. 2009). On the basis of the devices and type of jamming used by attacker the attack can be classified into four types (Xu et al. 2006; Vadlamani et al. 2016):

- Constant jamming: In this kind of attack, the attacker continuously produces radio signals to block legitimate users from using the communication channel.
- Deceptive jamming: In this approach, the attacker continuously sends data packets over the communication channel, putting the legitimate node in a perpetual state of busyness as it tries to continuously receive the packets.
- Random jamming: The random jamming takes into account the energy preservation by alternating between random sleep and generation of jamming signals. Different levels of effectiveness and energy conservation can be achieved by varying the sleep time and jamming time.
- Reactive jamming: In the reactive jamming the attacker observes the communication channel for any network activity and immediately transmits a signal to interfere with the ongoing communication.

According to Xu et al. (2015), the constant, deceptive and reactive attacks results in the falling of packet delivery to almost zero, but the jammers would exhaust their energy soon. The random jammer saves energy by sleeping.

- **Malicious Node Injection**

The attacker initiates this attack by inserting a malicious node between the authentic nodes of the IoT system. The malicious node can then be used to control operation and snoop on the data flowing between the authentic nodes (Illiano and Emil 2015).

- **Physical Damage**

This is a form of node tampering with a difference that the attacker tries to physically damage the devices of the IoT system to affect the availability of the service provided by the system.

- **Social Engineering**

A social engineering attack is a psychosomatic attack directed on human beings using smart devices to gain their trust via nefarious methods to collect some confidential information like passwords, credit card details, access permissions etc. These attacks are initiated through phone or online, so that offenders can hide their actual identities. For example, the attacker tries to convince the victim to disclose a credit card number over the phone or an attacker send multiple messages asking to click on the link of some spoofed webpage to get some sensitive information. The attackers are using this attack more frequently to initiate larger attacks because of the effectiveness of this attack.

Social engineering attacks can be generic attacks, fabricated for a wide-ranging of target, and targeted attacks specially designed for small group or an individual. Targeted attacks are often more effective than generic attacks as they are created for a smaller target group (Harris n.d.; Singla et al. 2022).

- **Sleep Deprivation Attack**

Usually most of the smart computing and sensing devices alternates between sleep mode and working mode exist without affecting the services in order to alleviate the battery consumption. The targets of this attack are battery operated devices, and the attacker introduces this attack by communicating with the target in a way that seems to be authentic; though, the intension of the communication is to keep the the target node in active state to consume its battery power as fast as possible. Further, it is challenging to detect this attack as it is initiated purely through the use of authentic looking communications (Pirretti et al. 2006). A similar type of attack is the barrage attack in which the attacker continuous bombards target nodes with control traffic packets. During barrage attack the targets spend somewhat more power but it is easy to detect.

17.6.2 Network Attacks

Due to the broadcast nature of the communication medium and the weaknesses of the network devices, IoT systems are vulnerable to network assaults. Attackers can launch network attacks against the IoT system network from any remote location. Network assaults can be broadly divided into passive and active attacks.

- **Traffic Analysis Attack**

Traffic analysis is done to capture and analyzing the data communication pattern in order to identify the IoT device, their activities and to collect the confidential information (Panagiotis et al. 2018; Goyal et al. 2020; Lilhore et al. 2022). An attacker first initiates device identification and activity recognition attacks (Hafeez et al. 2019) to observe the raffic pattern which can be used further to trigger bigger attacks. These attacks are passive in nature and are hard to discover.

- **RFID Unauthorized Access**

As there are no secured way to authenticate the access to the RFID systems, these can be accessed manipulated easily by any person. An attacker can take advantage of this vulnerability to read, update or delete the RFID data (Burmester and Medeiros 2007).

- **RFID Spoofing**

In RFID spoofing also called duplication of tags the attacker tries to analyze the sensitive information of a particular RFID system. After analysis the attacker uses the same information and format to generate a duplicate tag. Then the attacker can use this authentic looking fabricated tag to gain access of the system (Mitrokotsa et al. 2010).

- **RFID Cloning**

In IoT systems, a RFID cloning attack is to make one or more imitations of a genuine RFID tag, so that these imitations can be used to deceive the users for getting valid authorization. An attacker clones an RFID tag by replication the information of an original tag, to a blank new RFID tag. Though these RFID tags have duplicate data, the unique ID of the original tag is not replicated, which helps in tracing the compromised or cloned tag. Usually spoofing and cloning are done back to back, cloning duplicates the tags and spoofing is to acquire access permission on secured information (Smiley 2016).

- **Sinkhole Attack**

A sinkhole attack can be started by an attacker by hacking an authentic IoT network node or by introducing a rogue node into the network. The goal of the compromised nodes is to appear as the best route to the base station in an effort to entice network traffic away from practically all other nodes. The data can be altered by the sinkhole, which could jeopardizes the network security. Malicious nodes can potentially have an impact on the packet delivery ratio by sinking packets rather than sending them to their intended location. (Soni et al. 2013; Wallgren et al. 2013; Mayzaud et al. 2016; Mathew and Terence 2017).

- **Man in the Middle Attack**

It's a type of cyber security attack where the invader monitors the communication between authentic nodes. The attack takes place in between two legitimate parties, permitting the hackers to eavesdrop a communication between two nodes for which they are not authorized, that's why the name "man-in-the-middle (Cekerevac n.d.). The attacker compromises the security and privacy of the genuine nodes by observing, snooping and fabricating the communication between these nodes (Padhy et al. 2011). In such attacks the attacker does not necessarily need to be actually available near the targets, but this can be executed with the help of network communication protocols.

The techniques used for implementing MiTM attack include: Sniffing, Packet injection, Session hijacking and SSL stripping etc. and the various types of MiTM attack are: Rougue access point, ARP spoofing, Multicast DNS spoofing etc.

- **Denial of Service Attack**

It's a type of attack which can be used to restrict the network connection or IoT devices from providing its intended services to its users. Denial of Service attack is realized by inundating the communication channel or device of an IoT network with excessive surplus of activity that may incapacitate the network or device (Medaglia and Serbanati 2010). As most IoT devices are fabricated from inexpensive generic components from few manufacturers with some common default security vulnerabilities, the infrastructure for IoT is a latent advantage for the attackers. Attackers use Botnet also called zombie army, to launch denial of service attack. Botnet is collection of interconnected computers injected with malware. DoS attack can be categorized as outage attack, battery draining and sleep deprivation attack.

- **Routing Attack**

Secure routing is the key to the reliable operation of IoT applications; still we find instability and vulnerabilities in a numbers of routing protocols. Routing attacks target the packets routing information to carry out different attacks like modifying attack which alters the routing information to misroute the traffic or to create routing loops etc. (Goyal et al. 2021; Wu and Hu 2008).

- **Sybil Attack**

It's a type of masquerade attack in which an adversary node fabricates a number of false identities to control and compromise the authentic nodes of a system. The recent advances in IoT are providing benefits and opportunities but also increasing the security risks (Popli et al. 2021; Zhang et al. 2014; John et al. 2015). An attacker planning to initiate Sybil attack needs to fabricate links between the fake nodes or identities and the authentic nodes in a network. An attacker can generate node identities firstly by using real IDs in which an attacker creates multiple real IDs using a single device and secondly using virtual IDs which communicate with attackers' IDs only (Alharbi et al. 2018). The IoT under Sybil attacks may broadcast incorrect reports to nearby authentic nodes which unknowingly accepts and reacts accordingly for e.g. accepting a fake node as part of a route, or accepting multiple votes from fake node in IoT based voting system. Since most Sybil attackers behave as normal users, to trace a Sybil is extremely difficult (Pawar and Vanwari 2016).

- **Hello Flood Attack**

Usually nodes broadcast Hello packets over the network to show their existence to all the neighboring nodes. Any node that receives such Hello packets may adopt the sender as the neighbor and start communicating. The Hello flood attacks can be triggered by a malicious node by sending a Hello message for declaring itself as their neighbor over the network with such a power that it can be received by a maximum number of nodes and they select the sender as an authentic neighbor node and start communicated with that node and also updates its routing table (Hamid et al. 2006). As a result all nodes communicate with base station via this malicious node. This is how an attacker artifice of being a neighbor to other nodes and thus affects the existing routing mechanism.

- **Selective Forwarding Attack**

In this type of attack, malicious nodes may choose not to transmit particular packets in order to stop their propagation (Kakkar et al. 2022; Bysani and Turuk 2011). The rogue node may purposefully or randomly discard packets. The various varieties of selective forwarding attacks included of:

- Black hole Attack: The attacker selectively forbids packets from or headed toward a certain node. By dumping all the packets to the sink, this may be applied to the whole network.
- Neglect and Grey Attack: The rogue node delivers its own packet while at random times dropping others.
- By delaying the packet, the rogue node deceives the routing information.
- The last assault is a blind letter attack, in which a rogue node promises to send packets but then discards them.

- **Replay Attack**

A replay attack also called as playback attack is a sort of Man in the middle attack in which the adversary eavesdrops on a valid secure communication, intercepts the information and maliciously delayed or repeated. The reason why replay attack is dangerous is that the attackers can initiate this attack simply by resending the captured information without decrypting.. For e.g. if an infrared device is used to open a door then an attacker can unlock the door by recording and replaying the infrared modulation pattern (Coward 2017).

17.6.3 Software Attacks

In Software attacks an attacker exploits the application vulnerabilities in the system. The various types of software attacks are:

- **Virus and Wormhole attacks**

System can be infected by adversary by injecting malicious software or code into the system that can compromise the security and privacy of the system by exposing sensitive information or may results in denial of service. In a wormhole attack, two manipulative sensor nodes create a direct communication link to transfer packets between each other ignoring the intermediate nodes (Nagrath and Gupta 2011), to create an illusion of being shortest path in the network. Such shortcut in the network will attract the authentic nodes to include this path in their routing table. Then these malicious nodes keep on snooping and recording the network traffic. The attackers in the network create a tunnel to transfer recorded information of one position to another as shown in Fig. 11. The wormhole attack is very challenging as it can be implemented on any protocol and is also effective on encrypted data. Threats related to Wormhole attack includes: protocol failure, inserting erroneous routing information into the network etc.

- **Spyware and Adware**

Spyware is any type of malicious software that enables a hacker to gather private data from an intended victim's system through an infected device and transmits it to another system, including authentication passwords, network traffic, and internet usage patterns. The hazards posed by spyware have increased recently as a result of the Internet of Things' (IoT) expanding pervasiveness in seemingly all facets of business and daily life. It can be carried out at the application layer and aims to protect the IoT system's privacy. Adware is a software or application that automatically displays advertisements through pop-up windows. While they are not always malicious, but it can be designed to analyze the browsing behavior and collecting sensitive information, install malicious programs, and redirect user to insecure sites. Adware can be unwittingly downloaded by downloading freeware. Adware may slow down the networks and devices by running different programs in the background.

- **Trojan Horse**

A Trojan horse is a kind of malicious code designed to look authentic but pursues to deceive the user to load and execute the malware on their device. A Trojan can be used to inflict some harmful attack on the data or network. Unlike Viruses, Trojan needs someone to execute and replicate. For example, if you receive an email with some attachment coming from a cybercriminal using some fake authentic ID and downloading this attachment means installation of a malware on your device.

The common types of Trojan malware are:

- Password-stealing Trojans to retrieve and disclose system password.
- Backdoor Trojan to create a backdoor entry for third party to perform malicious activities on your system.
- DDoS attack Trojan to exhaust the resources by flooding the traffic.
- Downloader Trojan to download new harmful programs on infected system.
- Fake Antivirus Trojan which demands money to sense and eliminate threats.
- Game-thief Trojan pursues to steal account details of online gamers.
- Info stealer Trojan to get sensitive data from compromised system.
- Mail finder Trojan to collect email addresses stores in system.
- Ransom Trojan to seek a money to undo the harm done by the malware.
- Remote Access Trojan to access infected system through a remote connection.
- Rootkit Trojan to hide malicious objects.
- SMS Trojan to send and capture text messages of infected mobile.
- Trojan banker to retrieve bank details by observing online activity.

- **Phishing Attacks**

The goal of this attack is to extract confidential information deceitfully for individual gain. In phishing attack the attacker lures the authentic user to open infected email or webpage through which the attacker can gain access to confidential credentials of the user (Jagatic et al. 2007). A variation of phishing attack is spear phishing in which an attacker fabricates malicious communication for a victim after examining the victim's activities thoroughly (Rana et al. 2022a; William 2008).

- **Malicious Code Injection**

This type of attack can be initiated by infecting an authentic node by inserting a malicious code into it. The infected node can now be used to gain access of network information to carry out further attacks. Web-based vulnerability like XSS can be used by an attacker to attach a malicious code to a benign link. Clicking this authentic link executes the attacker's script on the local web browser and results in hijacking of web operation or even complete system (Zhang and Qu 2013; Yampolskiy et al. 2013). The common attacks include shell injection, running executable active-x scripts and HTML script injection etc. (Tobias et al. 2011).

- **Buffer Overflow**

A buffer is a piece of memory of specific size used to store information. Different applications use multiple buffers to store their data. Buffer overflow occurs when a program passes more data than the assigned capacity of the buffer and the surplus data overwrites the content of the next buffer. The attacker uses this vulnerability to overwrite the original instructions with malicious instruction (Scully 2017). Further, this attack enables an unauthorized user to gain administrator rights and execute malicious code (Zhu et al. 2011). IoT devices are prone to buffer overflow attack because of the following reasons (Desiner 2017):

- **Memory**: IoT devices use limited memory and can be overflowed very easily.
- **Language**: Most IoT programs are designed in C or C++ which don't support garbage collection and hence vulnerable to the buffer overflow attack. Also, the attackers can use pointers to interpret the location of critical code in memory.

17.6.4 Encryption Attacks

In these kinds of attacks the cyber assailants examine and infer the encryption keys to compromise the security and impose various attacks. The various attacks that fit in this category are:

- **Side Channel Attacks**

Side channel attack is not against any specific cryptosystem. This attack uses the weakness in implementation of the cryptosystem to retrieve security key of system rather than using software or encryption vulnerabilities. It measures execution related data like time, power usage and fault etc.to retrieve sensitive data like secret key etc. The types of side channel attacks are as follows-

- **Timing Attacks**—The attacker uses computation timing to initiate the attack.
- **Power Analysis Attacks**—The attacker uses power consumption as parameter to initiate the attack.
- **Fault analysis Attacks**—The attacker injects errors in the cryptosystem and analyze the output for retrieving information.

- **Cryptanalysis Attack**

Cryptanalysis is a process of retrieve original data from encrypted data. Here the basic goal of an attacker is to recover plain text from cipher text by penetrating the cryptosystem. An attacker can compromise the system if he somehow manages to extract secret decryption let from the cryptosystem. The classification of attacks on cryptosystems is as follows (Jing et al. 2014; Harris n.d.)

- Cipher text Only Attack: Here the attacker tries to interpret the plain text for a given set of cipher text. This is the main attack in modern cryptosystems.
- Known Plain text Attack: In this attack, the attacker knows a sample cipher text and plain text pair and attempts to decrypt the remaining cipher text. e.g. linear cryptanalysis.
- Chosen Plain text Attack: Here the attacker has the cipher text—plain text pair of his choice and tries to find out encryption key. e.g. Differential cryptanalysis.
- Dictionary attack: Here the attacker maintains a dictionary of known cipher text and its plain text and refers the same in future.
- Brute Force Attack: In this attack, the attacker attempts to crack the key by trying all possible keys.

17.7 Conclusion

The Internet of Things (IoT) is a significant technical advancement that has attracted both industry and academia. IoT is a collection of physically connected items, hence the security of the internet is greatly impacted by each connected object with a weak security system. Since the IoT devices were created with limited resources, implementing them could present more difficulties in the areas of security, interoperability, privacy, etc. In order to take appropriate action, a major security investigation is needed. We provide an overview of IoT in this chapter, covering its definition, architecture, attacks, and challenges with a focus on security in particular. By the investigation of various vulnerabilities, it is concluded that the major security risk is at perception layer owing to the limitations of resources and technology used by devices in this layer, followed by the network and the application layer.

References

Alharbi A, Zohdy M, Debnath D, Olawoyin R, Corser G (2018) Sybil attacks and defenses in internet of things and mobile social networks. Int J Comput Sci Issues 15(6):36–41. https://doi.org/10.5281/zenodo.2544625

Altium Desiner (2017) Altium. https://resources.altium.com/p/internet-of-things-security-vulnerabilities-all-about-buffer-overflow

Andrea I, Chrysostomou C, Hadjichristafi G (2015) Internet of things: security vulnerabilities and challenges. IEEE symposium on computer and communication, pp 180–187

Burmester M, Medeiros BD (2007) RFID security: attacks, countermeasures and challenges. The 5th RFID academic convocation, the RFID journal conference

Butun I, Osterberg P, Song H (2019) Security of the internet of things: vulnerabilities, attacks and countermeasures. IEEE Commun Surv Tutorials 20(10):1–25

Bysani LK, Turuk AK (2011) A survey on selective forwarding attack in wireless sensor networks. Proceedings of the international conference on devices and communications (ICDeCom), pp 1–5, https://doi.org/10.1109/ICDECOM.2011.5738547

Calihman A (2019) Architectural frameworks in the IoT civilization, www.netburner.com

Cekerevac Z, IoT and MITM attacks—security and economic risks MEST J 5(2):15–25, https://doi.org/10.12709/mest.05.05.02.03

Chang Z, Li S (2019) The IoT attack surface: threats and security solutions. Trend Micro. trendmicro.com/vinfo/in/security/news/internet-of-things/the-iot-attack-surface-threats-and-security-solutions

Coward C (2017) IoT devices may be susceptible to replay attacks with a raspberry Pi and RTL-SDR dongle. Hackster.io.https://www.hackster.io/news/iot-devices-may-be-susceptible-to-replay-attacks-with-a-raspberry-pi-and-rtl-sdr-dongle-de6eca268fbf

Dorai R, Kannan V (2011) SQL injection—database attack revolution and prevention. J Int'l Com L & Tech 6:224

Goyal N, Dave M, Verma AK (2020) SAPDA: secure authentication with protected data aggregation scheme for improving QoS in scalable and survivable UWSNs. Wirel Pers Commun 113(1):1–15

Goyal N, Sandhu JK, Verma L (2021) CDMA-based security against wormhole attack in underwater wireless sensor networks. In: Advances in communication and computational technology, pp 829–835. Springer, Singapore

Hafeez I, Antikainen M, Tarkoma S (2019) Protecting IoT-environments against traffic analysis attacks with traffic morphing. 2019 IEEE international conference on pervasive computing and communications workshops (PerCom Workshops). https://doi.org/10.1109/percomw.2019.8730787

Hamid MA, Mamun-Or-Rashid M, Hong CS (2006) Routing security in sensor network: hello flood attack and defense. IEEE ICNEWS, 2–4

Harris IG, Social Engineering attack on the Internet of Things. IEEE Internet of Things. https://iot.ieee.org/newsletter/september-2016/social-engineering-attacks-on-the-internet-of-things.html

Illiano VP, Emil CL (2015) Detecting malicious data injections in wireless sensor networks: a survey. ACM Comput Surv (CSUR) 48(2):24

Jagatic TN, Johnson NA, Jakobsson M, Menczer F (2007) Social phishing. Commun ACM 50(10):94–100

Jing Q, Vasilakos AV, Wan J, Lu J, Qiu D (2014) Security of the internet of things: perspectives and challenges. Wirel Netw 20(8):2481–2501
John R, Cherian JP, Kizhakkethottam JJ (2015) A survey of techniques to prevent sybil attacks. Proceedings of the international conference on soft-computing and networks security (ICSNS), pp 1–6, https://doi.org/10.1109/ICSNS.2015.7292385
Kakkar L, Gupta D, Tanwar S, Saxena S, Alsubhi K, Anand D, ... Goyal N (2022). A secure and efficient signature scheme for IoT in healthcare. CMC-Comput Mater Continua 73(3):6151–6168
Khullar V, Singh HP, Miro Y, Anand D, Mohamed, HG, Gupta D, ... Goyal N (2022) IoT fog-enabled multi-node centralized ecosystem for real time screening and monitoring of health information. Appl Sci 12(19):9845
Kranenburg, Van B (2012) IoT challenges. Commun Mobile Comput 1:9
Kumar A, Sharma S, Goyal N, Singh A, Cheng X, Singh P (2021) Secure and energy-efficient smart building architecture with emerging technology IoT. Comput Commun 176:207–217
Leloglu E (2017) A review of security concerns in internet of things. J Comput Commun 5(1):121–136
Li L (2012) Study on security architecture in the internet of things. International conference on measurement, information and control (MIC), pp 374–377
Lilhore UK, Imoize AL, Li CT, Simaiya S, Pani SK, Goyal N, ... Lee CC (2022) Design and Implementation of an ML and IoT based adaptive traffic-management system for smart cities. Sensors 22(8):2908
Lopez DD, Uribe MB, Cely CS, Torres AV, Guataquira NM, Castro SM, Nespoli PP, Marmol FG (2018) Shielding IoT against cyber attacks: an event based approach using SIEM. Hindawi Wirel Commun Mobile Comput 2018:1–18. https://doi.org/10.1155/2018/3029638
Mahmoud R, Yousuf T, Aloul F, Zualkernan I (2015) Internet of things (IoT) security: current status, challenges and prospective measures. Proceedings of the tenth international conference for internet technology and secured transactions (ICITST), pp 336–341
Mathew A, Terence JS (2017) A survey on various detection techniques of sinkhole attacks in WSN. Proceedings of the international conference on communication and signal processing (ICCSP), pp 1115–1119, https://doi.org/10.1109/ICCSP.2017.8286550
Mayzaud A, Badonnel R, Chrisment I (2016) A taxonomy of attacks in RPL-based internet of things. Int J Network Security 18(3):459–473
Medaglia CM, Serbanati A (2010) An overview of privacy and security issues in the internet of things. In The Internet of Things Springer, New York, pp 389–395
Mitrokotsa A, Rieback MR, Tanenbaum AS (2010) Classification of RFID attacks. Inf Syst Front 12:491–505, https://doi.org/10.1007/s10796-009-9210-z
Mpitziopoulos A, Gavalas D, Konstantopoulos C, Pantziou G (2009) A survey on jamming attacks and countermeasures in WSNs. Commun Surv Tutorials IEEE 11(4):42–56
Nagrath P, Gupta B (2011) Wormhole attacks in wireless adhoc networks and their counter measurements: a survey. Proceedings of the third international conference on electronics computer technology, 6, pp 245–250, https://doi.org/10.1109/ICECTECH.2011.5942091
Nguyen KT, Laurent M, Oualha N (2015) Survey on secure communication protocols for the internet of things. Ad Hoc Netw 32:17–31
Padhy RP, Patra MR, Satapathy SC (2011) Cloud computing: security issues and research challenges. Int J Comput Sci Inf Technol Security (IJCSITS) 1(2):136–146
Panagiotis I, Radoglou Grammatikis A, Panagiotis G, Sarigiannidis A, Moscholios ID (2018) Securing the internet of things: challenges, threats and solutions. Internet of Things 5(Elsevier), 41–70
Paul F (2019) Top 10 IoT vulnerabilities. Networkworld. https://www.networkworld.com/article/3332032/top-10-iot-vulnerabilities.html
Pawar S, Vanwari P (2016) Sybil attack in internet of things. Int J Eng Sci Innov Technol 5(4):96–105
Perrig A, Stankovic J, Wagner D (2004) Security in wireless sensor networks. Commun ACM 47(6):53–57

Pirretti M, Zhu S, Vijaykrishnan N, McDaniel P, Kandemir M, Brooks R (2006) The sleep deprivation attack in sensor networks: analysis and methods of defense. Int J Distrib Sensor Netw 2:267–287, https://doi.org/10.1080/15501320600642718

Popli R, Sethi M, Kansal I, Garg A, Goyal N (2021) Machine learning based security solutions in MANETs: State of the art approaches. In: Journal of physics: conference series (Vol 1950, No 1, p. 012070). IOP Publishing

Rana A, Sharma S, Nisar K, Ibrahim AAA, Dhawan S, Chowdhry B, ... Goyal N (2022a) The Rise of Blockchain internet of things (BIoT): secured, device-to-device architecture and simulation scenarios. Appl Sci 12(15):7694

Rana SK, Rana SK, Nisar K, Ag Ibrahim AA, Rana AK, Goyal N, Chawla P (2022b) Blockchain technology and artificial intelligence based decentralized access control model to enable secure interoperability for healthcare. Sustainability 14(15):9471

Saibabu G, Jain A, Sharma VK (2020) Security issues and challenges in IoT routing over wireless communication. Int J Innov Technol Exploring Eng 9(4):1572–1580, https://doi.org/10.35940/ijitee.D1797.029420

Scully P (2017) 5 Things to know about IoT security. DZone. https://dzone.com/articles/5-things-to-know-about-iot-security

Sharma S, Kumar A, Bhushan M, Goyal N, Iyer SS (2021) Is blockchain technology secure to work on?. In: Blockchain and AI technology in the industrial internet of things, pp 66–80. IGI Global

Singla D, Gupta D, Goyal N (2022) IoT based monitoring for the growth of basil using machine learning. In: 2022 10th international conference on reliability, infocom technologies and optimization (Trends and Future Directions) (ICRITO), pp 1–5. IEEE

Smiley S (2016) 7 types of security attacks on RFID systems. atlasRFISstore. https://www.atlasrfidstore.com/rfid-insider/7-types-security-attacks-rfid-systems

Soni V, Modi P, Chaudhri V (2013) Detecting sinkhole attack in wireless sensor network. Int J Appl Innov Eng Manag 2(2):29–32

Tait A (2017) 10 Internet of things security vulnerabilities. Learning Tree International. https://blog.learningtree.com/10-internet-of-things-security-vulnerabilities

Tech Enthusiast. Security in IoT-Security solutions for IoT communication protocols. CRYPTIOT. https://cryptiot.de/iot/security/security-solution-iot-com-protocol

Tobias H et al (2011) Security challenges in the IP-based internet of things. Wirel Pers Commun 61(3):527–542

Ullah A (2018) IoT: applications of RFID and Issues. Int J Internet Things Web Serv 3:1–5

Vadlamani S, Eksioglu B, Medal H, Nandi A (2016) Jamming attacks on wireless networks: a taxonomic survey. Int J Prod Econ 172:76–94, https://doi.org/10.1016/j.ijpe.2015.11.008

Wallgren L, Raza S, Voigt T (2013) Routing attacks and countermeasures in the RPL-based internet of things. Int J Distrib Sensors Netw 9(8), https://doi.org/10.1155/2013/794326

William S (2008) Computer security: principles and practice, Pearson Education India

Wu D, Hu G (2008) Research and improve on secure routing protocols in wireless sensor networks. 4th IEEE international conference on circuits and systems for communications, pp 853–856

Wu M, Lu TJ, Ling FY, Sun J, Du HY (2010) Research on the architecture of internet of things. 3rd International conference on advanced computer theory and engineering (ICACTE), vol 5. IEEE, pp 475–484

Xu W, Ma K, Trappe W, Zhang Y (2006) Jamming sensor networks: attack and defense strategies. IEEE Network 20(3):41–47, https://doi.org/10.1109/MNET.2006.1637931

Xu W, Trappe W, Zhang Y, Wood T (2015) The feasibility of launching and detecting jamming attacks in wireless networks. 6th ACM international symposium on Mobile ad hoc networking and computing, pp 46–57

Yampolskiy M, Horvath P, Koutsoukos XD, Xue Y, Sztipanovits J (2013) Taxonomy for description of cross-domain attacks on CPS. In: Proceedings of the 2nd ACM international conference on high confidence networked systems. ACM, pp 135–142

Zhang W, Qu B (2013) Security architecture of the internet of things oriented to perceptual layer. Int J Comput Consum Control (IJ3C) 2(2):37–45

Zhang Q, Wang X (2009) SQL injections through back-end of RFID system. In: 2009 international symposium on computer network and multimedia technology. CNMT 2009. IEEE, pp 1–4

Zhang K, Liang X, Lu R, Shen X (2014) Sybil attacks and their defenses in the internet of things. IEEE Internet Things J 1(5):372–383, https://doi.org/10.1109/JIOT.2014.2344013

Zhu B, Joseph A, Sastry S (2011) A taxonomy of cyber-attacks on SCADA systems. Internet of things (Ithings/CPSCom). International conference on and 4th international conference on Cyber, Physical and Social Computing. IEEE, pp 380–388

GPSR Compliance

The European Union's (EU) General Product Safety Regulation (GPSR) is a set of rules that requires consumer products to be safe and our obligations to ensure this.

If you have any concerns about our products, you can contact us on ProductSafety@springernature.com

In case Publisher is established outside the EU, the EU authorized representative is:

Springer Nature Customer Service Center GmbH
Europaplatz 3
69115 Heidelberg, Germany

Batch number: 10756376

Printed by Printforce, the Netherlands